Modeling Complex Engineering Structures

Other Titles of Interest

Engineering with the Spreadsheet: Structural Engineering Templates Using Excel
Craig T. Christy
Tools to quickly apply the powerful analytic capability of Microsoft Excel to structural engineering applications (CD-ROM included).
ASCE Press. ISBN 978-0-7844-0827-8

GIS Tools for Water, Wastewater, and Stormwater Systems
Uzair Shamsi
State-of-the-art information to develop GIS applications for water, wastewater, and stormwater systems.
ASCE Press. ISBN 978-0-7844-0573-4

Hydraulic Structures: Probabilistic Approaches to Maintenance
Walter O. Wunderlich
An overview of probabilistic methods and an introduction to the field of probability theory and its applications in engineering.
ASCE Press. ISBN 978-0-7844-0672-4

Numerical Methods in Geotechnical Engineering: Recent Developments
George M. Filz and D. Vaughan Griffiths (Editors)
Advanced numerical methods as they can be applied to a wide range of problems in geotechnical engineering.
ASCE Proceedings. ISBN 978-0-7844-0502-4

Principles of Applied Civil Engineering Design
Ying-Kit Choi
A pioneer reference detailing the guidelines, principles, and philosophy for designing heavy civil engineering projects.
ASCE Press. ISBN 978-0-7844-0712-7

Structural Safety and Its Quality Assurance
Bruce R. Ellingwood and Jun Kanda (Editors)
Structural safety problems from basic concept to design and construction.
ASCE Committee Report. ISBN 978-0-7844-0816-2

Modeling Complex Engineering Structures

Edited by
*Robert E. Melchers
and
Richard Hough*

Library of Congress Cataloging-in-Publication Data

Modeling complex engineering structures / edited by Robert E. Melchers and Richard Hough.
    p. cm.
Includes bibliographical references and index.
ISBN-13: 978-0-7844-0850-6
ISBN-10: 0-7844-0850-5
1. Structural engineering—Data processing. 2. Building materials—Computer simulation. 3. Structural analysis (Engineering)—Data processing. 4. Engineering models. I. Melchers, R. E. (Robert E.), 1945- II. Hough, R. (Richard)

TA640.M63 2007
624.101'13—dc22

2006026708

Published by American Society of Civil Engineers
1801 Alexander Bell Drive
Reston, Virginia 20191
www.pubs.asce.org

Cover images: structural model details, National Aquatics Center, Beijing—courtesy of Arup + PTW + CSCEC.

Contents

Preface

From time to time in the history of engineering, specialist engineers have paused to look over the fence at what their colleagues are up to in parallel disciplines. This has often led to useful insights and accelerated growth in one or the other discipline. For those of us interested in the modeling of structural systems, such a cross-discipline review is very timely at this point in our history, thanks to progress in the range, versatility, and computational power of analysis software during recent years.

The purpose of the present volume is to bring together structural analysis strategies from a wide range of current industries, providing readers with a rare opportunity to make comparisons and deductions from their own particular perspectives.

The level of sophistication available to structural system modelers has accelerated rapidly since the early days of electronic computing. This trend is likely to continue with the increased availability of higher speed vector-array machines and parallel-processing facilities. This availability, in turn, invites the development of new algorithms using improved numerical methods based on more detailed understanding of real material and structural behavior.

In parallel, more work is needed on defining appropriate criteria for structural performance and safety, including the measurement and codification of these ideas for more consistent outcomes. In most areas of structural analysis, issues remain to be tackled in the simulation of loading and environmental demands and in compatibility and consistency of modeling choices so that modeling efficiency and accuracy are both enhanced in the interest of improved life cycle performance of real structures.

The present volume provides an overview of many of these issues, typically through case studies. One volume cannot pursue each issue in depth, but it can stimulate cross-fertilization of modeling approaches, of computational techniques and system representation, and of ways of viewing satisfactory performance.

A discussion is also included, at a practical level, of concepts of uncertainty in system representation and the implications for performance and for safety.

The chapter authors all have a common goal of presenting a readable summary of the trends in their areas of expertise, usually by case studies. The frame of reference has been a reader with experience at student or practitioner level of at least one area of structural analysis. The chapter authors typically have provided references to more detailed works and expositions where these would be useful, and readers are encouraged to extend their exploration through these further sources.

Disclaimer

The editors accept no responsibility for any consequences arising from the use of any advice provided by chapter authors and expect all readers to apply their own due diligence in using such advice.

Acknowledgments

The editors appreciate the cooperation they have received from all the authors during the compilation of this volume. They also appreciate the confidence that ASCE Press has had in the project from the beginning and the efficient manner with which reviews, editing, and publication has been handled. Particular gratitude in this regard is expressed to Bernadette Capelle, ASCE's original acquisitions editor for this project, and to Betsy Kulamer.

Robert E. Melchers
Richard Hough
September 2006

Contributors

Tristram Carfrae is a principal of Arup and an Arup Fellow. He is based in Sydney and holds several Board and Board Executive positions throughout the international Arup Group. He has been active in many of the firm's technical groups and is now chair of the worldwide Design and Technical Executive. During his 20 years with Arup, Mr. Carfrae has applied innovative structural engineering in collaboration with many of the world's most famous architects on some of the world's best-known projects. He has received personal awards in the United Kingdom (UK) and Australia in recognition of his contributions to structural design and its integration into great architecture. Mr. Carfrae is also an adjunct professor at the University of New South Wales, Australia.

Paul Cross is an associate director of Arup, based in London. He graduated from Queen Mary College, London, and joined Arup in 1973. Paul's particular expertise is the structural analysis and design of tall buildings. He has developed systems to process data and results for very large analyses and to optimize the efficiency of structural systems. He has been responsible for the analysis of many major commercial and residential projects in the UK and elsewhere, including the headquarter buildings for Swiss Re and HSBC in London, Luk Yeung Sun Cheun and Exchange Square in Hong Kong, Century Tower in Tokyo, Torre de Collserola in Barcelona, and Tour Sans Fins in Paris.

Chun-Man Chan, Ph.D., is an associate professor in the Department of Civil Engineering at the Hong Kong University of Science and Technology. He obtained his B.Sc. in 1984 and M.Sc. in 1985 from the Massachusetts Institute of Technology and received his Ph.D. in 1993 from University of Waterloo, Canada. His research focuses on computer-aided design and structural optimization of tall buildings.

The effectiveness of his innovation on structural optimization has been demonstrated through actual applications to a number of high-profile building projects in Hong Kong. He has received research and teaching awards, including the 1998 ASCE State-of-the-Art Civil Engineering Award and the Teaching Excellence Appreciation Award from Hong Kong University of Science and Technology.

Bruce R. Ellingwood, Ph.D., received his undergraduate and graduate education at the University of Illinois at Urbana-Champaign. From 1972 to 1986, he held research positions with the U.S. Department of the Navy and the National Bureau of Standards. He served on the faculty of the Johns Hopkins University from 1986 to 2000 and was appointed College of Engineering Distinguished Professor at the Georgia Institute of Technology in 2000, the position he holds now. Dr. Ellingwood's professional interests in the field of structural reliability have focused on the analysis of structural loads and load combinations, performance of structures under occupancy, environmental and abnormal load conditions, and development of probability-based safety and serviceability criteria for design. He is the author of more than 250 technical publications and reports. He is recipient of numerous awards and is a member of the National Academy of Engineering.

Xiu-Li Guan works for Scott Wilson Pavement Engineering (SWPE) in Nottingham, England. Prior to joining SWPE, she was a postdoctoral research associate in the Department of Civil, Surveying, and Environmental Engineering at the University of Newcastle, Australia, working mainly in the field of structural engineering. Dr. Guan has some ten years of research and working experience in the field of civil and structural engineering, with major research expertise in the areas of structural reliability and risk assessment, finite element methods, and stress analysis.

Richard Hough is a principal in the Sydney office of Arup and Professor of Multi-Disciplinary Design at the University of New South Wales. Previously, he was a director in Arup's London office, where he led a structural engineering team that included the firm's lightweight structures group. He has been managing principal of Arup's California practice; has led many technical groups within Arup, including the Structural Coordination Group and the Seismic Skills Network; and is currently Design and Technical Leader for the Australasia Region. He has worked on some of Arup's most technically challenging projects, including many with longspan roofs. He is also a conjoint professor at the University of Newcastle, Australia.

Owen F. Hughes, Ph.D., is professor of ship structures, Department of Aerospace and Ocean Engineering, Virginia Polytechnic Institute. He is recognized internationally as a pioneer in the field of first-principles structural design, having been one of the first to achieve a synthesis of finite element analysis, ultimate strength analysis, and mathematical optimization. In doing so, he made several fundamental developments in all three areas. His book, *Ship Structural Design* (1993), presented an entirely new method for ship structural design. He also implemented the method in a computer program called Method for Analysis, Evaluation and STRuctural

Optimization (MAESTRO), which is now used by 13 navies, various structural safety authorities, and more than 80 structural designers and shipyards in Europe, North America, Asia, and Australia. Dr. Hughes has been NAVSEA Research Professor at the U.S. Naval Academy, chairman of the Society of Naval Architects and Marine Engineers (SNAME) Panel on Design Procedures and Philosophy, and chairman of the International Ship and Offshore Structures Congress (ISSC) Committee on Computer-Aided Design.

Tim Keer is a principal of Arup, responsible for the technical quality of services provided from Arup's Detroit office. These services mainly focus on analytical and design support to the automotive industry. Mr. Keer joined Arup in London in 1984 and was sponsored by the firm through studies at Cambridge University. He rejoined Arup in 1987 and worked in the Advanced Technology Group on a wide range of projects involving advanced analysis techniques and software development. In 1992, he transferred to the United States to open Arup's Detroit office. His particular interests lie in the development and application of techniques for nonlinear analysis and design.

Don Kelly, Ph.D., is a professor in the School of Mechanical and Manufacturing Engineering at the University of New South Wales. He has been a principal researcher and program leader in the Cooperative Research Centre for Advanced Composite Structures for 12 years, specializing in analysis and design for composite structures. He is known for his work in the development and application of finite element methodology and has published over 150 papers on finite element analysis, structural analysis, and design.

Robert E. Melchers, Ph.D., obtained his doctorate from Cambridge University in 1971. For some years afterward he was engaged in consulting and then joined the Department of Civil Engineering at Monash University. In 1986, he was appointed professor of civil engineering at the University of Newcastle, Australia. His main research areas are structural reliability theory applied to complex structures, structural performance, and structural deterioration. He is the author of several books and more than 300 technical papers. He is a fellow of the Australian Academy of Technological Sciences and Engineering.

Campbell Middleton, Ph.D., is a senior lecturer in structural engineering at the University of Cambridge, where he has been since 1989. Before that, he spent nearly ten years in bridge and highway construction and design in Australia and then worked for consulting engineers Arup in London. He holds an M.Sc. from Imperial College, London, and a Ph.D. from Cambridge. His main research interests cover the assessment of concrete bridges. He is consultant to a number of clients in the UK, Europe, Australia, and North America and is involved in the development of bridge codes. He chairs the UK Bridge Owners' Forum, established in 2000 by representatives of all the major national, regional, and private bridge-owning organizations in the UK to identify research needs and priorities for bridge infrastructure.

Torgeir Moan, Dr. Ing., has been a professor of marine structures since 1977. His main research interests are probabilistic modeling of structural load effects and resistance and also reliability and risk analysis, especially of complex marine structures. He has authored more than 250 scientific papers and delivered many invited plenary lectures in large international forums. He has supervised 40 doctoral students. Dr. Moan has contributed to the development of structural design standards in Norway and internationally and has been engaged in accident inquiries. He has been elected member of the three Norwegian academies of engineering, science, and letters (NTVA, DKNVS, and DNVA) and served as vice president of NTVA. He is an elected fellow of the Royal Academy of Engineering in the UK and a fellow in ASCE and International Association for Bridge and Structural Engineering (IABSE).

Dan J. Naus, Ph.D., received his doctorate from the University of Illinois in 1971. He is currently a Distinguished Research and Development Staff Member at the Oak Ridge National Laboratory in Oak Ridge, Tenn., where he has been employed for more than 29 years. His current research interest is managing the aging of nuclear power plant civil structures. He is the author of more than 275 technical papers and a fellow of the American Concrete Institute, ASCE, and International Union of Laboratories and Experts in Construction, Materials, Systems, and Structures.

Jeom Kee Paik, Dr. Eng., is professor of ship structural mechanics, Department of Naval Architecture and Ocean Engineering, at Pusan National University, Korea. He received his undergraduate degree from Pusan National University and his graduate degrees from Osaka University, Japan, all in naval architecture. He has 25 years of experience in teaching and research in the areas of limit state design, ultimate strength, impact mechanics (collision and grounding), age-related structural deterioration models, and structural reliability. He is the author of several books and more than 400 technical papers. He has been chairman of ISSC committees on collision and grounding and on condition assessment of aged ships, as well as convenor of Working Group for International Standards Organization (ISO) code development on strength assessment of ship structures. Dr. Paik is a Fellow and Council member of the Royal Institution of Naval Architects, UK.

Jack Pappin, Ph.D., is director of Arup, based in Hong Kong. After graduating in Australia, he completed a doctorate at Nottingham University, where he investigated granular road base course materials. He joined Arup Geotechnics in London in 1980 and transferred to Hong Kong in 1993. He is mainly involved with analysis and design of geotechnical problems with particular emphasis on basements, deep excavations and foundations, movement prediction, and developing numerical analytical techniques. He is responsible for seismic hazard assessment and geotechnical earthquake engineering in Ove Arup and Partners worldwide.

Murray L. Scott is the Chief Executive Officer of the Cooperative Research Centre for Advanced Composite Structures Ltd., which has its Head Office in Melbourne,

Australia. He has over 25 years experience in the field of aerospace engineering and has worked in both industry and academia, mainly in Australia, but also in Germany, Netherlands, UK, and the United States. He has made major contributions to the development of advanced composite materials and structures, particularly over the past 15 years. He is a Fellow of the Royal Aeronautical Society and the Institution of Engineers, Australia, and is an Adjunct Professor of RMIT University. From 2001 to 2003, he was the President of the International Committee on Composite Materials. In his current position, he is responsible for the leadership and management of 100 researchers across 20 organizations working in the field of design and manufacture of advanced fiber composite structures.

Richard Sturt is a director of Arup and the technical leader of the Midlands (UK) office of the Advanced Technology Group. He specializes in automotive computer-aided engineering analysis, including crashworthiness, occupant protection, stiffness, and dynamics. He plays a key role in the development of new analysis methods and software within his group and has contributed to development of tools in related areas, such as earthquake response of buildings, noise from railway viaducts, and train tunnel aerodynamics. He is a chartered mechanical engineer and a member of the Institute of Mechanical Engineers. Mr. Sturt graduated from Cambridge University in 1985 and is the author of numerous technical articles and conference papers.

Rodney Thomson, Ph.D., is a Senior Research Engineer at the Cooperative Research Centre for Advanced Composite Structures, based in Melbourne, Australia, responsible for analysis and design research activities. He has 15 years experience in the analysis and design of composite structures for aerospace, maritime and land transport applications. His research interests include computer-aided design, structural optimization, impact and crash simulation, and the development of improved methods for predicting failure of composite materials. He leads an ongoing international research project on modeling of defects in composite structures, and is actively involved in European Union 6th Framework Programs, developing new analysis tools that incorporate structural degradation. He has over 40 publications in the field of composite materials.

David Vesey is a director of Arup, based in the firm's London office, where he is responsible for building projects, carrying out design reviews, and mentoring and training young engineers. He has been responsible for structural analysis and design on several major tall buildings, in both steel and reinforced concrete. He conducted appraisals of a number of significant structures under repair or refurbishment and acted as expert witness on several occasions. Mr. Vesey maintains a special interest in monumental sculptures and railway infrastructure projects, as well as tall buildings. He is an examiner for the Institution of Structural Engineers Mutual Recognition assessment with the National Administration Board of Structural Engineers Registration (NABSER), China, and for the Chartered and Associate members exams.

Structural Modeling: An Overview

Robert E. Melchers and Richard Hough

1.1 Complexity in Structures

This book is concerned with the mathematical modeling and analysis of "complex structures." We will consider a structure to be complex when its analysis calls for modeling that is beyond the ordinary in terms of the required accuracy of representation, of geometry and connectivity, of material behavior, of loading environment, or of some combination of these. We will be concerned with the analysis of such structures for design purposes or for the assessment of the remaining life of existing structures.

This definition will lead into consideration of:

1. complexity of geometric representation, as might be indicated by a high number of components or elements and by a high level of statical indeterminacy (or, for some structures, kinematic freedom),
2. material and/or behavioral properties, which require detailed characterization through detailed mathematical modeling,
3. detailed representation of localized behavior, such as for high stress intensity regions,
4. accurate capturing of dynamic response or of geometric nonlinearity,
5. accurate modeling of the loading and operating environment,
6. accurate representation of boundary conditions including interactions across boundaries, and
7. refined performance assessment techniques.

Over recent decades, the pursuit of the theory of structures has tended to focus on the first and third items above. Some of the other issues have been left with rather less formal attention, even though they are essential to successful structural engineering and are well-known to those practicing in particular areas. This volume attempts to review the consideration of all these complexity issues, relative to widely varying applications.

To provide a contemporary overview of current developments and likely future trends in the modeling and assessment of complex structures, a selection of structural systems is considered. Widely different structural configurations, material properties, operating environments, and performance requirements are involved, including a number of applications not always thought of in terms of structural engineering. However, they share the same language, methods, and computational tools, and lead to related insights into structural behavior. Inviting comparisons between underlying methods and criteria across a variety of structural systems is a key intention of this book.

We will consider buildings and bridges, cars, ships, aircraft, nuclear and offshore structures, and structures inherent in geotechnics. All of these overlap in the modeling of the underlying structural systems and in the approaches and techniques. This is true also of the representation of environmental influences and of loading. Importantly, it is increasingly recognized that there should be similar confluences in safety and reliability assessment.

Tied up with reliability issues is, of course, the question of how complex, or detailed, an analysis is required. Several of the following chapters address the important question of choosing the level of analysis that is consistent with the overall needs and expectations of accuracy and reliability, and of balancing effort between different aspects of the model. There is often a risk of imbalance towards complexity of geometric representation. Better modeling of loading environments and material behavior, by comparison, requires substantial investment in research to improve our understanding of real-world phenomena. In seismic engineering, for instance, geotechnical engineers point out that our ability to model the nonlinear, dynamic behavior of building superstructures is disproportionate to our ability to model potential seismic signals arriving at the building site. The wide range of analyses considered here allows the reader to obtain a perspective on such strategically important relativities.

1.2 Historical Understanding of Structural Behavior

The remainder of this chapter considers the evolution of structural theory and its application to traditional civil engineering structures such as buildings and bridges. This strand of structural analysis and design is chosen because it has the longest history to consider—in terms of application, it goes back to the beginning of human settlement.

Historically, practical understanding of structural behavior developed quite separately from the theory of structures, and this itself developed separately from methods of calculation. There is a rich legacy of successful structures built entirely without calculations. There is evidence that before the Renaissance, structures were constructed using empirical rule books. These rule books reflected experience gained from successful (and unsuccessful) structures. Structural failure was not uncommon. The rule books essentially provided a theory of structural design, although not of the type now in use. Some survive from the building of medieval cathedrals. There is evidence also of rule books for the proportioning of wood and stone structures already some 600 years BC (Heyman 1996).

The rule books were the forerunners of what became our *codes of practice*—professionally agreed and accepted procedures known from experience to provide, usually, safe and reliable structures. Fortunately, the stresses in the early stone and wood structures were typically low, much below the crushing strength of the material. More important were matters of stability and ensuring that lines of force were contained within the structural form. From this perspective, geometry was an overriding requirement of structural theory. All this changed with the Industrial Revolution and the development of the new tension-capable materials of iron and later steel and with the subsequent introduction of reinforced concrete.

Theory, however, played little part in exploiting the new materials. The sizes of iron beams, columns, and ties were selected by using the results obtained by manufacturers from many tests on full-sized components (Addis 1994). Unfortunately, this does not readily allow extrapolation to new situations and conditions (Sibley and Walker 1977), and eventually it became necessary to base such extrapolation more formally on simplified mathematical models.

Interest in the bending capacity of beams preceded the Industrial Revolution. Galileo's (1638) hypothesis of the fracture of a wooden cantilever embedded in a brick wall by rotation at the support is immortalized in his famous sketch. His analysis contravened the requirements of equilibrium, but its publication led to experiments by the Frenchman Mariotte (1686), and eventually to the correct analysis for elastic beams by Parent (1713) deduced later also by Coulomb (1773). Already at this point there is a difference in emphasis between Galileo's concern with fracture (also shared by Coulomb) and the elastic theory being pursued by others (Heyman 1996). It is the latter theory that came to dominate subsequent thinking about structural behavior—a situation that prevailed until the mid-1900s when it was finally challenged by the plastic theory of structures (Baker et al. 1956).

Historical developments in elastic theory went largely unnoticed by contemporary engineers and builders, however. Design choices were still guided strongly by established "good practice." This consisted of accepted simplifications of structural behavior handed down by experienced engineers and recorded in volumes such as Reynolds and Kent (1936) and Russell and Dowell (1937). These contain a wealth of practical advice, accompanied by comments such as "experience shows." Importantly, the structures they considered were confined mainly to statically determinate systems for which graphical methods of frame analysis were highly suited (e.g., Grinter 1936). For statically indeterminate systems such as reinforced concrete plates, simplified methods were becoming available, sometimes derived from the growing body of theory, but more often the results from rather ad hoc experiments and "experience" were still preferred (Ferguson 1988).

1.3 The Dominance of Elastic Structural Analysis

The desire for greater spans and more economical use of structural and reinforcing steel led to the gradual introduction of statically indeterminate framing systems.

In parallel, the practical necessity for statically indeterminate frames as used for early aircraft spurred the development of a variety of simplified, analogous, and iterative methods, including the famous moment-distribution method of Hardy Cross. Soon there was an explosion of elegant solutions and solution techniques—many for very specific structural forms—with the only important practical constraint being the large number of variables for any but very simple structures (e.g., Matheson 1959; Norris and Wilbur 1960). Many of these developments were aimed at circumventing what was clearly a major obstacle—the large number of degrees of freedom for what we would now call stiffness methods or the similarly large number of redundants for flexibility methods.

Structural theory now had a theoretical basis in mathematical models, both of the structure and of the material(s) from which it was composed. However, it still could not be applied in practice to its fullest extent, despite the deployment of large teams of clerical assistants solving simultaneous equations and later inverting matrices as required for the design of more efficient and lighter designs. This limitation remained until the availability of high-speed computers.

Much attention came to be concentrated on the issue of computing power for increasingly larger elastic models, to the neglect of better models for material behavior and for representation of connectivities of members and of boundary conditions. Structural analysis came to be seen very much as having its respectability in the solution of complex elastic structural systems. Strong literature grew up around such solutions for a wide variety of systems—both an engineering literature and a mathematical one.

Elastic modeling had, and has, many attractions. It is simple, it allows superposition of partial solutions and any number of load cases, and for the relatively low stresses typical of earlier structures, it was also reasonably accurate. It became the currency of codes and the touchstone of practitioners for all kinds of applications. It still accounts for the vast majority of analyses carried out. This is appropriate in most cases, particularly in low-stress statical situations and serviceability analyses and low amplitude dynamic situations. It is generally easily understood and hence less vulnerable to oversight of computational errors and more fundamental errors of modeling.

Despite the attractiveness of elastic theory, research has shown repeatedly that for statically indeterminate structures the results from an elastic analysis usually are not good predictors of actual behavior. In part, this observation can be ascribed to the difficulties of modeling the structure, its details, and its supporting system. There are matters that are generally unknowable and can have a major influence on actual stress states, such as actual connection behavior or foundation movements.

Our stepwise progress away from purely elastic structural analysis began with the development of the theory of plasticity, which refocused on the problem originally of interest to Galileo—that of the fracture or ultimate strength or capacity of a structure. However, the theory of plasticity in its theoretical, analytical form is complex and unsuited to all but very simple problems. The major breakthrough was to simplify it to so-called rigid-ideal plastic theory. In this form it permitted a good first approximation for the prediction of collapse loads for structures composed of materials with a substantial yield plateau in their stress-strain behavior (Baker et al. 1956). However, it does not easily fit with elastic ideas. The transition from modeling structural behavior by elastic theory to modeling it by rigid-plastic theory remains discontinuous even today, although there are links that can be exploited (Melchers 1982).

For buildings, bridges, and large civil structures, it was the so-called ultimate strength rules of design codes that first brought ideas about nonlinear behavior into structural engineering practice. For reinforced concrete, for example, these rules allow the design of reinforcement in cross sections by plastic concepts (as elastic concepts are demonstrably erroneous) while the overall structural behavior is still generally modeled using elastic methods. Structural steel design codes are still restrictive on the use of plastic analysis, typically permitting cross-sectional design by plastic rules but limiting it for frame action.

A more recent trend is to nonlinear elasticity, with the potential to incorporate plasticity ideas as well. The step up to nonlinear analysis, in geometry or material properties, is a big step, with plenty of latent pitfalls. It has been most successful where it has been accompanied by adequate parallel effort on materials research, prototype testing, and data collection on loading and performance in-service, as later chapters indicate. For the moment, these real-world inputs still pose considerable challenges, and the use of structural engineering analysis techniques, elastic or otherwise, still requires plenty of insight and understanding—the "art" of structural engineering. Some of the issues involved are discussed in the next section.

1.4 Structural System Modeling: Materials, Connectivities, Boundary Conditions

High-quality material modeling is warranted in cases where small process and material savings can make big economies. Examples include weight-critical aerospace design, longspan bridge design, and cases where statutory detail design requirements must be met (irrespective of whether such analyses are realistic or not). High-quality modeling is also likely to be beneficial for assessments of existing structures, such as for life extension or forensic purposes, using actual or measured loading information and good quality estimates of boundary and material properties.

For example, for the sway dynamics of a reinforced concrete frame there is no point in simulating the effect of crack growth on member stiffness if no thought has gone into the flexibility of the beam-column intersections themselves. Nor is such detail of much use if, by comparison, there is insufficient information about the dynamic loading itself or the likely structural mass or structural damping or even foundation stiffness. As always, experience and even intuition must play a role in these choices.

Similarly, modeling the structural behavior of a steel structure under a fire scenario requires good quality information about the temperature dependency of steel. It also requires understanding of the progression of fire in the building and the resulting heat load.

Common ways of fabricating and assembling structural elements continue to be the subject of research. This improves understanding of the real behavior of different joint configurations, such as glued or cast, welded or fabricated, and under loading that is moderate or extreme, static or cyclic, fast or slow. There has been steady progression in our ability to model connection behavior, from traditional assumptions about pinned and fixed connections, to load capacity and load-deformation properties for particular forms of connection. Typically, such behavior is nonlinear. This will affect the performance of the structure and, perhaps more critically, its buckling or instability response.

The appropriate modeling of boundary conditions and their adequate calibration against real-world data continues to present challenges, particularly at soil-structure interfaces. Examples include slabs on soft ground, concrete arch-dams, and seismic response of structures.

Finally, we note that structural and material modeling is not usually a one-off effort. It is typical in design for the structural model to develop from very simple concepts, gradually maturing as the design develops. Typically there will be many iterations of progressively more detailed evaluations of a design concept. For example, modeling of dynamic response for a high-rise framed building might start with a single stick cantilever model, with judicious choice of interstory flexibility and mass distribution. When this indicates that the response is likely to be satisfactory, more detailed modeling will follow. This then requires sufficiently detailed information about the actual layout of the building and its proposed mass and stiffness distribution, perhaps as governed by architectural and other decisions.

1.5 Loading

For a long time understanding and appropriate representation of loads for structural analysis appeared to lag somewhat behind the power and capacity of analysis techniques. The pioneers of structural engineering, however, were as much concerned with loadings as with material strength and structural analysis. For example, actual weight measurements of groups of people were used to define a reasonable live load. Modern work on live loads has used essentially similar approaches and this has developed, in more recent times, into probabilistic models for live loads (e.g., Mitchell and Woodgate 1971; Chalk and Corotis 1980). Similarly, measurements made of the weights and impacts exerted by locomotives in the early days of railroad engineering and by transport vehicles for road bridges have led to much better understanding and modeling of loads for design purposes.

Evaluation of environmental loads such as wind, hurricane, snow, and earthquake loads was and is more difficult. In the Tay Bridge disaster of 1879 for example, there was much discussion about what should have been an appropriate lateral pressure to represent wind loading. Again, this simple level of wind modeling is now known to be simplistic. Deeper understanding of the mechanics of wind and the pressures exerted has indicated that a probabilistic framework is required for useful representation of wind load (Simiu and Scanlan 1978). Local vortex and eddy effects can lead to very high local wind pressures or suctions. Complex interplay between the exciting force system and the structural response is the focus of modeling aeroelastic admittance, affecting the design of large lightweight roofs and tall tower structures.

With the development of limit state design codes since the 1960s, loads came to be widely recognized as uncertain quantities, representable by probability distributions. For practical design, however, (traditional) nominal design values are still used but it is now recognized that these values have a finite and definable probability of being exceeded some (small) percentage of the time. More generally, loads need representing as stochastic quantities, varying in an uncertain but definable manner with time. This allows, for complex structures, dynamic analysis of likely behavior under extreme loads, such as frequency domain analysis or, with more effort, time-history analysis.

One area of great interest is the proper combination of loads. Theoretically it is a matter of the probabilities attached to the occurrence of various loads and the probability of their simultaneous occurrence at high levels (e.g., Melchers 1999). The basic ideas, however, have been intuitive for a long time. Thus Russell and Dowell (1937) noted that the permissible stresses could be reduced by a factor of 25% when "the stresses are induced by occasional wind forces only." They noted also moreover that snow loading could be neglected if one had already designed for high wind loads. It is unlikely that any modern design code would allow this. But their basic factor of safety was 4 reduced to 3 under wind—much higher than the equivalent factors implicit in modern design codes.

Developments in load definition could not have been achieved without a great deal of effort by often highly specialized groups, such as seismologists, atmospheric physicists, and hydraulic specialists. Although this has led to structural design process for major structures becoming more complex, the counterbalance is the availability of more effective software for data management.

1.6 Safety

Associated with structural analysis is the issue of performance evaluation. It is particularly critical for the design of unusual (often complex) structures when conventional code rules are not necessarily valid and for the assessment of existing structures for remaining life. In both cases, structural safety and performance are likely to be important, and economic considerations are usually also significant. Such assessment requires high-quality tools for structural analysis but also sound tools for making decisions about safety and performance. Such tools have not always been available.

What is at issue is the relationship between what the engineer estimates will be required of the structure and the ability of the structure to meet those expectations. This applies both to structures yet to be built (as in design) and to existing structures (as in assessment). The first uncertainty is knowing what is actually required over the future (or remaining) life of the structure. To meet this uncertainty, it has been traditional to apply a factor of safety to the stresses predicted under the adopted loading scenarios. Early factors were 4 or 5 or even greater. Choosing appropriate factors has vexed designers over a long period of time, as reported, for example, by Mayer

(1926), Pugsley (1955), and Julian (1957). These deliberations led eventually to what is now known as the theory of structural reliability that now underpins partial factor code formats for design codes in many countries. In principle, it is more rational than the earlier factors of safety, even if it is more complex to apply in practice.

Fundamentally, the same applies to the assessment of existing structures for remaining safe life. While cases of gross structural deterioration are easily handled, structures that appear sound but do not satisfy current design rules for safety are much more difficult. Probability methods can assist here also, using modern structural reliability theory and economic decision methods. These require good quality modeling of the structure, good quality representation of its properties and of the loading cases, and good quality estimates of the failure probabilities for the various failure modes. The estimated failure probabilities can then be combined with expected costs of failures or other consequences to determine appropriate courses of action (Faber 2000).

The combination of structural reliability theory with modern structural analysis tools can result in a very complex and highly demanding computational problem. This is illustrated in chapters 8, 9, and 10. Several simplifying approaches can be taken. All involve some degree of compromise in the estimated failure probability and in setting design factors and construction requirements for safety assurance. The bases for these approaches are reviewed briefly in chapter 10.

In principle, this brings together the various strands addressed in the other chapters: all engineering design sectors need to deal with the reliability of modeling and the consequences that follow. In each sector the choices can have significant social and economic effects.

1.7 Conclusion

The major challenge for modern designers at the cutting edge of competitive structural design is the adequate representation of the structure and its behavior under foreseeable loading and environmental scenarios. Simple adherence to design codes is unlikely to be sufficient. Nor are simple representations, with conservative assumptions, likely to be competitive. They are also unlikely to be defendable should structural failure occur.

In the chapters that follow, engineering experts from a range of specialist areas describe the approaches used in their field for representation of structures for analysis, either for initial design purposes or for reanalysis. The structures vary considerably in type, scale, and complexity and are subject to quite varied environmental and loading conditions. In all cases it is clear that there must be an intuitive understanding of structural behavior prior to representation in a mathematical model. Only then can the model represent reality in a sufficiently accurate, logical, and coherent manner.

References

Addis, W. (1994). *The art of the structural engineer*, Ellipsis, London.

Baker, J. F., Heyman, J., and Horne, M. (1956). *The steel skeleton*, Cambridge University Press, Cambridge, England.

Chalk, P. L., and Corotis, R. B. (1980). "Probability model for design live loads." *J. Struct. Engrg.*, 106(ST10), 2107–2033.

Coulomb, C. A. (1773). Essai sur une application des règles de *maximis & minimis* à quelques problèmes de statique, relatifs à l'architecture [Essay on an application of the maximum and minimum rules for a problem in statics in architecture], *Mémoires de Mathématique & de Physique, présentés à l'Académie Royale des Sciences par divers Savans, & lûs dans ses Assemblés*, 7, 343–382, Paris, 1776.

Faber, M. H. (2000). "Reliability based assessment of existing structures." *Progress in Struct. Engrg. and Mech.*, 2(2), 247–253.

Ferguson, P. M. (1988). *Reinforced concrete fundamentals*, 5th Ed., Wiley, New York.

Galileo, G. (1638). *Discorsi e Dimostrazioni Matematiche, intorno à due nuove scienze Attenenti alla Mecanica & i Movimenti Locali* [Dialogues concerning two new sciences], Louis Elsevier, Leiden. Facsimile reproduction, Brussels, 1966. Translated by Henry Crews and Alfonso de Salvio, Northwestern University Press, 1914.

Grinter, L. E. (1936). *Theory of modern steel structures*, Vol. 1, *Statically determinate structures*, MacMillan, New York.

Heyman, J. (1996). *Elements of the theory of structures*, Cambridge University Press, Cambridge, Mass.

Julian, O. G. (1957). "Synopsis of the first progress report of committee on safety factors." *J. Struct. Engrg.*, 83(ST4), 1316.1–1316.22.

Mariotte, E. (1686). *Traité du Mouvement des Eaux et des autres Corps Fluides* [Treatise on the movement of water and other fluid bodies], E. Michallet, Paris.

Matheson, L. (1959). *Hyperstatic structures*, Butterworth's, London.

Mayer, H. (1926). *Die Sicherheit der Bauwerke* [The safety of structures], Springer, New York, Berlin.

Melchers, R. E. (1982). "Deflection of statically indeterminate structures." *Int. J. Mech. Sci.*, 24(6), 341–347.

Melchers, R. E. (1999). *Structural reliability analysis and prediction*, 2nd Ed., Wiley, Chichester.

Mitchell, G. R., and Woodgate, R. W. (1971). "Floor loadings in office buildings—the results of a survey," *Rep. CP 3/71*, Building Research Station, Garston, Herts., UK.

Norris, C. H., and Wilbur, J. B. (1960). *Elementary structural analysis*, 2nd Ed., McGraw-Hill, New York.

Parent, A. (1713). *Essais et recherches de Mathématique et de Physique* [Essays on researches in mathematics and physics], J. de Nully, Paris.

Pugsley, A. C. (1955). "Report on structural safety." *The Struct. Engr.*, 33(5), 141–149.

Reynolds, T. J., and Kent, L. E. (1936). *Structural steelwork for building and architectural students*, The English Universities Press Ltd., London.

Russell, P., and Dowell, G. (1937). *Competitive design for steel structures*, Chapman & Hall, London.

Sibley, P. G., and Walker, A. C. (1977). "Structural accidents and their causes." *Proc. Inst. Civil Engrs., Part 1*, The Institution of Civil Engineers, London, 191–208.

Simiu, E., and Scanlan, R. U. (1978). *Wind effects on structures*, Wiley, New York.

2

High-Rise Buildings

Paul Cross, David Vesey, and Chun-Man Chan

2.1 Analysis Methods

This chapter considers the conceptual analysis and analytical techniques for high-rise buildings. It also describes analysis techniques used to optimize the structure and available floor space in tall buildings. The conceptual analysis and optimization methods are illustrated with case studies.

Analysis of a high-rise building is performed to check the response of the building to vertical and lateral loads. The prime concern is typically lateral analysis for wind and seismic loading. While there are a number of software packages designed specifically to analyze multistory buildings, this chapter will concentrate on methods that are appropriate for use with most general structural analysis software, in order to demonstrate and discuss techniques from first principles.

The analysis process involves a series of assumptions and approximations, both in the modeling of the building structure and the estimation and application of the loads. For example, it is common to use linear elastic software for analysis that assumes that the stiffness of the structure is based on its undeformed shape and that the materials behave elastically throughout the range of applied loading. With this type of analysis, the user can put different types of loading into separate load cases and subsequently factor and superimpose them to produce combinations and envelopes. For extreme loading under seismic conditions, nonlinear analysis may become appropriate, and this is discussed later.

For linear analysis, it is important to be aware of the underlying approximations and when they may not be appropriate. For example, while most design codes allow for a measure of second-order effects in their load factors, P-Delta effects (Fig. 2-1) can be significant in certain tall building structures.

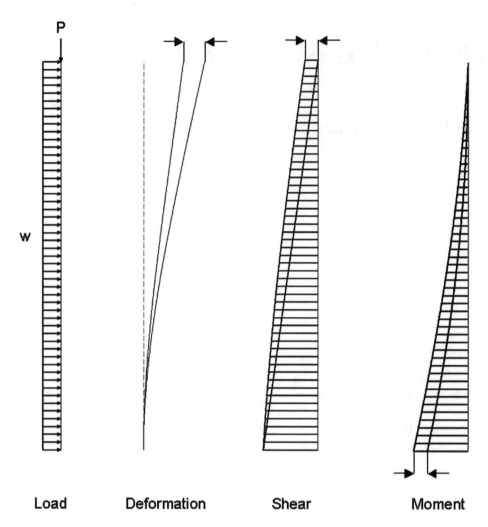

| Load | Deformation | Shear | Moment |

Figure 2-1. The P-Delta effect. Large displacements modify structural stiffness and generate additional forces and displacements.

When modeling the structure, it is important to avoid unnecessary complexity. For example, for a lateral-stability model, it is efficient to establish clearly the elements that form the building stability system and exclude those that should be designed to carry only gravity loads. If gravity beams and columns are included and modeled with moment connections, they will attract a small proportion of the lateral load and reduce that attracted by the real stability system. This is appropriate if the intention is to construct the connections that way, but they will of course be more expensive, and any future refurbishment of the building may be compromised as no beam or column that contributed to stability could easily be removed.

2.2 Appropriate Stability Systems for Tall Buildings

Stability systems for all tall buildings can be considered as vertical cantilevers, with both bending and shear flexibility and fixed at their bases (Fig. 2-2). Some examples of appropriate stability systems are illustrated in Figure 2-3.

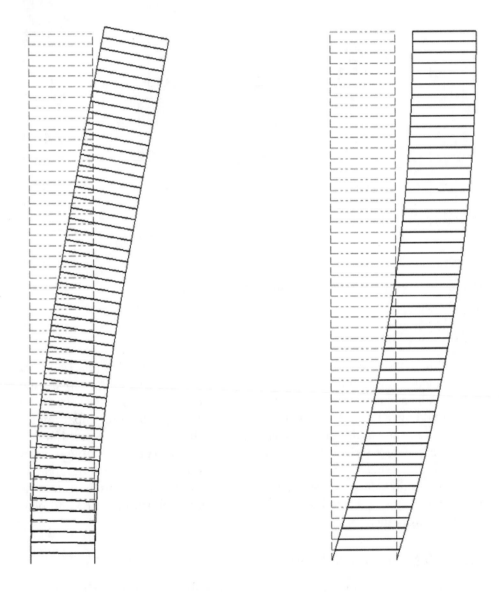

Bending   **Shear**

Figure 2-2. Deformed shapes of vertical cantilevers due to shear and bending.

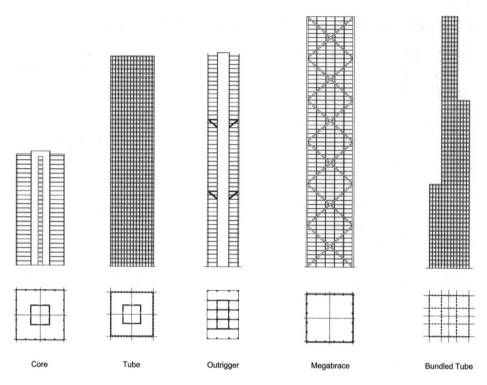

| Core | Tube | Outrigger | Megabrace | Bundled Tube |

Figure 2-3. Appropriate stability systems for tall buildings.

2.2.1 Core Structure

Tall buildings typically have one or more service cores housing elevators, staircases, toilets, plant rooms, and services risers. Cores have external and internal walls that can be concrete or braced steelwork, are commonly situated near the center of the building, and can resist lateral loads while having minimum impact on the architecture of the building. For taller buildings, the cores may not practically be capable of providing adequate strength and/or stiffness, so other solutions must be explored. Even when core-braced stability is appropriate, some architectural and planning requirements can create patterns of core wall penetrations, transitions in wall elements, or wall terminations such that quite sophisticated shell and plate analysis is still called for.

2.2.2 Outrigger-Braced Core

Some of the bending moment generated by lateral loads can be resisted as axial forces in external columns if they are connected to the core, at one or more levels, by outrigger beams. These are typically steel trusses, one or two stories deep, and are located at plant or lift transfer levels to minimize their impact on the architecture. They can reduce the bending moment on the core, and hence the deflection, by up to 50%. The core must still resist all of the shear. Additional columns can be mobilized

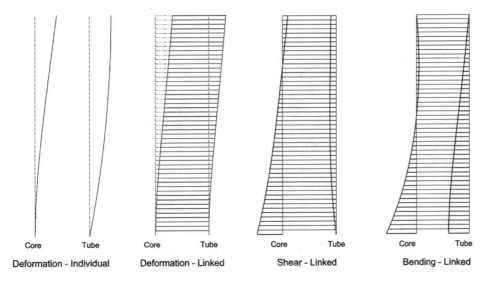

Core	Tube	Core	Tube	Core	Tube	Core	Tube
Deformation - Individual		Deformation - Linked		Shear - Linked		Bending - Linked	

Figure 2-4. Deformations and force distributions for a linked core and external tube system.

via external belt trusses allowing the number of outrigger trusses to be minimized. Outrigger systems are particular suitable for rectangular buildings where wind acting on the larger face of the building has to be resisted by minor-axis bending on the core. In this case, outriggers will typically only be required to assist the minor axis of the core, as bending about the major axis can be resisted by the core alone.

2.2.3 External Tube

The external columns can be connected to form a braced frame or a moment frame, which will resist a proportion of both bending moment and shear on the building. If we consider the external tube and the core as vertical cantilevers, the plan dimensions of the tube are significantly larger than the core, and the tube will have a relatively large bending stiffness. However, as it has much larger perforations (windows), it will be soft in shear. If we consider the deformed shapes of core and tube acting alone (Fig. 2-4), we can see that the tube will displace more at lower levels, so when core and tube are connected, the core will resist a larger proportion of the shear at lower levels than at the top. The two primary elements of core and tube can be tuned to work together with maximum efficiency.

Some other lateral load resisting systems are also illustrated in Figure 2-3.

2.3 Types of Analysis

A number of different types of analysis may be required on a tall building. A different model may be required for *serviceability* than that to assess *strength*. The serviceability model will be based on a best approximation of stiffness and in-service loading and may include nonstructural elements, whereas a more conservative approach is appropriate when assessing strength.

A *stability* analysis normally would only include the elements resisting lateral loads, but if a *gravity* strength analysis is also required to check nonlinear effects, it may be appropriate to enlarge the model.

A lateral *dynamic* analysis needs to model the total building mass, as the stability system must be designed to resist all forces arising from horizontal accelerations. A *static* gravity analysis need only consider mass directly supported by the stability system. However the total building mass must be included in the model for P-Delta and notional horizontal analyses as all destabilizing lateral loads must be resisted by the stability system.

A *modal* dynamic analysis will predict the natural modes of vibration of the structure, which can then be fed into a wind or seismic *response* analysis. Time-history modeling is appropriate for some seismic analysis, as discussed later.

In addition to overall models, which consider the behavior of the whole system, a number of detailed *substructure* analyses, which look in detail at components of the system, may be appropriate. Some examples of these are described later in this chapter.

2.4 General Modeling Techniques

2.4.1 2D or 3D?

It will normally be necessary to use a 3D model for overall analysis of tall buildings as the design of the stability system will often be governed by diagonal lateral loading and horizontal redistribution of vertical loads.

2.4.2 Rigid Diaphragms

For strength analyses, it is common to model only the significant stiffness components of the elements. For example, in a fully braced system only axial stiffness may be modeled, and in a moment frame it is common to neglect the minor bending and shear stiffness of steel sections and concrete walls. A rigid diaphragm is normally included at each floor level to maintain the plan geometry of the building and eliminate the need for minor-axis stiffness in the beams. In practice, the floors have neither infinite stiffness nor strength so care must be exercised when using this technique. For example, any element in the plane of the diaphragm will not deform axially, so if it forms part of a braced system, the stiffness of that system will be overestimated and the element forces will not be calculated (Fig. 2-5).

2.4.3 Building Mass

The total mass of each floor can be modeled with a single lumped mass connected to the rigid diaphragm and located at the center of mass of the floor. The I_{ZZ} mass term ($\Sigma m(x^2 + y^2)$) must be included if torsional modes are required in a dynamic

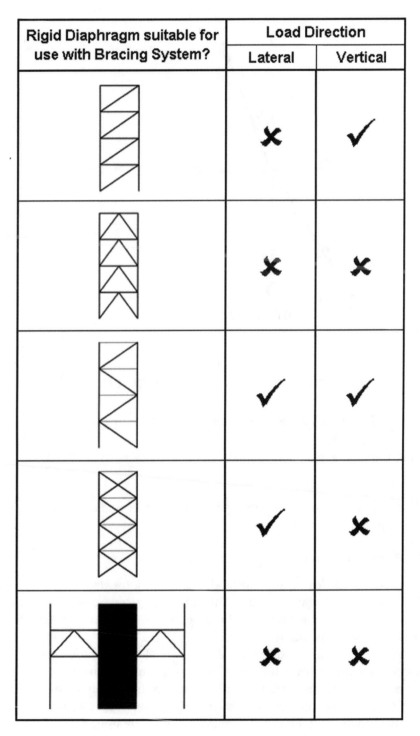

Rigid Diaphragm suitable for use with Bracing System?	Load Direction	
	Lateral	Vertical
	✗	✓
	✗	✗
	✓	✓
	✓	✗
	✗	✗

Figure 2-5. The suitability of rigid diaphragm modeling in conjunction with different types of bracing.

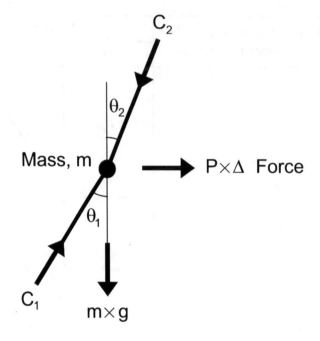

$$C_1 \times \cos \theta_1 = C_2 \times \cos \theta_2 + m \times g$$

$$P \times \Delta \ \text{Force} = C_1 \times \cos \theta_1 - C_2 \times \cos \theta_2$$

Figure 2-6. Mobilization of mass supported by nonstability elements in a P-Delta analysis.

analysis. The masses must be supported vertically either by nodal restraints or, for a P-Delta analysis, vertical struts pinned at each floor (Fig. 2-6).

2.4.4 Stiffness Corrections

It is common to use skeletal center line models based on gross elastic section properties when analyzing stability structures. This is normally satisfactory for estimating the force distribution in the model, but in some circumstances a more accurate estimate of element stiffness will be required.

For example, the stiffness of reinforced concrete elements will be increased by the presence of reinforcement and reduced by cracking of the concrete. The effect of cracking on section bending stiffness will depend on the magnitude of axial force and bending moment in the element, which may vary continuously along its length. Typically, the bending stiffness of beams may be factored by around 0.5 and the axial and bending stiffness of columns by 1.1.

Element stiffness may also be adjusted to allow for the finite size of the beam to wall/column joints. This is discussed in more detail in the case studies presented later.

It is, however, important to remember that all models will only be approximations of reality, and that for strength design any distribution of forces in equilibrium will suffice, provided the structure possesses the required level of ductility.

2.4.5 Foundations

While it is common simply to model encastre supports at the top of foundations, it is making the assumption that the stiffness of the foundations is infinite when compared to the stiffness of the structure. Differential settlements and rotations may redistribute forces in the frame, and overall foundation rotation from lateral loading will increase the horizontal displacement of the building and hence the magnitude of P-Delta effects. Sometimes it may be appropriate to model a horizontal restraint at ground level if the basement forms a stiff box, but this should not be done unless the stiffness and integrity of the load path can be ensured.

As the stiffness of foundations cannot accurately be predicted, these effects should be considered using both upper- and lower-bound estimates of each parameter.

2.4.6 Construction Sequence

When analyzing for the effects of vertical loads, it is often important to account for the sequence of construction. Generally, live loading can be applied to the whole structure but in some cases a temporary condition may arise that is more critical than the final state. For ductile structures, forces locked in by temporary conditions can be ignored for strength design as they can be redistributed at the limit state.

It is important to consider the effects of differential shortening of the vertical elements in a tall building as they can influence the tolerances required for false-floor and ceiling zones. These calculations can be particularly complex where the structure has a mixture of steel and concrete elements. While the steel will shorten elastically, the concrete movements will be time dependent due to creep and shrinkage effects. The control of differential shortening is particularly important for outrigger-braced core structures (described above), as there will be a tendency to redistribute gravity loads between core and columns.

2.5 Case Study 1: Concrete Core Analysis

This section will look in more detail at the analysis of a building with stability provided by a central, reinforced concrete core using, as an example, the HSBC UK Headquarters building at Canary Wharf, London. Images of the completed building and during construction are included in Figure 2-7.

2.5.1 Modeling

The *wide column analogy* method of analysis has been adopted to model the 3D-coupled shear wall structural system of the tower core.

Figure 2-7. HSBC UK HQ—Canary Wharf, London. Jump-form construction of core (left) and completed building (right). Photographs courtesy of CentralPhotography.com.

In this method, each wall panel is represented by an individual, vertical, two-node beam element at its centroid. These panels are connected to other panels by a system of horizontal, two-noded beams. Rigid beams, or *stiff-arms,* are modeled between the wall center and edges to ensure that plane sections remain plane for horizontal sections of each panel. Flexible beams representing the rectangular section of lintels span between the wall edges.

In practice, local deformations at the junctions of walls and lintel beams increase the effective span of the beams. This effect is simulated by reducing the bending and shear stiffness of the lintels (Fig. 2-8). In this case, the effective span, Le, was assumed to equal the clear span, L, plus the minimum of the lintel depth and 500 mm. The shear area was then factored by L/Le and the bending inertia by $(L/Le)^3$.

As described earlier, rigid horizontal diaphragms with nodes at the wind and floor mass centroids are included at each floor level.

Plots of the full 3D model are included in Figure 2-9, which shows all the elements skeletally and filled to represent the magnitude of the cross-section size. Some of the shear walls terminate at upper levels as the lower zone lifts terminate.

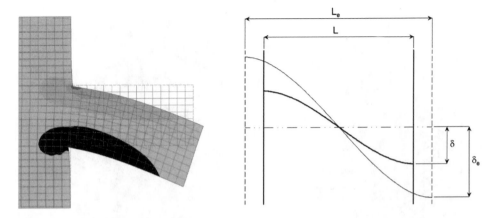

Figure 2-8. Modeling the effective length of spandrels in skeletal shear wall analysis.

Steel beams and columns are introduced to facilitate the application of dead and live loads at these floors. They are modeled as pin-ended and hence do not form part of the stability system.

2.5.2 Data Generation

The analysis data set for the 3D model (Fig. 2-9) was generated by spreadsheet from the single story *chassis* model shown in Figure 2-10.

X and Y coordinates were defined for each *chassis node* and heights for each story. The coordinate data for the full model was generated at all stories for each chassis node so that any point in the model could be defined by its chassis node and story number. Special W and M nodes were used to define the positions of wind and mass centroids whose coordinates could vary at each floor.

Topology, end fixities, and property numbers were defined for each *chassis element*. Property numbers were defined for each story in the building and could be left blank to omit the element at any level.

Loading data was described in terms of chassis node or element and the floor to which it was applied. Figure 2-11 shows a plot of the dead loading applied to an individual floor.

All other data modules were generated from chassis data in a similar way.

2.5.3 Analysis and Results

Static analysis for gravity, wind and notional lateral horizontal loads, and modal dynamic analyses were performed on the 3D model using the Arup program, General Structural Analysis (GSA).

Figure 2-12 shows the wind displacements and interstory drifts generated by characteristic wind loads. The maximum overall displacement ($H/2600$) and interstory

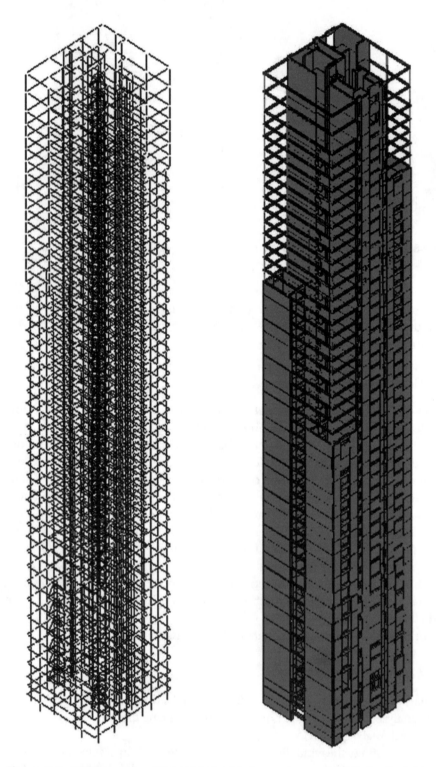

Figure 2-9. HSBC analysis—3D model of central core, represented skeletally (left) and with sections filled (right).

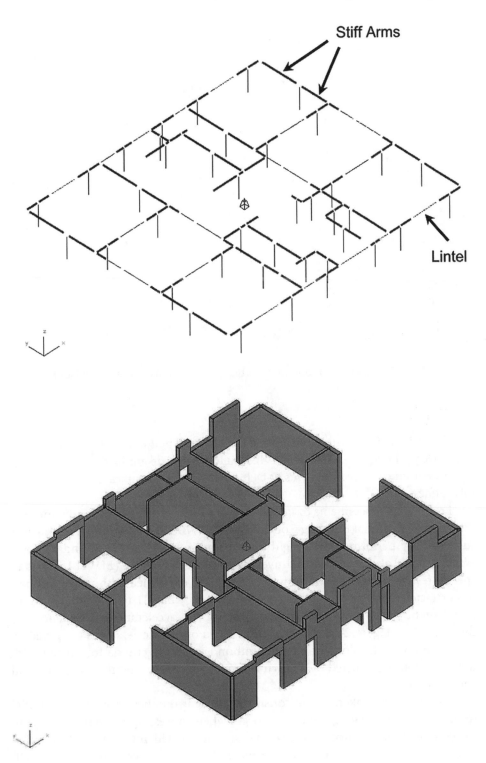

Figure 2-10. HSBC analysis—Illustration of shear wall modeling for a single story.

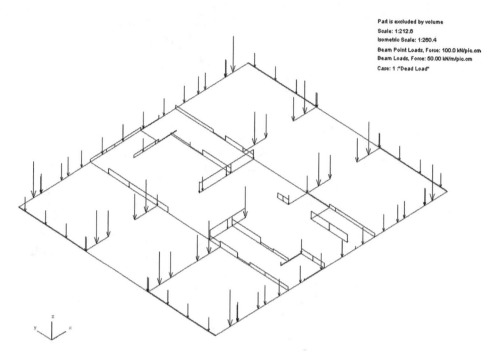

Figure 2-11. HSBC analysis—Diagram of vertical loading applied to a typical floor.

drift (h/2000), are well within the normal limits for tall buildings of H/500 and h/300 respectively. The size of the core on this building was determined by architectural rather than structural requirements. P-Delta effects are not significant on buildings of this stiffness.

Preliminary assessments of lateral accelerations were made using the method described in National Building Code of Canada (NBCC). Wind tunnel testing was subsequently performed on a rigid force balance model at the Boundary Layer Wind Tunnel Laboratory at the University of Western Ontario (BLWTL). The load spectra results from the wind tunnel testing were combined with modal properties extracted from the 3D analysis and processed by BLWTL to give predictions of pseudostatic peak wind forces on the building and accelerations experienced by the occupants. Figure 2-13 shows a comparison of the accelerations predicted by NBCC and BLWLTL plotted against common acceptance criteria for office buildings. The forces predicted by the wind tunnel tests were slightly less than local code requirements.

Advantage was taken of the consistent element numbering in the 3D model resulting from the data generation process when processing the results. Element forces were extracted from the 3D analyses and pasted into a spreadsheet that designed reinforcement for the core walls and beams in accordance with the UK concrete code. Results could be expressed in terms of chassis wall and beam numbers rather than analysis element numbers.

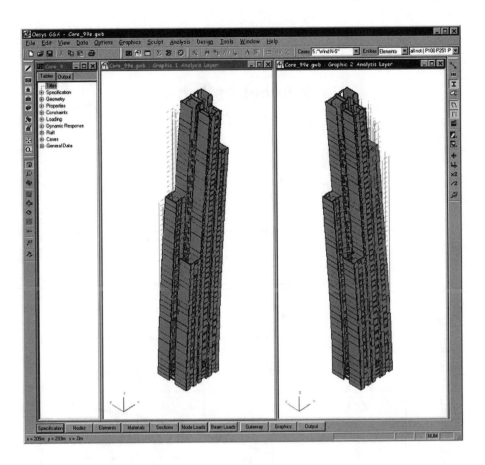

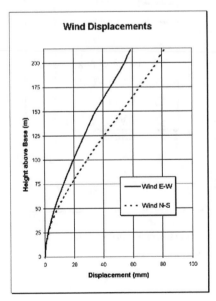

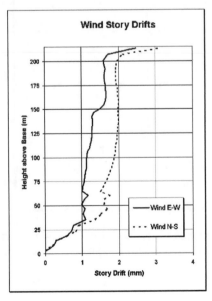

Figure 2-12. HSBC analysis results—Deformed shapes of 3D model (top) and graphs of floor displacements and story drifts (bottom).

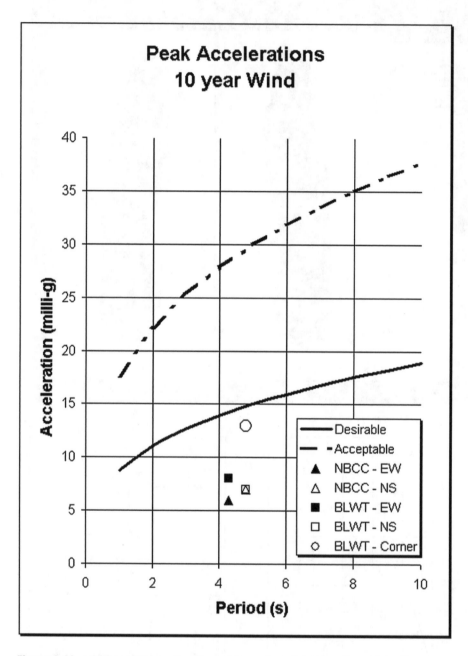

Figure 2-13. HSBC analysis results—Comparison of predicted maximum lateral accelerations with acceptance criteria for office buildings.

2.6 Case Study 2: External Tubes

In this example we will first look at two options for a building on St. Mary Axe in the City of London with stability provided by external tubes. In the first example, which was only taken to scheme stage, the tube was an irregular moment frame. In the second option, which has been constructed, the tube is formed by a cigar-shaped steel diagrid. We will also look at schemes for the London Millennium Tower and the Swiss Re London Headquarters.

2.6.1 3D Modeling

For the St. Mary Axe in the City of London the techniques used to model external tubes formed by a 3D moment frame are essentially the same as those required for 2D frames. As for shear wall structures, rigid diaphragms are normally used to represent the floors. Section properties are often adjusted to take account of the influence of the finite size of the junction between beams and columns on the overall stiffness of the frame. A center-line model will normally be adopted for the overall 3D analysis, with single, two-noded beam elements representing individual beams and columns connected at nodes at the center of each beam/column joint. The required element stiffness correction can be determined by subframes analysis of a single joint as will be shown later. While a 3D model will usually be required to determine individual element forces for final design, simpler models can be used earlier in the design process.

2.6.2 Single Stick Model

Consider the square external moment frame shown in Figure 2-14. We can calculate a bending and shear stiffness for a single stick cantilever model that will approximately represent the behavior of the whole frame. The tube can be represented by using separate effective wall thicknesses for bending and shear. The bending thickness, t_B, at a given point can be found by dividing the area of an individual column by the spacing. The shear thickness, t_S, is a function of the individual bending and shear stiffness of the beams and columns and can be considered as the thickness of a solid wall of equivalent stiffness to a beam/column subframe of the type illustrated. This representation of the tube makes a number of assumptions, including a point of contraflexure at the center of each beam and column.

Plane sections remain plane for bending. No warping restraint is considered for torsion.

2.6.3 Shear Lag

In practice, as the shear stiffness of a tube structure decreases, the effect of *shear lag* increases. The distribution of vertical stress at the base of the tube is not linear, resulting in peak stresses at the corners. However, this effect reduces above the base of the building as can be seen in Figure 2-15, which illustrates the column load

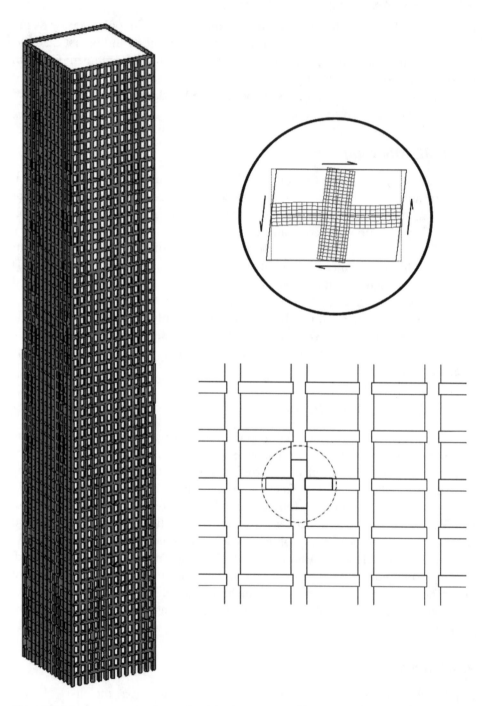

Figure 2-14. An example of an external tube structure illustrating shear panel behavior.

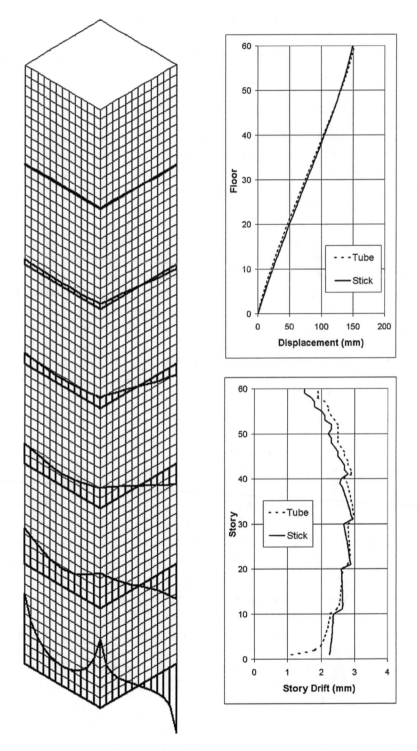

Figure 2-15. Shear lag variation with height in a tube structure (left) and comparison of results from single stick and 3D models (right).

distribution arising from lateral loading on a 3D tube model. Figure 2-15 also compares the deflected shape from a single stick model of a square external tube with that from a detailed 3D skeletal analysis. It can be seen that the results are generally within 5%.

2.6.4 London Millennium Tower

Similar techniques were used in developing the scheme for the proposed 92-story London Millennium Tower in the City of London. Here, lateral stability was provided by an external perforated tube structure around the irregular perimeter of the building. A preliminary 3D model of the external tube is shown in Figure 2-16 alongside an architect's image of the building.

Investigations were carried out for different column shapes, spacing, and beam depths in steel, concrete, and composite construction. In each case, an optimized solution was required. It would not have been practical to perform and interpret this number of analyses using 3D models, so spreadsheets were used to create single stick models from first principles to represent the overall behavior of the tube. The bending and shear thickness, t_B and t_S, are first calculated from the beam and column section properties as described above. Using the thin walled section theory, the Centroid and Principal Inertias (Iu, Iv) can be calculated from t_B, and the Shear Center, Shear Areas (Asu, Asv), and Torsion Stiffness (J) from t_S.

The single stick cantilever model was analyzed by spreadsheet to predict deflections from wind loading using the following procedure:

1. Calculate wind force applied at each level from pressure profile, width, and force coefficient.
2. Integrate loads to give shear force (Q).
3. Integrate shear force to give bending moment (M).
4. Integrate $M/(EI)$ to give rotation.
5. Integrate rotation to give bending displacement.
6. Integrate $Q/(G \times As)$ to give shear displacement.
7. Add shear and bending to give total displacement.

Typically 20% of the maximum overall deflection was due to twist (torsion), with the remainder split between flexure (bending) and sway (shear).

Spreadsheets were also developed to perform modal analysis and predict building accelerations using the NBCC method mentioned earlier.

The bending and shear stresses induced in the stick models were processed by the spreadsheet to give individual column axial forces, and beam and column bending moments. It was found that the element sizes required for strength did not provide adequate stiffness to satisfy the required stiffness and comfort criteria.

2.6.5 Tube Optimization

Studies were therefore required to determine the most economic way of providing the required stiffness.

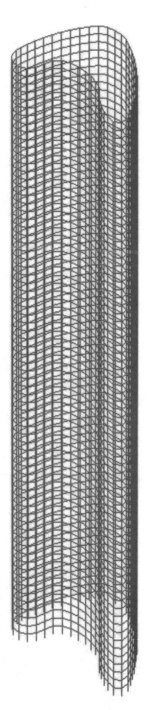

Figure 2-16. London Millennium Tower—Architectural image (left) and 3D analysis model (right). Photograph courtesy of Richard Davies/Foster and Partners.

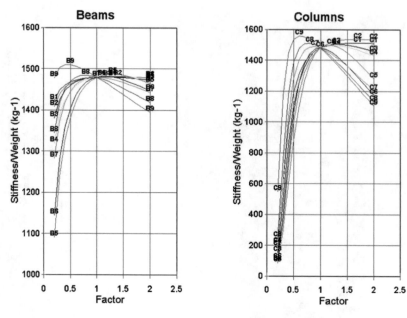

Uniform Member Sizes

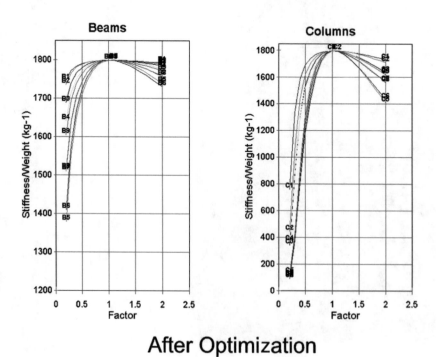

After Optimization

Figure 2-17. London Millennium Tower—Stiffness optimization of external tube structure based on a single stick model.

Figure 2-17 illustrates the results of an optimization study on a concrete scheme with circular columns and rectangular beams on a building split vertically into nine zones. Initially, a constant section is specified throughout the height of the building. Here the depth of the beams is fixed, but the beam width and column diameter is varied independently in each zone.

First, the beam width for zone 1 is factored by 0.2, and a deflection analysis is performed. The overall stiffness to weight ratio is calculated and the result is plotted on the graph. The procedure is repeated for factors up to a maximum value of 2, for the beams in the remaining zones, and by the column diameters in each zone. If the initial arrangement was the optimum, each curve would peak at a ratio of 1, as for each element any increase or reduction in dimension would result in a reduction in efficiency. Not surprisingly, the results indicate that the element sizes should be increased at the lower zones and decreased at the higher ones. Member sizes were adjusted until optimization was achieved, resulting in a concrete saving of 20%.

Similar studies were performed for steel schemes where the section depths and breadths were fixed and the web and flange thickness varied.

2.6.6 Substructure Analysis

Joint studies were performed using simple 2D FE models where the circular columns were represented by elements of varying thickness calculated to give the same area and inertia as the circle (Fig. 2-18). The stiffness predicted by the simple model was checked against that from the detailed 3D model shown, and found to be within 0.5%. Effective thicknesses for use in the single stick models were derived by comparing the lateral deflection from these analyses with that of a solid panel with unit thickness and the same overall dimensions.

Unfortunately, the project was stopped when it was felt that it would not ultimately receive planning permission.

2.6.7 Swiss Re London Headquarters

A new, significantly lower scheme at 180 m high, was then produced for the site. As can be seen from the construction image (Fig. 2-19), this was also based on an external tube structure, but in place of a moment frame there was a steel diagrid frame. The structural system arose from the architectural requirements of the façade, which expressed the position of light-wells that spiral up and around the perimeter of the building.

Before this final scheme evolved, a number of different options were considered. Some of these only used the perimeter to support vertical loading with the lateral loads being resisted by a central core. However, for the diagrid scheme, the structural sizes required to resist gravity loads already has adequate strength and stiffness for wind. The central core structure is therefore not required to contribute to lateral stability and is only braced for local and temporary stability.

With diagonal columns that resist both bending and shear, single stick representation of the tube would not have been simple. Also, stiffness optimization was

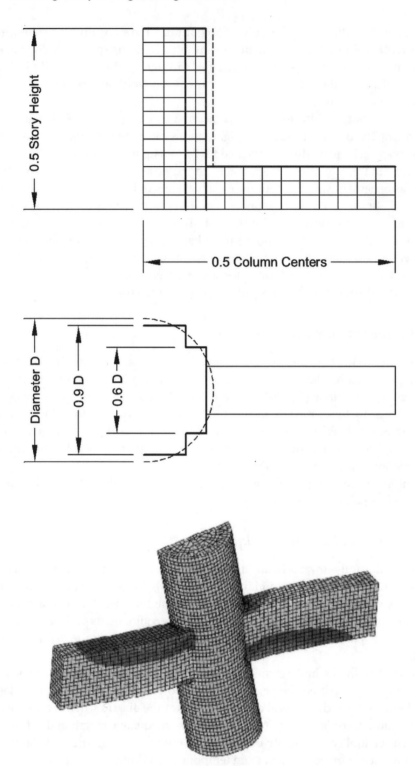

Figure 2-18. London Millennium Tower—2D and 3D analyses to determine the effect of the finite joint on shear stiffness of a typical panel.

Figure 2-19. Swiss Re London HQ under construction.

not required as members were governed by strength alone, so all analysis, static and dynamic, was performed on a 3D model. The diagrid was modeled with bar elements that included only axial stiffness. As significant diaphragm forces are generated by the diagrid, rigid links could not be used in the 3D analysis model of the external structure. Under symmetrical vertical loading, these diaphragm forces could be resisted, with equal efficiency, by either radial members or hoops shown in Figure 2-20. Radials, however, cannot resist circumferential shear at nodes arising from lateral or asymmetrical vertical loading, so hoops are required anyway. Figure 2-21 shows the results from two studies, one with infinitely stiff radials and one without radials. It can be seen that the increase in weight of steel required if the radials are omitted is minimal. It was therefore decided to minimize any reliance on radial action and design the hoops to resist all the loads. The radials were therefore modeled with minimal axial stiffness and used only to distribute lateral loads from a central node at each floor.

Forces required to restrain the Diagonal Columns may be provided by either Hoops or Radials

Hoop Forces are larger

Total Length of Radials is greater

For a given weight of Steel, Strength and Stiffness are identical

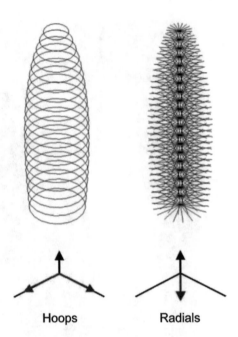

Hoops   Radials

Figure 2-20. Swiss Re—Options for lateral restraint of the diagrid.

Radials or Hoops

Rigid Radials

Hoops resist asymmetrical loading only

No Radials

Contribution of floors ignored

Hoops resist both symmetrical & asymmetrical loading

Hoops increased by 40 tonne

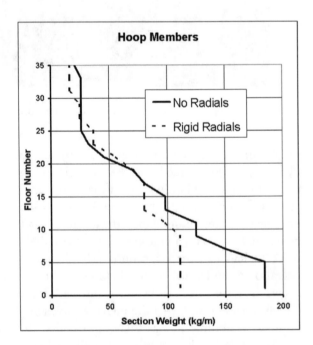

Figure 2-21. Swiss Re—Results of parametric study into hoop and radial diagrid restraint.

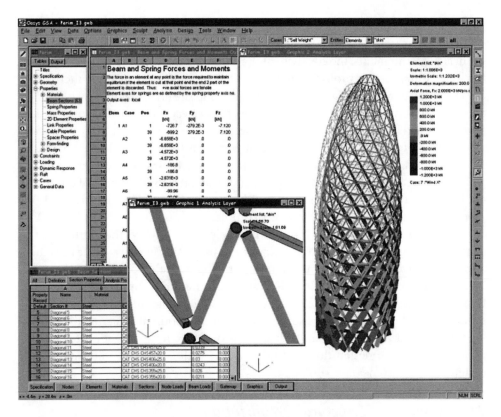

Figure 2-22. Swiss Re 3D Analysis model—Selected input and results.

An example of the results from the analysis is given in Figure 2-22, which shows axial force contours generated from wind loading. The results from individual load cases in the 3D analysis were exported to a design spreadsheet that included bending from intermediate loading.

2.7 A Systematic Optimization Method for High-Rise Buildings

2.7.1 Need for Optimization of High-Rise Structures

The quest for a way of performing a cost-effective design in a short period of time has led to the development of numerical optimization techniques for the structural design of tall buildings. The second half of this chapter presents an effective computational method for optimal element sizing of lateral stability models of high-rise structures.

Tall buildings are large-scale structures requiring enormous financial investment. One major role of the structural engineer is to develop an efficient structural system that is safe for its intended life and serviceable for its intended function. In addition, the design produced must be optimal in the sense of minimum construction cost, maximum structural efficiency, and maximum usable floor space.

Creation of the lateral stability system for a tall building is the first and most challenging task for a structural engineer. It requires a thorough understanding of the design requirements, the creativity of the engineer, and collaborative efforts of all the stakeholders of the project. Typically several preliminary structural alternatives for a building may be proposed at an early stage, and the final system is then decided after a careful evaluation of the cost-effectiveness of each system. This process has been discussed in the earlier part of this chapter.

Once the topology of the lateral stability system for a tall building is defined, the main effort is to size the structural members to satisfy all safety and serviceability performance criteria. Economically sizing of the lateral stability model of a modern tall building is generally not an easy task because such buildings consist of thousands of structural elements and are very complex in nature.

Even with the availability of today's finite element software, the optimal sizing of tall building structures is a rather labor-intensive and iterative process. Generally, an iterative, manual optimization is not sought due to time constraints and design budgetary considerations. It is desirable to have a computer-based resizing technique that automatically selects the most economical elements for the lateral stability model of tall buildings so as to result in the optimal design while satisfying all specified design criteria. With the aid of such an optimization technique, dramatic savings in design time can be achieved with significant improvement in design cost-effectiveness.

2.7.2 Concept of Structural Optimization

Structural optimization is an advanced computer-based technique that replaces the traditional trial-and-error design procedure by a systematic goal-oriented design synthesis process. Given a particular design problem, the main concept of optimization is the capability to seek numerically the optimum solution. The optimum solution here means that an objective function is to be minimized while satisfying all the specified design constraints.

For tall building design, the objective function usually represents the construction cost of the structure expressed in terms of the element sizes as design variables. The constraints may include multiple sets of design performance criteria such as the lateral stiffness requirements of a building, element strength criteria as specified in design codes, and the geometric sizing limitations defined by planning requirements.

The structural optimization methodology is viewed as a basic framework for computerization of the design process. Figure 2-23 illustrates the basic work flow of a structural optimization design process. The design synthesis process involves a coordinated application of structural analysis and design optimization. Given that the optimal design problem is explicitly formulated in terms of a set of design variables, the optimal design solution can be sought numerically by an iterative optimization technique while satisfying all specified design performance and element sizing requirements.

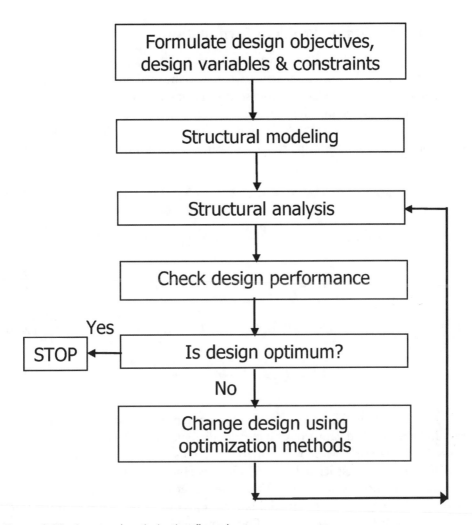

Figure 2-23. Structural optimization flow chart.

The optimization methodology is based on a rigorously derived Optimality Criteria (OC) method that has been shown to be most effective for large systems. The OC approach is an indirect method that first derives the stationary conditions at the optimum and then searches to the final design by applying recursive algorithms to satisfy the necessary optimality conditions. Although the earlier applications of OC were focused more on aerospace structures, recent research has shown that the OC approach is well suited for tall building design (Chan 1997; Chan et al. 1998; Chan 2001). Both 2D and 3D symmetrical and torsionally asymmetrical building structures can be considered under multiple static and dynamics loading conditions. In addition, the optimization technique is applicable to a wide spectrum of types and forms of large-scale tall building structures in which steel, concrete, or composite steel and concrete may be considered as construction materials.

2.8 Optimal Design Problem Formulation

Consider a general tall building structure having $i = 1, 2, .., N_S$ structural steel elements and $i = N_S + 1, \ldots, N$ concrete elements, where N is the total number of structural elements in the building. Taking the cross-sectional areas (A_i) of steel elements as design variables and assuming that the concrete elements have rectangular sections such that the width (B_i) and depth (D_i) are taken as design variables, the minimum structural cost design subject to multiple lateral stiffness constraints can be explicitly expressed as:

$$\text{Minimize:} \quad Cost = \sum_{i=1}^{N_s} (w_i \cdot A_i) + \sum_{i=N_s+1}^{N} (w_i \cdot B_i D_i) \tag{2-1a}$$

$$\text{Subject to:} \quad d_j = \frac{\delta_j - \delta_{j-1}}{h_j} \leq d_j^U \quad (j = 1, 2, .., M) \tag{2-1b}$$

$$A_i^L \leq A_i \leq A_i^U \quad (i = 1, 2, .., N_S) \tag{2-1c}$$

$$B_i^L \leq B_i \leq B_i^U \quad (i = N_S + 1, .., N) \tag{2-1d}$$

$$D_i^L \leq D_i \leq D_i^U \quad (i = N_S + 1, .., N) \tag{2-1e}$$

where w_i represents the cost coefficient for element i. Steel or concrete elements should have different values of unit cost coefficients that reflect the corresponding costs of the materials. Moreover, horizontal beams and vertical columns or walls may have different associated cost coefficients since they have different impacts on the use of floor space. Eq. 2-1b defines a set of multiple lateral drift constraints, where δ_j, δ_{j-1} = differential lateral deflections of two different levels, d_j, d_j^U = lateral drift and its corresponding allowable limit, and h_j = corresponding height between j and $j-1$ levels. Eqs. 2-1c through 2-1e define the sizing constraints where each element is specified to be within its corresponding lower- and upper-size bounds.

To facilitate a numerical solution of the design optimization problem, the implicit drift constraints must be formulated explicitly in terms of design variables A_i, B_i, and D_i. Adopting regressional relationships for linking cross-sectional properties as a function of cross section area A_i for steel sections (Chan 1997) and expressing concrete sectional properties in terms of B_i and D_i, the drift constraints can then be formulated explicitly, using the principal of virtual work, as:

$$d_j(A_i, B_i, D_i) = \sum_{i=1}^{N_s} \left(\frac{e_{ij}}{A_i} + e'_{ij} \right)$$

$$+ \sum_{i=N_s+1}^{N} \left(\frac{e_{0ij}}{B_i D_i} + \frac{e_{1ij}}{B_i D_i^3} + \frac{e_{2ij}}{B_i^3 D_i} \right) \quad (j = 1, 2, .., M) \tag{2-2}$$

where e_{ij} is the virtual strain energy coefficient, and e'_{ij} is the corresponding correction coefficient for steel elements; e_{0ij}, e_{1ij}, e_{2ij} are the respective virtual strain energy coefficients due to axial and shear forces, flexural moments, and torsional moments for concrete elements.

2.9 Optimization Technique

Once the explicit optimal design problem is formulated, a rigorously derived OC method can be employed to provide the solution. This method has been shown to be computationally very efficient for large-scale structures (Chan 1997). In this method, the constrained optimization problem is first transformed into an unconstrained optimization problem of a Lagrangian function that involves the objective function and the explicit drift constraints associated with corresponding Lagrange multipliers. Then, a set of the necessary OC for the optimal design is derived from the stationary conditions of the Lagrangian function. Based on the derived OC, a linear recursive relation can then be expressed respectively to resize the steel variable A_i and concrete design variables B_i and D_i as follows:

$$A_i^{v+1} = A_i^v \cdot \left\{ 1 + \frac{1}{\eta} \left(\sum_{j=1}^{M} \frac{\lambda_j}{w_i} \left(\frac{e_{ij}}{A_i^2} \right) - 1 \right) \right\}_v \text{ for active } A_i \qquad (2\text{-}3a)$$

$$B_i^{v+1} = B_i^v \cdot \left\{ 1 + \frac{1}{\eta} \left(\sum_{j=1}^{M} \frac{\lambda_j}{w_i} \left(\frac{e_{0ij}}{B_i^2 D_i^2} + \frac{e_{1ij}}{B_i^2 D_i^4} + \frac{3e_{2ij}}{B_i^4 D_i^2} \right) - 1 \right) \right\}_v \text{ for active } B_i \quad (2\text{-}3b)$$

$$D_i^{v+1} = D_i^v \cdot \left\{ 1 + \frac{1}{\eta} \left(\sum_{j=1}^{M} \frac{\lambda_j}{w_i} \left(\frac{e_{0ij}}{B_i^2 D_i^2} + \frac{3e_{1ij}}{B_i^2 D_i^4} + \frac{e_{2ij}}{B_i^4 D_i^2} \right) - 1 \right) \right\}_v \text{ for active } D_i \quad (2\text{-}3c)$$

where λ_j denotes the Lagrange multiplier for the corresponding jth drift constraint, v represents the current iteration number, and η is a relaxation parameter that can be adaptively adjusted to control the rate of convergence. Note that the resizing relations Eqs. 2-3a through 2-3c are only applicable for active members having their element sizes within their lower- and upper-size bounds. Any member reaching its size bounds during the recursive iteration is deemed an inactive member and has its size set to its corresponding size limit.

Before Eq. 2-3 can be used to resize the design variables A_i, B_i, and D_i, the Lagrange multipliers λ_j must first be determined. Considering the sensitivity of the

kth drift constraint due to the changes in the design variables, one can derive a set of M simultaneous equations to solve M number of λ_j as:

$$\sum_{j=1}^{M} \lambda_j^v \cdot \left\{ \begin{array}{l} \sum_{i=1}^{NA} \left(\dfrac{e_{ij}}{w_i A_i^3} \right) + \\[2ex] \sum_{i=N_s+1}^{NB} \dfrac{1}{w_i B_i^3 D_i^3} \left(e_{0ij} + \dfrac{e_{1ij}}{D_i^2} + \dfrac{3e_{2ij}}{B_i^2} \right) \cdot \left(e_{0ik} + \dfrac{e_{1ik}}{D_i^2} + \dfrac{3e_{2ik}}{B_i^2} \right) + \\[2ex] \sum_{i=N_s+1}^{ND} \dfrac{1}{w_i B_i^3 D_i^3} \left(e_{0ij} + \dfrac{3e_{1ij}}{D_i^2} + \dfrac{e_{2ij}}{B_i^2} \right) \cdot \left(e_{0ik} + \dfrac{3e_{1ik}}{D_i^2} + \dfrac{e_{2ik}}{B_i^2} \right) \end{array} \right\}_v \quad (2\text{-}4)$$

$$= -\eta \left(d_k^U - d_k^v \right) + \sum_{i=1}^{NA} \left(\dfrac{e_{ik}}{A_i^3} \right)_v + \sum_{i=N_s+1}^{NB} \dfrac{1}{B_i D_i} \left(e_{0ik} + \dfrac{e_{1ik}}{D_i^2} + \dfrac{3e_{2ik}}{B_i^2} \right)_v$$

$$+ \sum_{i=N_s+1}^{ND} \dfrac{1}{B_i D_i} \left(e_{0ik} + \dfrac{3e_{1ik}}{D_i^2} + \dfrac{e_{2ik}}{B_i^2} \right)_v \quad (k = 1,2,...,M)$$

where NA = number of active steel sizing variables, NB, ND = number of respective active width- and depth-sizing variables for concrete elements. Eq. 2-3 for the sizing variables and Eq. 2-4 for the Lagrange multipliers form the basis of the iterative OC method for the solution of the design problem. Having the current design variables A_i^v, B_i^v, and D_i^v, the corresponding λ_j^v values are readily determined by solving the simultaneous equations Eq. 2-4. Having the current values of λ_j^v, the new set of design variables A_i^{v+1}, B_i^{v+1}, and D_i^{v+1} can then be obtained by the respective recursive relations Eqs. 2-3a, 2-3b, and 2-3c. By successively applying the above recursive optimization algorithm until convergence occurs, the continuous optimal solution for the drift optimization is then found.

2.10 Optimization Case Studies

A number of prominent buildings in Hong Kong have been optimized by the optimization technique developed. Two of these projects are selected to illustrate the practicality and effectiveness of the analytical optimization approach for tall building design.

2.10.1 Case 1: Two International Finance Center, Hong Kong Station

At a height of 420 m, the 88-story tower Two International Finance Center (IFC) will be the fifth tallest building in the world when complete. The tower comprises a reinforced concrete core with composite perimeter megacolumns and large steel outriggers at three levels. Figures 2-24 through 2-26 present an exterior view, an elevation view of the lateral stability system, and a typical floor plan of the tower. With an overall aspect ratio of 7.6:1 and the severe wind conditions in Hong Kong, the structural design of the tower is primarily governed by the lateral serviceability stiffness criteria.

Figure 2-24. Two International Finance Center (IFC). Photograph courtesy of Ove Arup & Partners (Hong Kong) Ltd.

In order to investigate the optimal proportions of the structural elements of this hybrid, mixed steel and concrete structure, two key design objectives were defined at the outset of the optimization study. The first objective was to minimize the overall structural cost of the building, while the second objective was to minimize the overall structural cost involving both the material cost and the cost associated with floor area occupied by vertical structural elements. Construction-unit cost coefficients and a unit price of floor area were adopted for various structural elements based on estimated market values at the time of the study.

Both orthogonal and diagonal wind loading conditions obtained from wind tunnel tests were considered in this study. Under each wind condition, the lateral top deflection was limited to $H/400$ and the interstory drift was limited to $h/300$, where H is the overall height of the building above the pile cap level and h is the

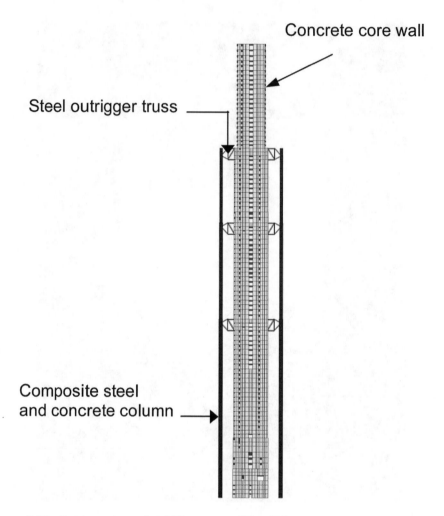

Concrete core wall

Steel outrigger truss

Composite steel and concrete column

Figure 2-25. Outrigger-braced stability system of Two IFC.

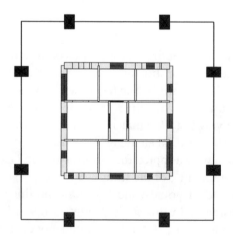

Figure 2-26. Standard floor plan for Two IFC.

story height. It should be noted that overall top drift and interstory drift performance criteria were the major design issues adopted for the purpose of this study at a very early stage in the design. Design considerations addressed subsequently also included the issue of acceleration performance criteria.

For constructability reasons, core wall panels, as well as the megacolumns, on opposite faces were grouped symmetrically to have the same sizes with respect to each perpendicular direction. All elements were to be sized within practical lower- and upper-size bound limits. The lower-size bound limits correspond typically to the minimum strength-based sizes while the upper-size bound limits represent the largest buildable dimensions for structural elements. Along the height of the building, the megacolumns were sized in four different zones. Variations in wall thickness were considered whenever it was deemed appropriate, but were not allowed to have more than one variation for every eight to ten stories. Lintel beams in the core were sized to have their width to be the same as the thickness of adjacently connected walls and their depth to maintain minimum headroom. The outrigger steel members were modeled as truss elements while the concrete lintel beams were modeled as frame elements and the shear walls as shell elements. Each composite megacolumn mobilized by the outriggers is idealized as a pair of two axially jointed truss elements representing respectively the steel and concrete materials involved in the composite megacolumns.

The lateral deflections and interstory drift ratios of the building under the most critical diagonal wind loading condition are shown in Figure 2-27. Initially, some violations have been found in both the overall lateral deflection and interstory drift ratios near the top of the building. In terms of interstory drift ratios, all stories above the 70th level are found to exceed the limit of 1/300, the most critical violation at the 87th level with a drift ratio of 1/234 (i.e., 28.2% violation). For the overall building deflection, an initial violation of 11.5% in the top deflection has been found. At the optimum, both the top deflection and interstory drift constraints are found simultaneously active with their most critical values reaching their allowable limits. In any event, the optimization ensures that the final design must satisfy the specified drift performance requirements for the building.

Figure 2-28 presents the design history of the optimization process for the building. The results of normalized structural costs (with respect to the initial cost) are presented for each design cycle, which includes the numerical computation of one structural reanalysis and one resizing optimization process. Rapid and steady solution convergence is obtained in a few design cycles for the two optimization runs considered in this paper. In other words, with the time required for the order of only a few additional analyses of the structure, the optimal distribution of element sizes of the building has been achieved by the optimization technique. In addition to quick solution convergence, the final design is found to be more economical than the initial design. With the complexity and scale of the 88-story composite steel and concrete building, such results demonstrate clearly the efficiency and effectiveness of the optimal design technique. It is believed that the results could not have been easily achieved by the traditional trial-and-error design method.

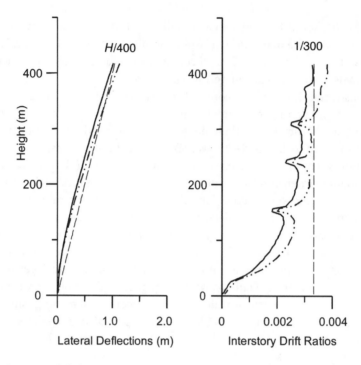

Figure 2-27. Lateral drift response under the most critical wind condition.

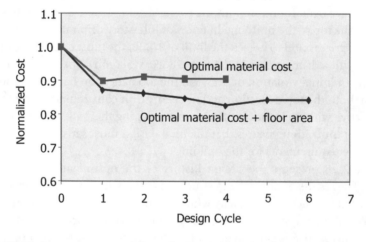

Figure 2-28. Design convergence history for Two International Finance Center.

Table 2-1 presents a summary breakdown of the nondimensionalized cost distribution of various key structural components of the tower for the two optimization runs. When the impact of floor area on the benefits of revenue is considered in the overall cost optimization run, the optimization recommends a substantial increase in the steel materials in the outriggers since the outrigger materials, being located within mechanical and refuge floor zones, do not occupy usable floor space. A close look at the proportions of the composite megacolumns indicates that the steel

Table 2-1. Nondimensional Cost Comparison with Respect to the Total Cost for Run (1)

Elements	Run (1) Minimum Material Cost	Run (2) Minimum Overall Cost	% Difference
Core	0.368	0.364	−1.0
Megacolumns	0.478	0.496	+39.0
Top Outrigger	0.041	0.057	+42.0
Middle Outrigger	0.052	0.074	+3.7
Bottom Outrigger	0.061	0.090	+47.5
Total Material Cost	1.000	1.081	+8.1
Floor Area Occupied	0	−0.172	−17.2
Total Cost	1.00	0.909	−9.1

quantities in these columns are also increased. Although the outcome of increasing the more expensive steel material results in an 8.1% increase in the total material cost for the structure, the savings in floor areas by the reduction of the concrete in the core wall and the composite megacolumns achieve approximately a 9.1% reduction in the total overall cost. The optimization run (2), while taking into account the benefit of usable floor area savings, did identify zones within the tower where the wall thicknesses and the megacolumn sections could be reduced. In fact, much of the usable floor area saving is found on the more valuable upper levels of the building.

2.10.2 Case 2: Harbourside, Kowloon Station Package Four Development

The Harbourside Development comprises three 69-story residential towers linked vertically to form a wall-like building on top of the Kowloon Station of the Airport Railway. Upon completion, the development will stand 240 m above street level and provide a total residential gross floor area of some 129,000 m^2. Figures 2-29 and 2-30 present an artist's impression of the development and the typical floor layouts of the towers.

The lateral load resistance of the three towers is basically provided by a reinforced-concrete-coupled shear wall system. The structural systems for the three towers are essentially identical, except that the two side towers have some minor variations in their end walls. Lintel beams connect the shear walls on each floor level wherever possible to utilize the total dimension of the building to resist wind loads. As shown in Figure 2-29, the three towers are linked together to work as an integrated unit at all levels between the two refuge floors and the twelve levels below the four top duplex floors. All structural shear walls are supported on the transfer plate, which transmits loads from columns and core walls down to ground levels.

The effectiveness of the lateral load resisting system depends on many factors such as the configuration of the structural form, the variable element sizes of the shear walls and beams, the lateral stiffness of the supporting podium structure, and

Figure 2-29. The Harbourside. Photograph courtesy of Ove Arup & Partners (Hong Kong) Ltd.

the various grades of concrete materials for the structural elements. In this case study, the main aim was to determine the optimal element sizes for the lateral stability system necessary to produce a cost-effective structure with maximized usable floor area subject to allowable top drift constraints of $H/500$ and interstory drift constraints of $h/300$ under all wind load combinations.

The effects of 24 wind load combinations derived from the wind tunnel test were analyzed. Under these combined wind conditions, the linked towers were found to sway and twist so that one end of the building deflects more than the opposite end. A close scrutiny of the drift response under the most critical wind load combination indicates that approximately 34% of the top drift value is contributed by the torsional component of the wind load. Such a peculiar lateral drift behavior imposes a greater challenge to the task of optimizing the structure with strongly coupled swaying and torsional effects.

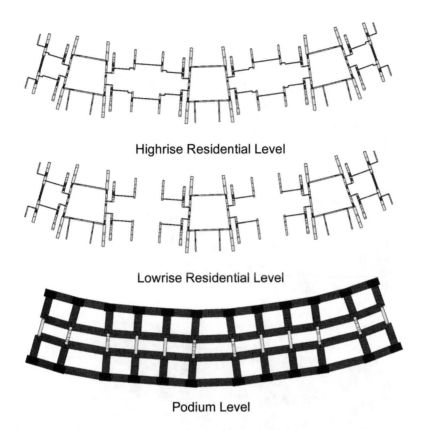

Highrise Residential Level

Lowrise Residential Level

Podium Level

Figure 2-30. Standard floor plans for the Harbourside.

Before the formal analytical optimization is performed, the computer-based optimization technique must be able to identify the cost-efficiency of each structural element that contributes to the lateral stiffness of the building. Figure 2-31 presents graphically the relative cost-efficiency of the structural elements in terms of percentage of virtual strain energy densities. The darker shading represents elements of higher efficiency, whereas the lighter shading represents less efficient elements. Using the rigorously derived OC technique, structural elements can then be resized in such a way that the more efficient elements are increased in size and the less efficient elements are downsized so as to result in minimum cost and maximum usable floor area. The scale of redistribution can be set within parametric limits based on the minimum required dimension for strength and the maximum practical size from aesthetic and spatial consideration.

The major findings of the optimization are highlighted in the following sections.

2.10.2.1 On the podium structure. Stiffening the podium columns has been found to significantly improve the overturning resistance at the lower part of the building where the wind moment is the largest, thus resulting in a maximum reduction in the thickness of the wall elements on the residential levels of the tower structure.

The optimization recommends reducing the transfer plate thickness to its minimum strength-based size because it has been found to be used primarily in

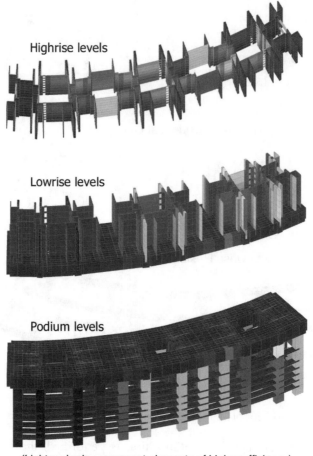

Highrise levels

Lowrise levels

Podium levels

(Lighter shades represent elements of higher efficiency)

Figure 2-31. Identification of efficient structural elements for the Harbourside.

transmitting gravity loads from the towers and it is structurally inefficient in providing lateral resistance to the building.

2.10.2.2 On the tower structure. Structural walls of the lower levels located at the outermost locations of the building plan are found to be the most efficient elements in resisting wind loads and therefore are recommended to be thickened to their maximum allowable dimensions. The thickening of these wall elements causes a reduction in the thickness of the less efficient walls, resulting in a net gain in the usable floor area of flat units.

The back walls and coupling beams on the upper levels are found to be efficient elements to link the three towers and form an effective closed section to better resist wind-induced torsion on the upper levels. From the structural standpoint, improving the torsional rigidity rather than the bending and shear properties of the building on the upper floor levels leads to a more efficient structure.

In order to fully utilize the outermost structural walls to resist wind loads, the major lintel beams around the main core and connecting the end wall are recommended to be widened to their optimum dimensions. For constructability reasons, these beams are, however, not allowed to be wider than two times the thickness of their adjacently aligned walls.

The optimization recommends dividing the structural elements into seven zones to allow for a more refined breakdown of element variations along the building height. This recommendation results in a more gradual transition of wall size from maximum dimensions in the lower zones to minimum strength-based sizes in the upper zones and also allows for more floor area to be saved in the upper levels of the building.

Compared to the initial design, the final design resulted in significant savings in construction materials and usable floor area. In reducing the concrete volume, the quantity of steel reinforcement may also be reduced. For this project, over 10,000 m^3 of concrete and 2,800 tonnes of reinforcing steel (equal to 11% of the total material quantity for the lateral stability system) have been saved and an increased usable floor area of 2,600 m^2 has been achieved. On average, there is approximately 2.5 m^2 increase in usable floor area for each flat unit, which is approximately equivalent to 2.2% increase in the floor efficiency ratio.

2.11 Conclusion

The effectiveness of the computer-based structural optimization technique has already been verified by a number of tall building projects in Hong Kong. In general, savings of 10% to 30% in structural material cost and 1% to 5% increase in usable floor space have been consistently achieved. While it is understood that some savings could be realized through a process of trial-and-error, the formal structural optimization technique provides a systematic, goal-driven approach to achieve the best values of tall building projects at the lowest overall cost.

Acknowledgments

Arup were consulting engineers for the projects described in the case studies. For the case studies in the second half of the chapter, acknowledgments are given to the Hong Kong University of Science and Technology, the Hong Kong Mass Transit Railway Corporation, Sun Hung Kai Properties Ltd., Hang Lung Project Management Ltd., and Ir K K Kwan, Ir Kelvin Lam, Ir Paul Tsang, Ir Alexis Lee, and Dr Craig Gibbons of Arup.

The computer-aided optimization technique presented in this chapter is part of the research program being carried out in the Department of Civil Engineering at the Hong Kong University of Science and Technology. The work was partly funded by the Research Grant Council of Hong Kong under project no. HKUST6226/98E. Special thanks are also due to the Central Waterfront Property Project Management Co. Ltd. (the developer of Two IFC) and the Hang Lung Group Ltd. (the developer of Harbourside) for taking the lead to support the use of the emerging optimization technique in their building developments.

References

Chan, C.-M. (1997). "How to optimise tall steel building frameworks." *Guide to structural optimisation*, ASCE, Reston, Va., 165–196.

Chan, C.-M. (2001). "Optimal lateral stiffness design of tall buildings of mixed steel and concrete construction." *J. Struct. Des. of Tall Buildings*, 10, 155–177.

Chan, C.-M., Gibbons, C., and MacArthur, J. (1998). "Structural optimisation of the North East Tower, Hong Kong Station." *Proc., 5th Int. Conf. on Tall Buildings*, China Translation and Printing Services, Hong Kong, 319–324.

3

Lightweight Longspan Roofs

Richard Hough and Tristram Carfrae

3.1 Roof Types

Some of the most complex and sophisticated man-made structures have been created for the purpose of enclosing large-volume spaces. The enclosing structures need to be as light in weight as possible, not just to reduce the significant costs of supply and erection, but also because of the magnifying effect on material quantities that follows from the fact that the major single load is usually the self-weight of the structural material itself.

Advanced analysis techniques are likely to assist in reducing self-weight, assuming they succeed in representing the actual behavior of the structure more closely, so the material optimization process can be carried out in confidence.

This chapter considers two main categories of longspan roofs, depending on whether the structure relies principally on compression or on tension to deliver its loads to ground. In the first case, analysis issues tend to center on the member design implications of nonlinear effects leading to instability. In the second case, understanding is needed of analytical methods for form-finding and shape-control, and the beneficial effects of nonlinearity in providing improved stiffness in tension systems.

Computational methods are discussed by reference to case studies of some prominent recent examples of both compression-based and tension-based roof systems around the world.

3.2 Compression Roofs

3.2.1 Free Arches and Vaults

The simplest longspan compression roofs are arches or unrestrained vaults, and they are prolific in the history of longspan roofs, in timber, steel, stone, and concrete. The

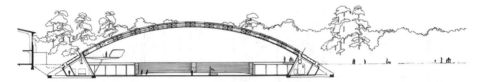

Figure 3-1. Space-frame vault for athletics hall, Frankfurt (project). Illustration by Foster and Partners.

lightweight, arch roof enclosing the world's largest airship hangar is an example of such a vault system, and its analysis, design, and construction are described in Janner et al. (2001).

One of the most elegant and slender space-frame vault designs was by architect Sir Norman Foster, for the Athletics Hall in Frankfurt, Figure 3-1, with a span of 70 m and a ratio of span to arch thickness of 60. The structural geometry comprises diagonals at 60 degrees to the long sides, stabilized by horizontal longitudinal members. The double-layer truss design featured 114 mm diameter tubes for chords that spring from concrete blade abutments tied through the ground slab. To maximize transparency and access through the end glass walls, no bracing structure was proposed in these walls, so they did not participate in stabilizing the roof. Therefore the vault structure behaved as a "free" vault without additional stiffening, and analysis proceeded on the basis of an equivalent two-dimensional arch (Zunz et al. 1985).

As in most iterative design processes, analysis began with simplified models that became more complex as the design converged. Thus nonlinear effects were dealt with initially in an arch-beam model where beam second moment of area was decreased to simulate shear flexibility in the arch-truss, depending on the moment/shear relationship around the arch. The slenderness of the arch led to a first eigenvalue with pinned bases of only 2.6 times the dead plus full snow load, so fixed bases were adopted, raising the factor to 6.0.

To quantify nonlinear effects for member design purposes, and to check overall factors against instability, a segmental beam-arch model was subjected to a series of nonlinear analyses at progressively larger loads until failure to converge was first detected at a load factor of 5.0. Both symmetrical and asymmetrical snow loads were considered, as were full detailed member properties and real rotational and translational spring stiffnesses at abutments. At service load levels, final moments at convergence of the nonlinear analyses were everywhere less than 5% higher than after the first iteration cycle, so nonlinear effects were incorporated globally for member design purposes by adding a 5% margin to linear analyses.

Enveloping nonlinear effects by an overall percent increase in design forces is a simple way of retaining linear analysis as the basic method for design purposes. There are other, more sophisticated and more accurate, ways of retaining a linear analysis basis, which are discussed later. Given that roof structures may require consideration of 50 to 100 different combination load cases, moving from a linear to a nonlinear basis is a major increase in time and effort. It is rarely justified if nonlinear amplification for typical member forces is in the 5% to 20% range. This is often the

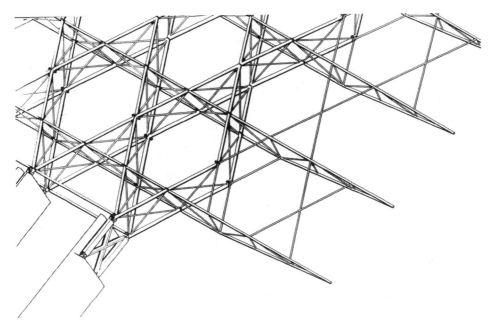

Figure 3-2. Space-frame vault for Athletics Hall, Frankfurt (project), showing member geometry.

case for roof structures, as nonlinearity beyond that implies levels of flexibility that start to transgress more conventional serviceability criteria.

Where nodes are closely spaced in complex roof space frames, as in Figure 3-2, member slenderness can be quite low, and achieving member squash load at ultimate can become a reasonable expectation. A crucial distinction can then be made between equilibrium requirements and forced displacements that generate member forces and moments not required for equilibrium, such as foundation movements and thermal effects. Nonequilibrium effects are then sometimes ignored for design purposes, provided the member and joint strains arising from equilibrium and nonequilibrium effects together do not cause unacceptable local distress like rupture and crack growth or serviceability problems.

Of course, if nonlinear analysis is avoided, it is still important to allow for the force demand placed on those elements that are restraining the more slender and less stable elements, and some approaches are discussed later.

3.2.2 Restrained Arches

If the entire volume beneath the arch or vault does not have to be kept free of structure for functional or even aesthetic reasons, then completely different opportunities open up for controlling the stability of the arch, and also for its analysis.

For the arch-roof over the station at Chur, Switzerland (Fig. 3-3), the bottom tension rod ties the feet of the tubular section steel arch members, and the rest of the tension rod array serves to control the shape of the arch under downward load cases.

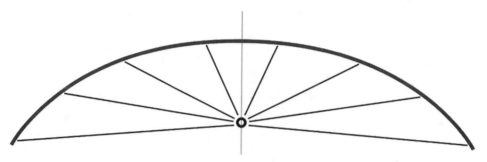

Figure 3-3. Cable-stayed arch for station roof in Chur, Switzerland.

The system performs like a bicycle wheel, though without the prestress opportunity. That is, no inward arch curvature can occur without outward curvature at another point, and that is prevented by the tension restraints.

Nonlinear dynamic relaxation routines deal particularly effectively with the system's behavior using multiple nodes along each tension rod. Distributing the rods' mass between the nodes allows self-weight curvature of the rods to be simulated at different tensions, and so the increasing nonlinearity of the rods' axial stiffness approaching zero tension can also be captured. Arch curvature changes, and hence moments, are sensitive to these effects, but once understood, the system proves remarkably robust and has been used in a range of lightweight roof applications (see Hughes 1993; Rice 1993). A particular application is discussed in Le Bourva and Wernick's work (1995), involving two arch-spans with stabilizing ties more external to the arch than internal, converging at a high point between the two arches.

Other examples where dynamic relaxation analysis has been used to good effect for justification of tension elements as part of a stability system include the Chek Lap Kok air terminal roof (Foster et al. 1999), and the Pavilion of the Future in Seville (Rice and Lenczner 1992). The latter has the extra advantage that the compression in the circular stone arch actually arises from the radial ties anchored on a smaller, parallel, circular tie. The system is therefore self-stabilizing in that any shape change the arch might try to undergo is more than resisted by the inner tension circle, which would need to undergo a greater shape change. This self-correcting feature meant that most of the analysis work could be kept linear.

Another particular situation arises where the arch members in a vault perceive a shear stiffness in the vault surface that offers restraint at least to the first few unrestrained buckling modes. Such shear stiffness is easily provided by tension cross bracing in the reticulated vault surface but must of course itself find restraint at the vault ends, by attachment to end wall diaphragms for instance. An application of such a "vault-shell" structure is described in the work of Bailey et al. (1999) and shown in Figure 3-4. Eigenvalue analysis of the timber arching elements was used to identify Euler compression loads, which were used to deduce effective lengths for design purposes. These were found to be about 25% of the arch length, or about half of the unrestrained first mode length. Some extra stiffening was also obtained from continuity of longitudinal members in the framing.

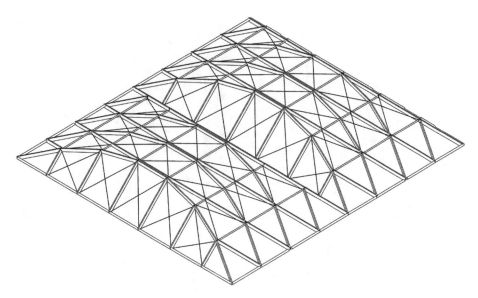

Figure 3-4. RAS Exhibition Halls, Sydney. Framing for timber grid-shell vaults.

3.2.3 A Longspan Retractable Segmental Tied Arch Roof

The retractable roof for the Miller Park baseball stadium in Milwaukee comprises seven arch segments spanning about 600 feet. The segments rely on arch action, and some of the arching elements are quite slender. For a structure of this type and of this magnitude, it was important to choose a reliable analysis method for dealing with second order effects arising from slenderness.

The structure is described in Figures 3-5 and 3-6. The outermost roof segments are fixed. The typical moving segments have a highly asymmetric structure, with a 60 in. ✕ 30 in. box section arch on the outer edge of the deck structure and a horizontal tie on the inner edge. Given the asymmetry, all parts are interactive under load. In particular, the upstand arch along the outer edge of the deck relies for its stability on the stiffness and strength of the remainder of the framing, particularly the W40 radial upstand hangers that connect directly to it.

A calculation method was needed that would check the strength of the compression chords themselves, as well as predict the restraint forces that would be imposed back onto the hangers and the deck. The results of such a full analysis of buckling effects would then need to be combined with the results of all the usual load cases to check the combined-stress ratios in each individual member.

Consideration was first given to the likely nature of buckling in the members. In general terms, when a slender member experiences failure due to a compressive load, the failure arises from the combined effects of the axial stress and the effects of additional bending moments. Sources of this bending include not only the primary loads, but also the effects of initial imperfections in the geometry of the members. All these moments are magnified by P-Delta effects arising from the compression.

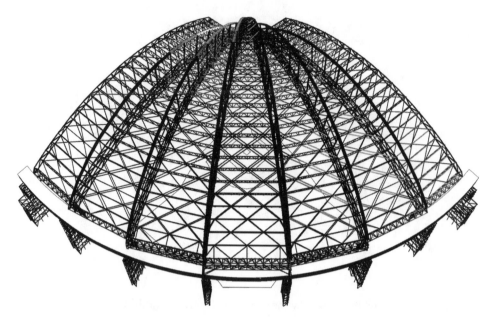

Figure 3-5. Miller Park, Milwaukee. Overall roof geometry.

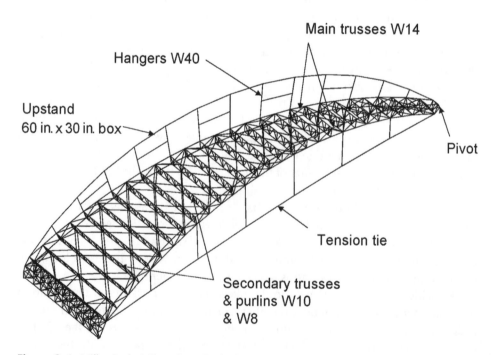

Figure 3-6. Miller Park, Milwaukee. Typical roof segment.

3.2.3.1 Initial modeling. Reduction in allowable axial stress is the method usually adopted by design codes to account for the effects of the bending moments due to imperfections rather than requiring that they be calculated explicitly. Column curves are typically provided that propose a lower axial stress capacity at higher slenderness. To allow use of the code equations for preliminary calculations, an effective length was estimated for each upstand arch. This was taken to be the length of a simple pin-ended strut, considered to have the same axial capacity as the real member assisted by its various real restraining elements.

To estimate this equivalent pin-ended strut length, analysis was first carried out to determine the compression in the chord under a load causing overall elastic buckling. This was then related to the Euler load of an equivalent pin-ended compression member. This required first an elastic buckling analysis of the structure under the greatest downward load case. The compression in the upstand was then deduced, at a load that just caused buckling (F_{cr}), and so the effective length of the pseudomember was calculated such that its Euler load was equal to F_{cr}. This method indicated that effective lengths for lateral buckling of the upstand chords were approximately three bays. Using this effective length, member strengths were then checked, based on code methods, along with the bending moments due to applied loads. The attraction of this preliminary method was its speed, and it was also used for check comparisons with the results of more rigorous methods applied later.

This so-called effective length method does not, of course, take account of any effects from higher modes of buckling nor does it deal directly with the additional moments arising from imperfections. This means the extra restraint forces required of the hanger and deck structure to restrain the compression chord were not directly calculated; they had to be allowed for empirically at this stage.

3.2.3.2 Detailed modeling. The detailed method relied on direct calculation of the additional bending moments caused by imperfections, then combining these with all other design effects, and then checking against the full axial and bending strengths of the members. This was in lieu of the usual code method of requiring reduced axial strength to allow for the effects of additional buckling moments. At the time of calculating the additional moments in the chords themselves, the restraint forces in the hanger and the deck arising as a consequence of buckling effects were also generated and could be checked for those members after further combination with the effects of the main loadcases.

A question implied by the chosen method is what distribution of imperfections should be assumed? The conservative answer adopted was to base the distribution on the natural buckling mode shapes of the structure in order to provoke the buckling mode directly. Structural analysis program GSA (licensed by Oasys Ltd.) was used to calculate these distributions by eigenvalue analysis. A preselected number of critical buckling modes are calculated by the program, based on a specified load case. A mode shape is produced for each mode, scaled to unit deflection, plus a set of forces and moments related to achieving that mode shape. The load factor is also calculated, being the factor that the applied load has to be multiplied by to generate instability in that mode. Typically, the load case selected was the largest downward load case, which caused maximum compression in the upstand chord.

Despite the apparent complexity of the method, there were some aspects of the roof's behavior that led to simplifications. For example, the ratio of compressions between different members was approximately constant, even though the actual compressions varied with different load cases. A consequence was that the mode shapes were effectively constant for all load cases, and advantage was taken of this.

The moments generated by the eigenvalue method described above could have been used directly after appropriate scaling. It was decided, however, to represent the mode shapes by specific applied lateral loads on the upstand chord, scaled and distributed to generate the necessary moments in the members. The advantage of this procedure was that these forces could then go directly into the main SAP 90 (Computers and Structures, Inc.) models, where all stress checks were being carried out. Another advantage was improved understanding of the contribution of different members to the restraint of buckling effects. Typically 10 to 15 modes were generated for each panel, and for each element final results were based on the mode causing the greatest stresses. Three other issues needed addressing to finally apply the method. These were the size to assume for the initial imperfections, the amount these imperfections are magnified by the P-Delta effects of the axial loads, and the amount of lateral force needed to represent the effects of the imperfections.

Regarding the size of imperfections, the intention was to reproduce the implied requirements of design codes. Taking a simple strut, compression failure can be considered to occur when

$$F/F_c + FsA/M_c = 1 \qquad\qquad (3\text{-}1)$$

where
F = ultimate axial load
F_c = axial capacity of a short section (squash load)
s = magnitude of imperfection
A = amplification factor ($= 1/(1\text{-}F/F_{cr})$)
F_{cr} = $\pi^2 EI/L^2$
M_c = local section moment capacity

A represents the additional moments due to the imperfections. Solving the equation for F, assuming a given yield strength and set of section properties, allows a curve of axial strength versus slenderness to be plotted. Different curves are obtained for different values of imperfection, s. Typical code-based column curves put $s = L \times k$, where L = length and k is a constant ranging between zero and 0.004. Code column equations based on the Perry-Robertson formulae, such as those in British Standard BS5950, can be expressed this way for example.

The curves tend towards a simple Euler curve as k tends to zero, although with a cutoff when the yield stress F_y is reached. For $k = 0.004$, the curve becomes close to the BS5950 curve with a Perry factor of 5.5. This corresponds to welded box elements such as those in the Miller Park upstand arch chords and also corresponds to an imperfection of length $L/250$. The formulae outlined in the US LRFD code (Load and Resistance Factor Design Specification for Structural Steel Buildings of the American Institute of Steel Construction, Inc. [AISC]) do not exactly match this method. An initial imperfection of $L/250$ gives a close approximation slightly on the safe side for the range of stresses under consideration, so this was the assumption adopted. Length L was chosen as the greatest length between points of contraflexure for each mode, due to bending in that mode.

The next question to be addressed -was the magnification of these imperfections by P-Delta effects from axial loads. Axial compression in a simple strut acts to amplify any initial imperfections by the factor $1/(1\text{-}F/F_{cr})$. For a structural system as a whole, this factor becomes $1/(1-1/Q)$, where Q is the ratio of the elastic critical load to the actual load, which can be calculated by the buckling analysis mentioned before. If elastic buckling occurs at three times the factored loads for a particular buckling mode for example, then the amplification factor will be 1.5, and an initial imperfection of $L/250$ will be magnified to $1.5 \times L/250$ along with consequent bending effects. The amplification factor will therefore vary for each mode.

Finally it was necessary to calculate the effective lateral force that would represent the above effects of an imperfection. Bending and shear forces within a simple axially loaded element undergoing a sinusoidal deformation, can be represented exactly by those in a similar straight element with a lateral load of sinusoidal distribution. The size of the imperfection determines the total magnitude of the lateral load W. For example, taking a simple strut with an initial sinusoidal imperfection of $L/250$, and an axial force F, bending will be the same as in a straight strut with the same axial load, and a sinusoidal lateral load of $W = 2.5\% \times F$. This value is then increased by the amplification factor described above. Checking test cases indicated that this assumption could also be applied to larger structures with intermediate restraints.

3.2.3.3 Results of the analyses. Figures 3-7a through 3-7f show a typical set of results for the first three modes of a typical retractable panel. On the left are the buckled mode shapes and on the right, the distributions of equivalent applied forces used in generating the mode shapes. The load factors that were obtained for each mode are set out in Table 3-1. They relate the elastic critical load for that mode, to the factored applied load.

To calculate the distributions for the equivalent lateral loads, offsets were measured from a straight line joining points of zero moment to the deflected position of the chords. The shape of the load distribution obtained in this way was then factored so that the total applied load in the largest half wave of the deflected shape was equal to

$$W = 0.025 \, F \times 1/(1-1/Q).$$

The outcome of the procedure was then to generate stresses equivalent to the effects of a length $L/250$ initial imperfection, magnified by the P-Delta amplification factor. For further reassurance, the moments obtained this way were checked against those generated by scaling the results of the buckling analysis, and close agreement was obtained.

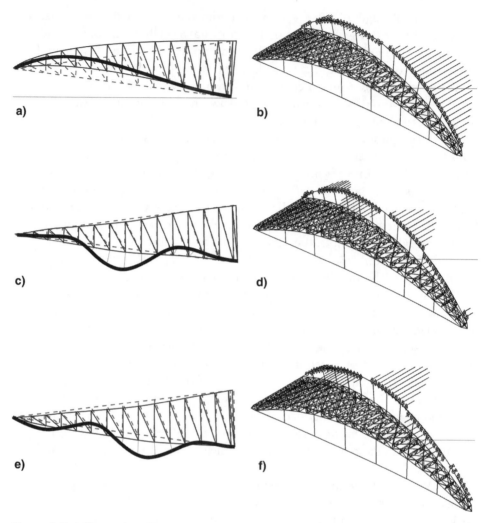

Figure 3-7. Miller Park, Milwaukee. Three instability mode shapes (a, c, and e, left), and equivalent applied forces (b, d, and f, right).

Table 3-1. Miller Park, Milwaukee. Load Factors for Elastic Critical Loads

	Mode 1	Mode 2	Mode 3	Mode 4	Mode 5
Panel 3	2.89	3.51	3.59	5.35	5.55
Panel 2	2.62	3.29	3.35	5.11	6.38
Panel 1	2.53	2.91	3.33	3.69	3.77

3.2.3.4 Combination of results. For uploading into the main SAP 90 models, two load cases were input for each mode shape, representing both positive and negatives values for the mode. Partial member strength utilization ratios were calculated to LRFD methods, representing section usage after application of the buckling load cases. The primary load combinations such as wind, snow, and temperature were then analyzed, including P-Delta effects to take account of the amplification of moments due to these primary loads. Total strength utilization ratios were then obtained by adding utilizations due to buckling to those from primary loads. So 20 buckling cases and 100 primary cases, for example, could be enveloped without needing to analyze 20 × 100 = 2,000 load combinations.

The strength demands predicted for the upstand chords were very close to those predicted by the initial more approximate calculations. About 15% to 20% of the upstand strength demand was due to buckling effects, arising mainly from the lowest modes. The whole process of developing the detailed calculation method was of major assistance in achieving a robust conceptual understanding of the likely instability effects and of the way they would be resisted.

3.2.4 Computational Methods for Buckling Analysis

The eigenvalue method used above for the Miller Park roof is one of three main methods for inclusion of P-Delta effects in the analysis of compression frameworks likely to exhibit nonlinearity.

The theoretical basis of the method relies on small deformations and on linear material behavior. In the overall structural stiffness equation

$$\{F\} = [K]\{\Delta\}, \qquad\qquad (3\text{-}2)$$

the stiffness matrix is taken to comprise two parts, as follows:

$$[K] = [K]_l + [K]_g \qquad\qquad (3\text{-}3)$$

where $[K]_l$ is the conventional linear stiffness matrix (no consideration of member axial forces), and $[K]_g$ is the geometric stiffness matrix that depends on the structural geometry and member axial forces.

The geometric stiffness matrix is independent of member material properties but depends on the choice of a particular load case and hence on a particular ser of member forces. Assuming that the load system causing buckling is related to the initial imposed load system by the buckling load factor λ, then at buckling, Eq. 3-2 becomes

$$\lambda\{F\} = ([K]_l + \lambda[K]_g)\{\Delta\}. \qquad\qquad (3\text{-}4)$$

At buckling, the structure's stiffness becomes zero, so the determinant of $([K]_l + \lambda[K]_g)$ disappears, which allows calculation of λ, the buckling load factor. The buckling load capacity provides an upper bound on loading. The analysis does not provide deflection versus load because the method assumes buckling capacity is unrelated to deflection.

By introducing a stiffness component that is dependent on nodal deflections $\{\Delta\}$, the displacement behavior of the structure can be tracked up to the buckling load. In this approach,

$$[K] = [K]_l + [K]_g + [K]_\Delta \qquad (3\text{-}5)$$

where $[K]_\Delta$ depends on nodal deflections. Eq. 3-2 becomes nonlinear (as $[K]$ now depends on $\{\Delta\}$), so some form of iterative solution is needed. Initial geometric imperfections may be incorporated to simulate potential real conditions more closely.

The third and most comprehensive method addresses the extra complication of potential nonlinear material behavior at larger displacements. This method provides the most accurate prediction of behavior approaching failure, provided the assumed nonlinear material characteristics are realistic enough to capture the real-world effects important to the case in hand. These may include elastic/plastic and strain hardening phases; effects of cracking; residual stresses from material production, fabrication, and installation; and the effect of multidirectional stress interactions on local failure criteria. Potential initial geometric imperfections from fabrication and erection also need to be modeled. Applied accurately, and with conservative assumptions about the pattern of initial imperfections (geometric and material), the method provides a lower bound on load capacity. The three methods of buckling analysis described above bound the structure's behavior as indicated in Figure 3-8.

For the Miller Park roof structure, there was a typically long list of load case combinations to be considered. Load case dependent nonlinear analysis would have denied the use of linear load case superposition and created an unrealistically lengthy analysis program, so an adaptation of eigenvalue analysis was used as

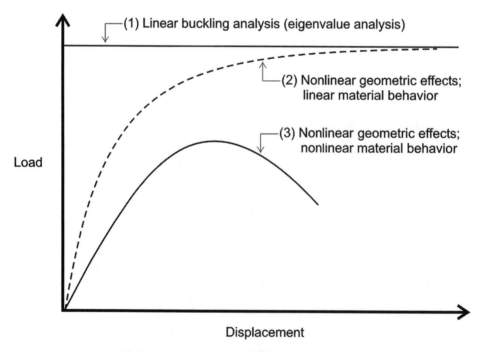

Figure 3-8. Typical load-displacement curves for different methods of modeling instability.

described. Explicit nonlinear analysis was adopted for snap-through buckling for the velodrome roof described later however.

3.2.5 Use of Strain Energy Density

While the eigenvalue analysis method is effective and popular for simulating nonlinear behavior of compression structures, it relies on correct estimation of the (half) wavelength of a given buckling mode shape. This then allows a corresponding pattern of assumed initial imperfections to be incorporated, for estimation of worst-case stress conditions.

The half wavelength in question is the one exhibited by the primary buckling members. That is the set of members whose initial imperfections will precipitate the buckling and then be affected by it.

For simple compression structures, the members in question and the corresponding wavelengths are usually obvious. For more complex systems, the primary buckling members may turn out to involve the bracing system as well, and the appropriate wavelength may be embedded in the buckled shape of the bracing system, rather than just the compression members themselves.

Plotting the structure's strain energy density (moment x curvature or axial force x extension divided by cross-sectional area, per meter length) gives an important clue for determining which is in fact the primary buckling member set, as strain energy is a measure of member participation in resistance to a particular mode shape.

3.2.6 Analysis of Domes—Origins

Analysis of domes and other shell structures has engaged the attention of mathematicians, scientists, and engineers for many centuries. The development of the equation of a catenary by Bernoulli, Huygens, and Leibnitz in the late seventeenth century could be considered the starting point. Parent, de la Hire, and Gregory then explored the equilibrium of vaults through mathematics. Poleni progressed the theory by his study of St. Peter's dome in Rome. He applied Newton's principle to deduce an explanation of the relation between thrust lines and the shape of a suspended chain, then extended his experiments to consider chains carrying varying masses to simulate the weight of different segments in a vault. In the late nineteenth century, Gaudi used this work as a basis for suspended models of complex vault structures. He was seeking thrust line geometries that would deliver compression without the penalties of bending and shear and without recreating the buttress methods used by the builders of Gothic cathedrals. His methods were in effect iterative equilibrium methods, deducing relations between the intended mass of constructed elements and the geometric configurations needed to keep them in equilibrium.

With the arrival of reinforced concrete, domes and vaults started to become thinner and lighter, and analytical methods became more important to their success. A large body of classical mathematical methods was generated to describe their behavior. As formwork and construction costs began to favor prefabricated, reticulated grid-shells, analogies could be made between continuous and skeletal shells for analytical purposes.

This was particularly useful for considering instability issues, before computer power grew to provide feasible iterative methods. Accounting for framework joint stiffness by shell analogy was never easy, however, and local snap-through failures of shallow reticulated domes were not unknown.

3.2.7 Analysis of a High-Rise Timber Grid-Shell Dome

The 100 m diameter dome for the New South Wales Royal Agricultural Society's exhibition facility in Sydney uses glued-laminated (glulam) plantation pine in the arching direction, with containment by circular tube steel sections in the circumferential direction (Bailey et al. 1999).

Rise from ring beam to crown is 35 m. The timber members are constant in depth (800 mm), vary in width (135 mm at crown, 280 mm at base), and are jointed by steel end-plate bearing details using grouted, end-grain bolting into the glulam sections.

The stability of domes against first-mode buckling (usually asymmetric, analogous to an arch) depends critically on the shear stiffness available to the shell framework, so a triangulated framing pattern was chosen at the outset (Fig. 3-9) even though it would be more challenging of construction, in terms of tolerance, than a simpler radial/circumferential pattern. The end-plate jointing detail was simple to fabricate and also useful in providing local moment continuity hence extra stiffness for stability, both local and global.

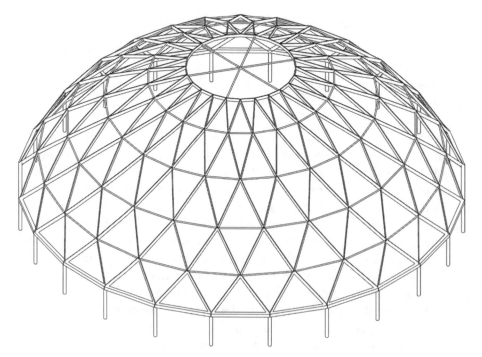

Figure 3-9. RAS Exhibition Halls, Sydney. Framing for timber grid-shell dome.

Buckling modes and mode shapes were first predicted by eigenvalue methods, similar to those described for the Miller Park structure, for several different loading configurations. This allowed determination of the local member compressions at critical major axis buckling loads, hence deduction of an effective length, based on Euler theory for pin-ended members. A view was taken on the appropriate E value to be used for the timber, depending on the long- and short-term components in the particular load case. The curvature and shear stiffness of the structure were both high enough to constrain effective lengths to between one and two individual member lengths.

Stability of the glulam sections in simultaneous negative bending (bottom edge compression) was considered by analysis of an individual member torsionally restrained at intervals by (flexible) purlins attached rigidly to its top edge. The results of this analysis were also interpreted in terms of effective length, so a combined axial force/moment check could then be carried out to code-based methods.

The third instability mode affecting member design was translational instability about the member weak axis. This was of some interest because of the potential for an entire ring of members at one circumference level to buckle simultaneously under a similar member compression. The purlins, at about 1.2 m centers along each member, were available for restraint. Their resistance resided in their attachment to a shear-stiff metal deck cladding system, plus their moment-capable connections to the glulam members themselves. Analysis indicated load factors against this mode were sufficiently high for it not to be an issue for member design.

An interesting analytical issue for any longspan roof is the extent to which temporary stresses occurring during erection will become preserved or "locked-in" to the completed structure. The mistake is often made of ignoring such residual stresses and assuming that the analysis can be carried out as if the structure were brought into existence in its final form instantaneously.

Many of the initial proposals for erection of the dome would have been quite disadvantageous in this respect, particularly those that relied on significant simply supported spans between lines of temporary props under member self-weight, with the prospect of the self-weight moments becoming built-in after installation of circumferential members and depropping.

The erection method finally adopted solved that and several problems to do with speed, access, and safety very neatly. The dome was in fact erected progressively from the crown outwards to the ring beam by upward jacking of the partly completed structure after installation of each successive ring of members. While analysis indicated there would be no consequential residual stresses arising from the method, there was one significant analytical consequence. Each time the partly completed dome was landed on the ground slab ready for another lift, the outermost circumferential tie was required to perform as a ring beam, with temporarily very high tension. So reanalysis of the structure was required for four extra configurations, which generally proved noncritical, except for the need to strengthen some joints.

3.2.8 Analysis of a Low-Rise Steel Grid-Shell Dome

The Sydney Olympic Velodrome roof is a shallow, steel-framed, grid-shell structure about 130 m × 100 m, with just 13 m rise, set out on the surface of a torus. The triangulated framing comprises a 900 mm × 400 mm (approximately) continuous steel box section in the short direction, with 457 mm diameter tubular members diagonally, pinned at their intersections with the box section (Fig. 3-10).

Although similar in span to the timber dome described above, the very low rise to span ratio for the velodrome roof led to much higher member compressions and greater sensitivity to nonlinear effects. A series of different analyses were carried out to build up confidence in the roof's stability under a wide range of load cases.

While the continuous box sections were of great use during erection, it is the smaller circular tube sections running diagonally that carry the large part of the gravity loads. This is because the elliptical plan shape of the ring beam, laid on a torus surface, features higher curvatures, hence greater resistance to thrust, at the four corners of the ellipse where most of the diagonals terminate in the chosen framing pattern. Various linear analyses were carried out first to test the stiffness of the frame, particularly under asymmetrical loads. The decision was made to maximize the bending stiffness of the box member by requiring fully rigid splices, while allowing pin-ended joints for the intermediate diagonal members.

While the low rise and high stresses implied that nonlinear effects would be consequential, direct nonlinear analysis was not chosen as the basic analysis tool because of the very large number of separate analyses required for the wide range of design load cases. A more pragmatic approach was adopted, with a mixture of methods, with the buckling characteristics of the box section member and the tube members considered separately at the outset, and with design rules deduced for each.

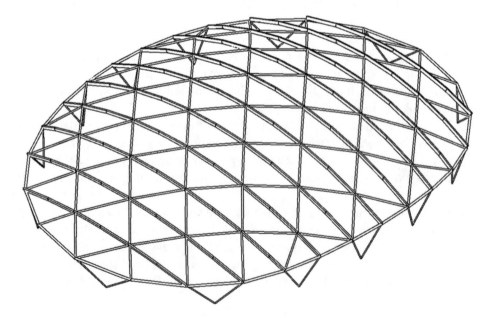

Figure 3-10. Sydney Olympic Velodrome. Steel grid-shell roof structure.

For the box sections, the out-of-surface elastic critical buckling load was esti-mated by eigenvalue analysis using a special load case that induced compression in the box sections while avoiding compression in the tube diagonals. This avoided the problem of the diagonals dominating the eigenvalue solution because of their greater slenderness. The major axis effective length obtained for the box section was slightly greater than the node to node length. A corollary was that the box section design was therefore controlled by minor-axis stiffness.

For a line of tube diagonals, restraint is provided by the intersecting box section members, which can be represented as elastic supports. Analysis of this reduced sys-tem, with careful choice of restraint stiffness, led to the conclusion that the critical mode was individual tube buckling, node to node, and that the box sections were stiff enough to ensure this.

Restraint forces imposed on the systems restraining each primary compression member were estimated by applying restraint force patch loads for each element in iso-lation. The corresponding maximum level of utilization for each section property was then taken as an initial capacity reduction across the property group. Lateral torsional buckling was also considered. The box section perceives direct torsional restraint only at its ends, 100 m apart. Translational restraint at the intermediate framework nodes represent a strong constraint on intermediate torsional shape change however, and analysis was used to test for the effective torsional stiffness available at any cross sec-tion, to allow a further, though small, design capacity reduction.

Snap-through buckling is a highly nonlinear effect and was analyzed separately using FABLON (owned by Oasys Ltd.), an in-house, nonlinear, dynamic relaxation routine. The curvature, frame shear stiffness, and box-section bending stiffness all combined to generate enough local stiffness so that a ductile failure of the ring beam occurred before any snap-through modes.

As for the timber dome, much thought went into suitable erection methods and the likely consequences of locked-in residual stresses they might generate. The box sec-tion was intended as an erection opportunity, and modeling assumptions were made about the number of intermediate props that might be used along its (max.) 100 m length. Each gave rise to different self-weight residual bending stress distributions.

The contractor, however, elected to avoid propping altogether by lifting two half-length box members with two cranes and performing a bolted splice in the air. The arch was then stabilized with some tube diagonals and released against temporary external props buttressing the ring beam at each box section location. The result was very small residual bending stresses, although the degree of control over the exact geometry of the arches was probably not as great as in a propped erection. Devia-tions from the theoretical position of the erected structure were analyzed as a final condition, to simulate actual geometry, and there were no unacceptable outcomes.

3.3 Tension Roofs

3.3.1 Introduction

When we move from compression structures to tension structures, our analytical concerns change. Analysis of compression structures is dominated by concerns about stability both local and global, whereas tension structure analysis is dominated by shape determination, optimum use of curvature and pretension, and the strength and stiffness of the whole system including abutments and restraints.

Typically, we need to carry out analysis to determine a doubly curved shape that is competent to carry potentially reversible loads with a good degree of efficiency. The structure will normally need pretension both to stiffen up the response to uniform load and, unless shear stiffness is available within the surface, to provide stiffness against nonuniform loads.

3.3.2 A Doubly Curved Cable Net

The Aurora Place development in Sydney comprises two buildings set each side of a public plaza. Environmental wind testing demonstrated that the plaza would be unusable for most of the year without a roof to protect patrons from the prevailing wind, which is directed down the facade of the taller office building.

The roof was conceived as a doubly curved anticlastic cable net supporting a flat glass canopy beneath (Fig. 3-11). One set of cables is draped between the two buildings; the other set is stretched downward between front and back anchorage cables.

Analytically, the first step in the shape-finding process was to choose a suitable cable grid. Geometric constraints included the need to connect at right angles to the

Figure 3-11. Aurora Place, Sydney. Doubly curved, single-layer cable net structure with glass cladding.

curved building facades so that thin plates could slip through the curtain wall mullions and the need to restrict the maximum glass dimension to about 2 m. There was also the opportunity to reference tension nets that occur in nature, such as spider webs. The plan form that developed from these drivers is shown in Figure 3-12.

The preliminary shape-forming process used a grid of beam elements, fixed at each end perpendicular to the façade, but with no axial stiffness. This was then analyzed using the in-house, nonlinear, dynamic relaxation program FABLON. The fabric form-finding feature in this program seeks iteratively to space nodes equally along each string of nodes, to achieve an even and flowing geometry. To account for some unacceptably irregular glass shapes and some anomalies near to the anchorages, the grid generation process was completed with assistance from a CAD package.

Once the cable (and therefore the glass) layout was established, the next step in the shape-finding process was to determine a set of heights for the nodes such that the resulting cable net was in equilibrium under the chosen set of pretensions. The target conditions for the set of pretensions were that none of the cables lost its tension under the most adverse combination of service loads and that no pretension remained under the most adverse combination of ultimate loads. This ideal situation allowed the maximum beneficial stiffness to be gained from the pretension without any penalty on strength requirements.

The height of anchor locations was chosen first, with due account of the reaction capability of the adjacent buildings. An iterative series of form-finding analyses was

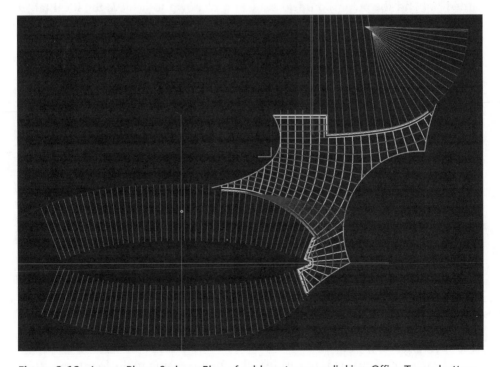

Figure 3-12. Aurora Place, Sydney. Plan of cable net canopy linking Office Tower bottom left, to Residential Tower top right.

then performed. The first determined the height of the nodes under a given pretension field. The second determined the appropriate pretension field for the shape. This pair of analyses was repeated iteratively until the change in either shape or pretension was small enough to be insignificant.

To determine the shape, the nodes were all fixed in plan but released vertically (using a modeling device sometimes referred to as *greasy poles*). The elements were given no axial stiffness but assigned a constant force or pretension. The resulting FABLON analysis provided an equilibrium position in which all vertical forces were resolved. Horizontal equilibrium was enforced by the greasy poles.

To determine the pretension field, the greasy poles were removed and the elements given their correct (or indeed any arbitrary but fixed) stiffness and the same pretension force set as in the shape-forming analysis. The result of this FABLON analysis was a slight change in shape, but more importantly, a set of forces that were in equilibrium, in all directions, for that shape.

Iterating these two analyses, feeding the shape from the first into the second and the tensions from the second back to the first, eventually led to a stable shape. The only catch was a tendency for pretension to leak from the system during the second analysis. The small change in shape during this step tended to reduce the overall pretension field. This was easily accounted for by setting one element as a control element with no axial stiffness. The force in this element did not change and therefore controlled the pretension field throughout.

Once the shape was confirmed, analysis under various load combinations was carried out and the pretension field scaled up or down to achieve the ideal target conditions of tension everywhere under all service conditions but no pretension remaining under the most onerous ultimate conditions.

3.3.3 A Cable-Stayed Catenary

The design of the roof structure for the City of Manchester Stadium explores the idea of divorcing the roof cladding surface from the supporting structure. The cladding surface is a slightly distorted section cut from a cylinder whose axis is parallel to the pitch centerline. This surface had insufficient curvature to work successfully as a compression vault such as those described earlier. Instead, it was decided to hang the surface from a network of cables supported by 12 masts distributed around the perimeter of the stadium. The masts were, in turn, restrained by pairs of backstays, plunging dramatically to the ground to form a gigantic loggia outside the stadium. One of the earlier analysis models is shown in Figure 3-13.

The cable network consists of a set of forestays, one per rafter, anchored on four looping catenary cables that are held down in turn by four corner ties, at the four corners of the stadium. This catenary system has a loose fit with the forestays and the predetermined roof surface, while keeping sufficient curvature in the catenary for reasonable efficiency.

The catenary's function is to ensure sufficient pretension in the forestays so they can resist wind uplift. It also distributes patch loads over several rafters and forestays

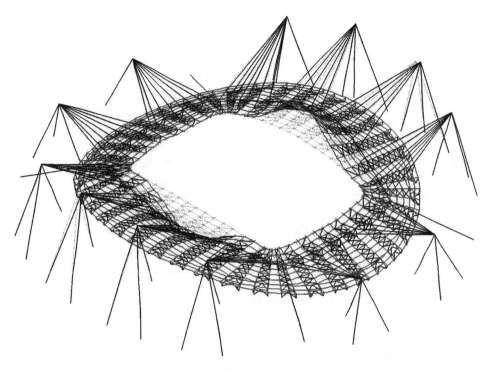

Figure 3-13. Manchester Stadium. Magnified deflections of rafters under nonuniform loading.

by means of the nonlinear tension stiffening effect of catenary prestress. It also resolves the lateral forces where forestays and catenary intersect.

Preliminary analyses were carried out to test the overall behavior of the system, particularly the compatibility of the relatively soft catenary and the relatively stiff forestay structures. As expected, after pretensioning of the forestays, the force level in the catenary (about 3,500 kN) hardly alters when the roof is loaded. According to the nonlinear analyses carried out, this high pretension provides a nonlinear stiffness that is useful in damping relative rafter movements under uneven roof loading, such as those shown magnified in Figure 3-13.

The first step of the form-finding analysis was to determine the optimum shape for the cable network. Objectives were:

1. to ensure that the forestays remained taut under all loading conditions, which prevents the masts from rotating backwards under their own weight and causing undesirable bending in the backstay rods,
2. to gain sufficient curvature in the catenary for structural efficiency,
3. to obtain acceptable angles of altitude and azimuth for the geometry of the corner tie-downs,
4. to control the length of the struts that connect the catenary to the rafters, and
5. to support the rafters near the inner edge of the roof so that the rafter size was not dominated by cantilever effects.

The shape-finding process followed a similar iterative strategy as that for the Aurora Place cable net. Firstly the catenary shapes were established within their predetermined planes under a given set of pretensions. The ratio of catenary pretensions had to result in the correct corner tie azimuth at their intersections and strike the right balance between objectives two and four above. The forestay pretensions were set equal and opposite to the compression generated by the maximum upward load case.

The iterations were on three levels. The first dealt with the intention to keep the catenary/forestay nodes in the same vertical plane as the rafters. The second dealt with the fact that every different possible forestay/catenary geometry developed a different set of forestay forces under load. So for each shape the forestay forces were reset and the shape refound. The third layer of iteration was needed to ensure that the catenary forces resolved at the corner tie intersection into a force in the desired direction. If not, then an entire catenary/forestay pretension set had to be scaled up or down and/or the intersection point had to be moved, which affected the catenary/forestay plane and the rafter support points.

Once the shape had been found and the pretension determined, the majority of the analysis under load was reasonably straightforward and was performed using linear analysis. The major nonlinear effects were those mentioned earlier—the tension-stiffening effect of the catenary under uneven loads and the bending behavior of the backstay rods under low levels of axial force.

Far more analytical effort was involved in modeling the erection process. The best strategy for this process often involves starting with the self-weight plus pretension geometry and then taking it apart analytically, in a reversal of the erection process. This is a complex and highly nonlinear process. The model needs sufficient detail to properly model the self-weight sag of all tension elements. All connection details must have their proper weight assigned. And the stiffness of all members, particularly cables, needs to be properly modeled.

Two different strategies are available for this unloading process. One involves the progressive removal of members from the model and the continual resetting of the initial geometry and pretension. The second starts from the same starting point (the end point of the real erection sequence) but involves setting more and more members to zero weight and stiffness, thereby effectively eliminating them from the analysis. So even for a relatively simple cable network, the shape-finding process can be quite complex, and the simulation of erection even more so!

3.3.4 Dynamic Relaxation

Form-finding and subsequent load case analysis of prestressed cable and cable net structures can be carried out by a range of iterative calculation methods.

The method used for both of the tension structures described above was dynamic relaxation rather than a stiffness matrix method. This method avoids the need for special requirements for numerical stability, such as a nonsingular stiffness matrix, and so allows a wider range of movement-sensitive structures to be analyzed. It is also more computationally efficient.

Dynamic relaxation uses iterative analysis to converge on the at-rest equilibrium condition of a loaded structure by first allocating mass and inertia to each free node according to the node's translational and rotational stiffnesses. External loads are then applied suddenly, and the system of nodes begins to oscillate about its equilibrium geometry according to the basic equations of dynamics.

The motion of each node is traced through small time increments until it comes to rest. Damping applied to the system may be viscous, proportional to the product of modal velocity and mass, with convergence achieved soonest by critically damping the lowest mode of vibration. Alternatively, kinetic damping may be applied, and this is often more effective where large locally unbalanced forces exist initially. Day (1965) and Barnes (1994) describe the method in more detail.

3.4 Conclusion

There is a range of computational methods available to the analyst and designer of lightweight longspan roofs, requiring more or less computational effort and time. As illustrated by reference to a range of worldwide case studies, the most appropriate method will depend on the level of modeling accuracy required, given the various demands of safety, serviceability, material consumption, ease of fabrication and erection, and aesthetic intent.

Acknowledgments

Arup, as consulting engineer for all the case study projects described, is gratefully acknowledged.

John Hewitt is also acknowledged for his assistance with the description of the Miller Park Stadium roof analysis and design, as well as Peter Bailey for assistance with the description of the Sydney Olympic Velodrome roof analysis and design.

The Architects for the Miller Park Stadium were HKS, NBBJ, Epstein Uhen; for the RAS Exhibition Halls, Ancher Mortlock Woolley; for the Sydney Olympic Velodrome, Paul Ryder with DesignInc; for Aurora Place, Renzo Piano Building Workshop; for the City of Manchester Stadium, Arup Associates.

References

Bailey, P., Carfrae, T., Hough, R., and Stevenson, P. (1999). "Exhibition halls for the new Sydney showground." *Arup J.*, 34(N2), 30–35.

Barnes, M. (1994). "Form and stress engineering of tension structures." *Struct. Engrg. Rev.*, 6(N3–4), 175–202.

Day, A. S. (1965). "An introduction to dynamic relaxation." *The Engr.*, January 29, 1965, 218–221.

Foster, A., Gibbons, C., Manning, M., and Scott, D. (1999). "The Chek Lap Kok airport terminal building." *Arup J.*, 34(N1), 4–11.

Hughes, A. (1993). "Chur Station roof." *Arup J.*, 28(N2), 3–7.

Janner, M., Lutz, R., Moerland, P., and Simmonds, T. (2001). "World's largest self-supporting enclosure." *Arup J.*, 36(N2), 24–31.

Le Bourva, S., and Wernick, J. (1995). "Euralille: The TGV station roof." *Arup J.*, 30(N2), 3–6.

Rice, P., and Lenczner, A. (1992). "Pabellon del Futuro, Expo '92, Seville." *Arup J.*, 27(N3), 20–23.

Rice, P. (1993). An *engineer imagines*, Artemis, London.

Zunz, G. J., Hough, R., and Banfi, M. J. (1985). "Analysis and design of a prefabricated three-way trussed grid vault." *Analysis, design and construction of braced barrel vaults*, Z. S. Makowski, ed., Elsevier, Barking, 367–381.

4

Bridge Assessment

Campbell Middleton

4.1 Challenges in Bridge Engineering

Our ability to design spectacular new structures of increasing complexity has been revolutionized over the last two decades of the twentieth century, thanks mainly to the development of numerical methods such as finite element analysis in line with dramatic advances in the processing speed of computers.

One of the great achievements of the last few years has been the successful construction of several longspan suspension bridges such as the Akashi Straits Bridge in Japan in 1998 with a span of 1,991 m, and the Great Belt East Bridge in Denmark, also completed in 1998, at 1,624 m. Other bridge technologies have also made dramatic advances. For example, in 1995 the Pont de Normandie in France took the then span record for cable-stayed bridges, which was held by the Yangpu Bridge in China (completed in 1993), from 602 m up to 856 m—an increase of over 40% in a single step! This record was then increased to 890 m in 1999 with the completion of the Tatara Bridge in Japan. Another important advance has been the development of new materials such as fiber-reinforced plastics (FRPs) as the search for more durable structures accelerates. For example, in the UK an 8 m span lift bridge made entirely from fiber-reinforced plastic and capable of carrying full highway loading (40 tonne) was built in 1994 over a canal at Bond's Mill. In the United States various forms of FRP deck replacement schemes have been tested and similar trials are in progress in a number of other countries.

But many challenges still remain for bridge engineers. The Millennium Bridge in London, a 325 m span steel pedestrian bridge, was to be a landmark for the new millennium and was conceived as a "blade of light" cutting across the Thames River. However just two days after its opening in June 2000 it was closed again due to the excessive sway experienced on the opening day by the estimated 100,000 users. A substantial retrofitting program was instigated to install vibration dampers to deal with the lateral movement induced by the synchronous steps of a large number of pedestrians (now referred to as synchronous lateral excitation) (see www.arup.com/millenniumbridge). This led to a total rethink and revision of the provisions for pedestrian loading in the British code. This proved an expensive and timely reminder of the need for engineers to continue to evolve our understanding of the interaction of imposed loads and the structural behavior of bridges. Another problem is the wind or rain induced vibration of stay cables on large bridges that remains a frustrating reminder of the difficulty faced by engineers in modeling satisfactorily the complex way in which the environment and structures interact. The earthquake-induced failures of so many structures in the last 20 years in disasters in countries such as Japan, China, Turkey, India, Pakistan, Taiwan, and the United States provide further reminders of the limits on our ability to predict, understand and tame the great and uncertain forces of nature.

So what are the major challenges facing bridge engineers in the twenty-first century? Are the complexities of designing new bridges really the source of the greatest challenge for bridge engineers today? The public simply demand that our bridges are "safe" and that they perform the function for which they were intended, that is, to keep the traffic on our highways flowing. Historically the structural engineering profession has an exceptional record of designing structures with a very low risk of collapse. This is, in part at least, the result of the conservativeness built in to most designs although this level of conservativeness can vary considerably depending on the degree of refinement used for the analysis. It also shows that there is considerable benefit from designing conservatively as structures will often be called upon to perform different functions from those originally intended and be subjected to unforeseen loads over their service life. These reserves of strength often enable such changes to be accommodated.

It would thus seem prudent to continue our current conservative practices for the design of new structures, as these have proved so successful in the past. In fact there is an argument for investing more in reserves of strength and improved durability at the design stage to reduce the need for strengthening, maintenance and repairs through the lifetime of the structure. The traffic delay costs and disruption caused by works associated with strengthening or repairing a bridge can swamp the actual cost of the repairs themselves or even, in some cases, the cost involved in completely replacing the structure. Perhaps simple measures such as ensuring all bridge decks are waterproofed, as is standard practice in the UK and many continental countries, or increasing the cover to reinforcement in concrete deck slabs by 5 or even 10 mm to reduce the susceptibility of the reinforcement to corrosion, might pay off in the long run many times over. Many bridge-owning authorities have adopted a much stronger emphasis on durability and whole-life cost in recent years. For example, in the UK specific durability requirements are outlined in the Highways Agency's advice note on the Design for Durability of Concrete Structures (BA57/90).

The answer to where the greatest challenge lies may be found in the results of national bridge evaluation surveys that identify the current condition, load-carrying capacity and functional performance of the *existing* bridge infrastructure around the world. Many countries have instigated national bridge assessment programs aimed at verifying that each nation's bridges will be safe for the demands of the new millennium. The U.S. National Bridge Inventory (NBI), which contains details of over 600,000 bridges, reported in 2000 that 25% of the bridges in the United States are structurally deficient and/or functionally obsolete. The cost of road and bridge improvements to overcome these problems has been estimated to be over US$200 billion (McClure 2000). US$23 billion has been allocated to the Highway Bridge Replacement and Rehabilitation Program (HBRRP) for the period 1997 to 2003 alone. In the UK, with a bridge population of around 150,000, estimates of the number of deficient bridges vary from around 10% on the national highway network to over 40% for some counties, which have responsibility for secondary and minor road bridges. The cost of rehabilitating or replacing these bridges has been estimated to be of the order of £4 billion (US$6 billion) (Middleton 1997).

Thus it can be argued that the greatest challenge facing bridge engineers today is the efficient management of this vast stock of existing bridges. Firstly we must ensure that those deemed inadequate are made safe as soon and as economically as possible, and secondly we must maintain the bridge stock in a satisfactory condition and not allow it to deteriorate and become deficient in the future. To achieve these goals there is an economic imperative for bridge engineers to employ the very best methodologies for assessing the load-carrying capacity of our existing bridge stock.

This fundamental shift in emphasis from the design of new structures to the *assessment, maintenance,* and *repair* of *existing* bridges requires bridge engineers to adopt a quite different philosophy and rationale in their thinking. Engineers are now asked to reappraise the bridge population and predict its response under what are often quite different load conditions to those envisaged in the original design and with limited information on the state of the material components.

Many bridges fail to meet functional requirements such as width whilst others suffer from serviceability problems or deterioration. Nevertheless a great many have failed assessment due to the predicted load capacity of the structure being deemed inadequate for the current code loading requirements. For example, the NBI 2000 report lists 55,000 bridges with an intolerable structural evaluation under U.S. Federal guidelines.

The major reason for so many theoretical failures is the dramatic increase in the weight and volume of heavy truck traffic over the last 50 years. Bridge designers of yesteryear could not reasonably have foreseen these increases and, as a result, our bridges were designed to lower loading standards and are inadequate for current load requirements. In addition the design standards and detailing requirements have also changed over recent years as research has provided a better understanding of structural behavior. For example, research in the UK in the earlier 1970s led to a better understanding of shear and changes were made to the design code requirements for shear links in reinforced concrete beams. As a result, few beam-and-slab concrete bridges built in the UK prior to this date comply with the current design standard in shear.

So when we come to the assessment of existing structures we must decide whether our conservative design methodology remains appropriate. Can we afford to condemn thousands of our bridges on the basis of the conventional approach to analysis? It appears we are faced with two choices—either condemn these structures and strengthen or replace them, or else undertake a more refined analysis. In such circumstances we need to reappraise the meaning of the word "failure" and investigate in more depth what our structural analysis modeling actually tells us about the likely behavior of a particular structure.

With such high costs involved, bridge engineers must justify the level of safety adopted for bridge assessment by examining the options available, and the potential risks involved, in reducing this level perhaps even down to the "do nothing" option for bridges with low theoretical load ratings.

We must also ask what are the most important categories of existing bridges on which resources should be concentrated? It is estimated that somewhere around 80% to 90% of all bridges are in the short to medium span range. (Here short span is defined to be less than 20 m and medium span to be between 20 m and 50 m.) In the UK around 65% of bridges are less than 15 m in span. Although not perhaps the most high profile structures in terms of aesthetics or public perception, these short to medium span bridges form the backbone of our infrastructure networks. Despite their relatively short span they still present great challenges, and there is potentially enormous economic benefit to be obtained from adopting a more rational and advanced approach to modeling and analyzing them. Although there are endless possible configurations of bridge forms and component materials, concrete bridges (reinforced and/or prestressed) in the form of slabs or slab-on-girder bridges form one of the most common and important categories of bridges and therefore this chapter will concentrate on how we can improve our ability to assess more realistically the strength and safety of existing short to medium span concrete bridge decks.

The real emphasis in the twenty-first century will be on maintaining the functionality, condition and safety of the existing infrastructure for the changing demands and conditions of the coming decades rather than refining the methods used to design new bridges.

4.2 Structural Analysis

4.2.1 The Analysis Credibility Gap

There is a growing realization that the analytical techniques developed for design are in many cases unable to accurately model the actual structural behavior of existing bridges. As a result, assessments often significantly underestimate the actual load capacity of bridges. This discrepancy between theoretical predictions and reality has been highlighted by the number of bridges that have failed their assessment even though the assessing engineers' experience and intuitive feelings tell them that the bridges are capable of safely carrying significantly higher loads. There are examples cited of bridges that have regularly carried abnormal vehicles weighing 180 tonnes without distress being assessed to have an ultimate load capacity of 7.5 tonnes (Brodie 1997). In many cases, these failed bridges exhibit no outward signs of distress. Although this does not, in itself, imply that failure may not be imminent, it

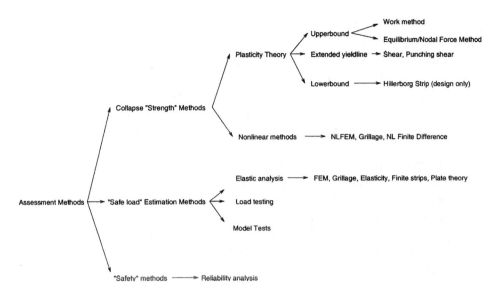

Figure 4-1. Analysis methods for assessing concrete bridges.

is likely that some evidence of damage or significant deformation will precede collapse. This brings into question the appropriateness of using elastic analysis for the determination of ultimate strength for many types of bridge.

Clearly it is important that engineers carefully evaluate the various methods of analysis available to them to ensure the most realistic and relevant ones are used for determining the strength of bridges (Fig. 4-1). To be overly conservative could result in expensive and often unwarranted remedial action such as replacement or strengthening being undertaken. Even placing some form of traffic or weight restriction on a strategic bridge can impose a substantial economic burden on a community.

4.2.2 Elastic Analysis—The Conventional Approach to Assessing Load Capacity

Current codes of practice are written with the implicit assumption that the design and assessment of bridges will usually be undertaken using linear elastic analysis techniques. Elastic theory is well established and understood, is supported by many computer software packages, and has been found most satisfactory for the design of bridges. As a lower bound method the engineer can be confident that the analysis method should be conservative and hence safe. This approach is quite understandable for design where a certain degree of conservativeness costs relatively little. However despite the enormous cost implications, elastic analysis methods are also relied upon as the primary analysis tool for assessing the ultimate strength of existing bridge decks and failure of an individual section or element is used as the criterion to determine structural adequacy even though there is a wealth of evidence from model experiments and full-scale load tests to show that bridge superstructures, and in particular concrete bridge decks, are often able to carry loads well in excess of the theoretical capacity calculated using elastic techniques. In practice, once an individual section of deck has reached ultimate or yield, the failure must develop into a full-collapse mechanism before the structure will actually fall down. In most bridges there is sufficient redundancy and ductility such that the consequences of a single element yielding are likely to be small and may only affect the serviceability of the structure. If one accepts that serviceability criteria do not govern, and collapse is the criterion on which to base the assessment, such conservativeness is difficult to justify.

In the research environment, where the best possible predictive methods are sought to model the actual behavior of bridges, researchers have, almost without exception, used yield-line theory, and in more recent years nonlinear finite element methods, to predict the flexural collapse behavior of concrete bridge decks. The question is, do these provide a practical alternative to elastic analysis for assessing ultimate strength and could they be adopted widely in practice.

Although current codes are predominantly based on the premise that elastic methods will be used for analysis, most codes allow for some redistribution and the use of other forms of assessment such as plastic methods, nonlinear finite elements or load testing in appropriate circumstances. For example, in the UK the assessment code explicitly states that "nonlinear and plastic methods of analysis (e.g., yield-line methods for slabs) may be used with the agreement of the Overseeing Organisation" (BD44/95, Clause 4.4.3). The bridge design code also permits these methods to be used stating that "plastic or yield-line methods may be adopted when appropriate to the form of construction" (BS5400: Part 1:1988, Clause 7.3.1). In the United States, Clause 4.4 of the American Association of State Highway and Transportation Officials Load and Resistance Factor Design (AASHTO LRFD) design code states that yield-line analysis may be used and Clause 4.5.2.3 refers to provisions for inelastic behavior.

4.2.3 Refined Elastic Analysis

Structural analysis is primarily aimed at determining how the stress resultants from the applied loads are distributed throughout the component parts of a bridge. In the case of a well-defined problem we can certainly obtain a unique solution to Navier's equations of elasticity and the result should be the same whether it is derived by slide rule or supercomputer. The majority of design (and analysis) is performed using one of the myriads of elastic analysis programs currently available. Some of these are easier to use than others, but the fundamental principle on which they are all based is the same. The programs use the equations of equilibrium, compatibility, and knowledge of the material behavior (constitutive properties) to solve the equations of elasticity and hence find the distribution of load effects throughout the structure. They require the user to define the applied loads, the boundary restraints, and also to make some assumption about the initial state of stress in the elements (which is usually assumed to be zero). Hence a unique set of stress resultants acting on the elements of the structure is calculated. These stresses are then compared with the strengths of the corresponding components.

As computers become ever faster, finer mesh or grillage sizes and more detailed analyses can now be performed economically. This permits greater flexibility for the designer, and sensitivity analyses of alternate schemes to optimize a design become feasible. But with regards to the fundamental analysis itself, it could be argued that much of the recent development work on computer programs concentrates on simplifying the input and output of information by improving the graphical user interface rather than developing the intrinsic analysis engine itself. The real limits on the analysis relate to the difficulty faced in defining the stiffness of the elements of the structure, the degree of restraint provided by the boundary supports and the initial state of stress in the structure.

So can we improve our current elastic analysis methods and thus obtain higher-rated load capacities? The key issue to recognize here is that, although the solution to a particular elastic problem will be unique, this solution does not necessarily reflect the actual state of stress in the structure being modeled. In practice, it almost certainly will not give the correct stress distribution primarily because the initial state of stress of the components and the boundary conditions in reality cannot be known in the vast majority of cases.

4.2.4 The Lower Bound Theorem—The Use of Cracked Section/Variable Stiffness Analysis

The Lower Bound Theorem of Plasticity underpins all the structural analysis we do; it is the application of this theorem that allows elastic analysis methods to be adopted for design and assessment even if the derived elastic solution is not the actual stress distribution in the structure. It is easy to forget this when so many sophisticated elastic analysis programs churn out stresses at all points in a complex structure with such impressive speed. But what is the meaning of these numbers and in what context are they valid?

Inherent in the Lower Bound Theorem is the assumption that structures, and elements within structures, are in general ductile and hence can yield and allow redistribution of loads to other locations within the structure. The reason this is essential is the fact that the initial level of stress in any real structure is not only unknown but also unknowable. In general, structural analysis is based on the assumption that a structure is initially stress free whereas in practice this will rarely be the case. All structures have locked-in stresses due to the initial fabrication or construction methods (e.g., rolling of steel beams locks in stresses within the beams), imperfections and tolerances in the building and connecting components, uncertain external loading at the time at which the connection to the structure is made, thermal effects from ambient conditions and so on.

The Lower Bound Theorem tells us that our structure will be safe provided a distribution of stress resultants can be found that is still in equilibrium with the applied loads and that nowhere exceeds the yield capacity of the material.

Thus there is scope to modify or optimize the load rating of a bridge deck by the judicious choice of member stiffnesses in, for example, a grillage analysis. This approach is quite widely adopted by bridge engineers in the UK but seems to be less favored in the United States. Some guidance on suitable grillage geometries such as spacing and orientation of members, and the selection of member properties is given in a report by West (1973) and textbooks by Hambly (1991) and Barker and Puckett (1997). The grillage analysis method is ideally suited to many common forms of bridge deck configuration, in particular slab and beam-and-slab, composite slab on steel, or concrete girder bridges. The key parameters to consider are the geometrical configuration and relative stiffness of the grillage members.

4.2.5 Inelastic Rating Procedures in the United States

Inelastic rating procedures for steel beam and girder bridges are examined in the National Cooperative Highway Research Program (NCHRP) Report 352 (Galambos et al. 1993), which recommends a shakedown limit state for rating this type of bridge. This takes into account plastic hinge formation in continuous bridges and the effects of repeated heavy loadings on the ultimate load capacity. It still doesn't represent a full mechanism analysis and assumes failure will be by progressive formation of hinges across midspan. This work is an important step towards the realization that bridge assessment engineers need to look beyond the elastic limit and first yield of an element for the definition of failure.

4.2.6 Compressive Membrane Action (Arching Action)

There is also potential to increase load ratings by taking into account the effect of rigid support restraints. Compressive membrane action, also referred to as *arching action,* can lead to significantly enhanced load capacity and yet it is usually ignored in assessment. The analytical tools exist in modern programs for modeling the support stiffnesses. The difficulty, however, lies in quantifying the magnitude of these restraint stiffnesses. If known, we could allow for them; however, in most practical applications these fall into the unknowable category in the same way as the initial state of stress in a structure. They do, however, give some reassurance to the assessor that untapped reserves of strength are likely to exist. This can be particularly helpful when employing upper-bound methods such as yield-line analysis. The Canadian and Ontario bridge codes include an empirical design method for deck slabs in composite girder bridges. This takes advantage of the membrane enhancement present in such structural elements and results in less reinforcement than would be obtained using elastic analysis. Other codes have yet to follow this lead.

4.2.7 Denton's Equations

An extremely useful computational tool that can potentially increase the assessed capacity of concrete bridge decks was recently developed by Denton and Burgoyne (1996). In the UK, the Wood-Armer equations (Wood 1968; Armer 1968) are used extensively for the design of bridge decks since they produce an optimal reinforcement configuration. However these same design equations are also often used for determining the moment fields when assessing existing bridges. In cases where the reinforcement in the actual bridge is not optimally distributed the predicted capacity can be very conservative. Denton recognized this and derived modified equations that provide a better estimate of the load capacity of the bridge. These equations have been incorporated as a subroutine in the SAM-LEAP5 structural analysis program, which is commonly used in the UK. An example is given in Denton's paper where an increased capacity rating of the order of 30% has been obtained by employing Denton's equations. This approach has now been very widely adopted in the UK and has resulted in very significant savings.

The other alternative is to consider inelastic procedures such as nonlinear finite element analysis and yield-line analysis.

4.2.8 Nonlinear Finite Element Analysis (NLFE)

Many researchers have applied nonlinear finite element techniques to predict the full load history response and ultimate capacity of concrete structures. There are many commercial programs available that can allow for both geometrical and material nonlinearities (e.g., ABAQUS and DIANA). For example, Shahrooz et al. (1994) analyzed a three-span skewed reinforced concrete bridge that was loaded to failure in the field using NLFE. In this research the sensitivity of the results to the support restraint was examined. Taplin and Al-Mahaidi (2000) also used NLFE to predict the ultimate shear capacity of a three-span T-beam bridge that was tested to failure in Australia. Cope (1987) used NLFE to examine a skewed reinforced concrete bridge. He predicted that there would be high stresses in the obtuse corner of the deck slab. This was subsequently verified during a site investigation that found extensive cracking under the asphalt surfacing in this region. Kotsovos and Pavlovic (1995) have used NLFE to study reinforced and prestressed concrete beams and, in particular, shear failure mechanisms. Collins and Mitchell (1991) have also extensively applied NLFE methods to various reinforced concrete structures including complex offshore oil platforms. Wills and Crisfield (1989) analyzed a half-scale prestressed concrete beam-and-slab bridge using the UK Transport Research Laboratory's in-house NLFE research program, however, this analysis predicted a collapse load 25% above that observed in the testing.

Although NLFE methods have developed to a sophisticated state, their applicability is severely limited by their high cost in computing time and the advanced level of expertise required to use them. In addition, the technique is load-history dependent and very sensitive to the choice of material parameters, in particular the assumed tensile strength and shear retention characteristics of the concrete. Another disadvantage often cited is that finite element programs (both linear and nonlinear) usually generate a large amount of output data, and it is often difficult to verify the results using some simple form of hand calculation. In practice it is necessary to calibrate the model parameters used in an NLFE program against test results from similar structural forms before it can then be used with any confidence in a predictive capacity.

There are a large number of research groups around the world working on the development of NLFE methods for the analysis of concrete structures. However the reality is that it remains most suited to in-depth, specialized assessments of major structures or for laboratory research, and it is not presently considered to be a practical option for use in assessing large numbers of existing bridges. This situation could well change in the future as computing developments continuously result in decreasing costs and greater speed with NLFE programs, although the sensitivity of results, need for calibration, and specialized expertise required are still likely to limit their application.

4.2.9 Plastic Collapse Analysis: The Yield-Line Method

The other analytical alternative is to use plastic collapse analysis or yield-line methods for assessing the strength of concrete bridges. Yield-line analysis considers the global collapse of a concrete slab rather than the failure of a single element thus utilizing the full distributed strength capacity of a structure. As a result it is usually significantly less conservative than elastic methods. Up until the development of nonlinear finite element methods in the 1960s and 1970s, researchers investigating the ultimate strength of concrete slabs almost exclusively adopted this approach as the best available theoretical method for predicting flexural strength. Over the last 80 years many large- and small-scale models of slabs and bridges have been tested. Several full-sized highway bridges have also been tested to destruction. These experiments have consistently shown that plastic collapse or yield-line theory is an extremely powerful and accurate tool for predicting the ultimate flexural strength of concrete slab, and many beam-and-slab, structures (Clark 1984).

Although most engineers are taught yield-line theory during their training, it has been, until recently, rarely used for the assessment of concrete bridges. Even though nearly all design and assessment codes permit the use of plastic methods, there exists a somewhat paradoxical situation in that yield-line methods have been used extensively in research but have not, as yet, been widely adopted in engineering practice although this position is quickly changing within the UK. A possible explanation for this is that traditional "hand" yield-line analysis methods can be extremely tedious and, without some form of computer program to facilitate the analysis, they are impractical to apply to anything but the simplest slab geometry, reinforcement, loading and failure mode configurations. In addition, as an upper-bound technique, there is always a degree of uncertainty that the critical failure mode has been found. The usual concern has been that one has to laboriously check a number of possible failure mechanisms for many different load cases and even then, other more intricate geometries might be possible. A further concern is whether or not the bridge has sufficient ductility to justify the assumptions inherent in yield-line theory (Beeby 1997). Incorporating a ductility check is somewhat difficult using conventional yield-line methods.

Although good predictions of ultimate shear strength can be obtained using plasticity theory for reinforced concrete members that contain sufficient shear reinforcement to ensure ductile behavior, further research is required to validate its use for the assessment of shear strength in general practice (Ibell et al. 1997). As a result, shear capacity must still be checked using the conventional elastic approach. Fortunately shear tends not to be the primary cause of failure in most concrete slab bridges, and thus yield-line analysis is well suited to structures that have been found to be adequate in shear but have failed in flexure.

Although yield-line analysis is an upper-bound method, there is a wealth of experience available from the literature on the types of failure mechanisms likely to form under typical highway loading configurations (Clark 1984). In addition, significant reserves of strength are found in many bridges resulting from compressive membrane or arching action and, to a lesser extent, work hardening of the reinforcing steel. This evidence supports the view that, for appropriate types of concrete bridge decks, the method can be applied with confidence provided the limitations of the technique are well understood by the assessing engineer.

The Highways Agency in the UK recognized the potential for applying yield-line analysis to concrete bridge assessment. Whereas elastic analysis programs are widely available, it was found that there were no generally available yield-line analysis programs in widespread use anywhere in the world. The few programs referred to in the literature tend to be very specific and restricted in their applicability to quite simple structural configurations. As a result, the Highways Agency commissioned the Department of Engineering at Cambridge University to develop such a program. This project resulted in a novel collapse analysis program called COncrete BRidge ASsessment (COBRAS), which was validated against model tests, theoretical solutions, and the Transport Research Laboratory's NLFE research program. The development of this yield-line program has provided a very powerful tool with which plastic collapse analyses of concrete bridges can be undertaken for a wide selection of possible failure modes and assessment load cases.

4.3 Generalized Analysis Method for Yield-Line Analysis

The breakthrough that allows a generalized yield-line solution scheme to be computerized is the realization that the yield-line problem can be reduced to what is fundamentally a problem of geometry. Although with the benefit of hindsight this might appear trivial, it must be recognized that no general solution scheme has previously been developed. Using developments in computer graphics and solid modeling theory, an analysis technique has been developed that creates a 3D "picture" of the bridge. This is used to derive all the required geometrical relationships for the failure mechanisms while incorporating features describing the component material properties and the applied loads.

Perhaps the most significant feature of this modeling technique is its ability to analyze rigorously realistic configurations of loading, bridge geometry, support fixity and failure mechanisms without the need to derive mathematical expressions describing the inter-relationship between these parameters. Multilayered, banded and curtailed reinforcement layers can be included. It is also possible to make some provision for the effects of steel corrosion and concrete deterioration.

The analyst selects from a predefined library of failure mechanisms, as shown schematically in Figure 4-2, and the program then iterates through a large number of possible

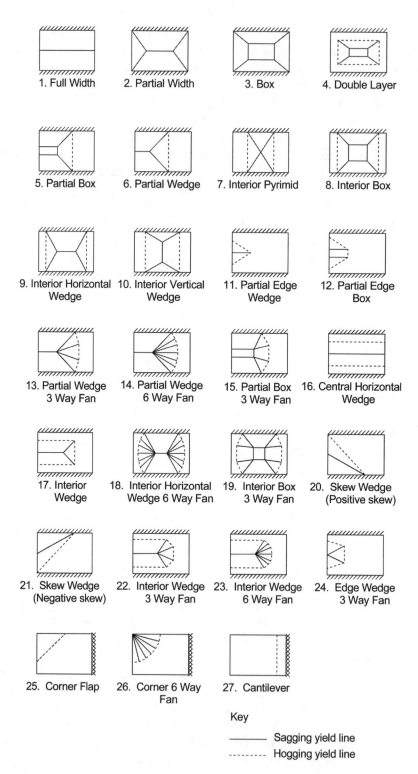

Figure 4-2. Schematic diagram of failure mechanism library.

geometries for each mechanism in search of the lowest, and hence critical, failure mode. As a result, structures that were hitherto impractical to assess by hand can now be analyzed automatically. With a modern portable computer a typical concrete bridge assessment can be performed in a couple of minutes.

The new method revolves around performing the following five tasks:

1. Modeling the bridge and its structural components
2. Modeling the applied live loads
3. Modeling the failure mechanisms
4. Optimizing the failure mode geometry
5. Calculating the ultimate strength and factor of safety (F.O.S.)

4.3.1 Modeling the Bridge and Its Structural Components

The fundamental parameters governing the collapse behavior of a concrete bridge are the geometry in plan, the support fixity, the cross-sectional dimensions, the concrete strength and density, the details of the various layers of steel reinforcement, and the applied loading. The modeling method developed permits each of these features to be separately specified and then merged together using computer graphics solid modeling techniques to form a single bridge structure model. For example, a layer of reinforcing steel can be defined by the outline plan of the bars and also properties such as area of steel (per meter width), yield strength, effective depth, and orientation in plan.

Many other parameters can also be represented in this same way. For example parameters such as strength reduction factors to allow for deterioration in the concrete and/or steel and membrane strength enhancement factors can be defined and merged with the bridge model.

This process of combining all the components together uses principles from set-theory, and the actual merging of component parts is performed using a generalized 3D solid modeling package specifically written for this purpose. The final model represents the entire bridge and incorporates all the required analysis parameters of material components and geometry.

4.3.2 Building the Applied Load Models

Complex loading combinations allowing, for example, for lane loads, vehicle or individual wheel loads, or line (knife-edge) loads can also be represented in terms of regions on which a given intensity of load acts. Since the magnitude and position of applied live loading is independent of the structural components of the bridge, the various load cases to be assessed are assembled in the computer in the specified location on top of the structure model of the bridge deck. However they are not combined with the bridge model. In this way a separate, independent, graphical representation of each load case is stored in the computer and allows the engineer to represent any desired combination or complexity of loading on the structure.

4.3.3 Modeling the Failure Mechanisms

The generalized analysis method generates 3D polyhedral failure models, which are 3D pictures of each of the chosen yield-line failure mechanisms, and stores these in the computer. These solid failure models provide all the required geometric information needed for the work calculation in the yield-line analysis.

One of the major strengths of this approach is that the failure modes are described totally independently of the loading and the structural properties of the bridge, depending only on the shape of the bridge perimeter taken from the boundary representing the plan of the bridge deck.

By incorporating an extensive library of predefined yield-line patterns within the program, the user can easily choose a selection of collapse modes for assessment. Figure 4-2 shows the library of standard failure modes currently incorporated into the COBRAS package. This includes a selection of some of the most commonly reported failure modes for bridge slabs and also some complex fan mechanisms. The library can easily be extended to any practical failure mechanism geometry if required.

By merging the three models representing the structure, the loading, and the failure mechanism, a single 3D solid model or picture of the entire bridge in its collapsed state is produced and stored in a data structure within the computer. This merging of the three component models to form a solid bridge model is accomplished using a solid modeling package developed specifically for this purpose.

Contained within this new solid bridge model are full details of all the information necessary to describe the structural parameters of the material components and dimensions of the bridge, the external loading acting on the bridge, and the required geometric information needed for the collapse mechanism analysis using the work method. This includes the location and length of all the yield lines, the details of abutment fixity at each of the boundaries, and the relative rotations between adjacent rigid plate elements of the failure surface.

4.3.4 Optimization of Failure Mode Geometry

By changing the position of some of the vertices of the solid bridge model, a rapid step-like iteration of the failure mode geometry can be performed (Fig. 4-3). By using this procedure for all the different failure mechanisms chosen from the library of failure modes, a search can be made for the critical global collapse mechanism with the lowest factor of safety.

The significant advantage of this approach lies in its speed and simplicity as it avoids the necessity to derive expressions for the work equation or undertake an often-difficult partial differentiation calculation to obtain an estimate of the critical failure mode geometry. With a computer, a large number of iterations can be examined quickly, thus ensuring that the critical geometry for the particular mode is found to within the accuracy dictated by the selected iteration step size.

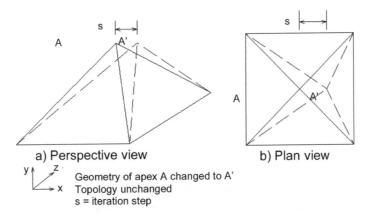

a) Perspective view   b) Plan view

Geometry of apex A changed to A'
Topology unchanged
s = iteration step

Figure 4-3. Iteration of the failure mechanism geometry.

4.3.5 Calculating the Ultimate Strength and Factor of Safety

The load capacity of a bridge is determined using the yield-line method in which a global factor of safety (F.O.S.) is derived. Having applied a given assessment load to a postulated failure mechanism and derived an F.O.S., the parameters defining the failure mode geometry are varied to minimize the F.O.S. and hence determine the load capacity of the bridge. The F.O.S. at collapse is given by the ratio of the calculated strength to the effects of the applied loading.

$$Factor\ of\ Safety\ (F.O.S.)\ /\ \frac{Strength}{Load\ effects}$$

In the work method this equation can be expressed in terms of the energy dissipated in the yield lines or plastic zones (ED) and the work done by the applied loads and the self-weight of the structure (WD).

$$F.O.S.\ /\ \frac{ED}{WL}$$

where

    ED = energy dissipated in the yield lines
    WD = work done by the loads (i.e., self-weight, superimposed dead load, and live load).

An alternative measure of safety, commonly used in nonlinear finite element analyses and bridge assessments, is the *live load factor* or λ-factor, which is defined as the ratio of the applied live load required to cause failure to the initially specified assessment live load, with the self-weight and superimposed dead load remaining constant.

$$Live\ Load\ Factor\ (LLF\ or\ \lambda)\ /\ \frac{Failure\ live\ load}{Assessment\ live\ load}.$$

In terms of the work method used here, the F.O.S. at collapse is 1.0.

$$F.O.S. = \frac{ED}{WD} = \frac{ED}{\lambda \cdot WD_{LL} + WD_{SW} + WD_{SDL}} = 1.0, \text{ and}$$

$$\lambda = \frac{ED - WD_{SW} - WD_{SDL}}{WD_{LL}}$$

where

WD_{SW} = work done by the self-weight of the bridge

WD_{SDL} = work done by the superimposed dead loads

WD_{LL} = work done by the specified live loading.

These two measures of safety, F.O.S., and λ, will usually have different magnitudes. As the work done by the self-weight and superimposed dead load increases as a proportion of the total work, the difference between the two measures widens. This is likely to become more pronounced with larger span bridges where self-weight dominates the loading.

The default setting on the COBRAS package determines the ultimate capacity of the structure by minimizing the F.O.S., but there is an option for the user to undertake the optimization by minimizing the live load factor (λ-factor). In the program, the values of both the F.O.S., and the load factor are determined. The critical failure mode found by minimizing one of these factors will not necessarily be equivalent to that obtained by minimizing the other.

4.4 Other Features of the Generalized Yield-Line Analysis Method

4.4.1 Generalized Moment Capacity Program

Because all the components of the bridge deck are incorporated in the bridge structure model, the rigorous theoretical moment capacity of the actual concrete section about the axis of the yield line allowing for all the orientations, depths, and types of reinforcement that cross the selected yield line can be calculated. In contrast, when attempting such calculations by hand it is usual to simplify the analysis by adopting Johansen's stepped yield criterion or the affinity theorems to account for orthotropic reinforcement layouts. Such simplifications are not necessary in the computerized approach.

4.4.2 Ductility

Since most code measures of ductility, defined in terms of rotation capacity of a section, are related to the geometry and material components along the yield lines of the structure, each of which is fully defined in the solid bridge model, the rotation capacity at all yield-line sections can be checked directly. The method does not ensure ductility is available; however, it does enable all the yield-line sections to be checked for compliance and a warning is given if any section does not satisfy user defined limits on neutral axis depths, and percentage of steel reinforcement.

The limits currently checked are:

1. The percentage of steel crossing all yield lines to warn of heavily reinforced sections: A warning is given if the steel reinforcement percentage, $\rho + 3.5\%$.
2. The neutral axis depth, x: A warning is given if this exceeds 40% of the overall depth of the slab, D (i.e., if $x + 0.4 \times D$).

4.4.3 Geometric Compatibility of Failure Mechanism

The 3D solid bridge model representation of the collapsed shape of the structure provides a simple method for checking geometric compatibility. For the solid shape to form a valid failure mechanism, all the faces of the solid must be planar. A routine incorporated in the program checks the planarity of all the faces of each of the chosen failure modes and prevents analysis of an incompatible mechanism. Thus one of the primary requirements of yield-line theory can be automatically checked.

4.4.4 Reinforcement Corrosion and Concrete Deterioration

Allowance can be made for deterioration in the steel reinforcement due to corrosion or a reduction in the concrete strength. An affected region is defined by its geometric location and the magnitude of the deterioration factor appropriate to the materials in the region specified. For reinforcement, a factor on the area of steel is used to model the effects of corrosion. Thus, any reinforcement within the corrosion-affected zone has the area of steel reduced by the corrosion reduction factor that must be selected by the assessor. In an identical manner, areas of concrete with reduced strength are identified and a concrete deterioration factor applied. This is recognized to be a simplistic approach to the problem of deterioration and makes no allowance for potential problems from loss of bond, delamination, or spalling; however, it does provide a means by which some measure of the effects of deterioration on flexural strength can be made.

4.4.5 Membrane Action in Concrete Bridges

It is well known that in many practical situations there is often some restraint to the lateral expansion of a slab deflecting under external loading. This restraint can be provided by edge beams, diaphragms, or the supports to the slab and results in internal arching action within the depth of the slab that can significantly enhance both the flexural moment capacity of the structure as well as its resistance to punching shear under concentrated loading.

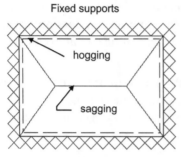

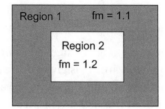

a) Restrained slab with
   uniformly distributed loading

b) Feature regions for
   membrane enhancement

Figure 4-4. Example regions for membrane enhancement factor.

There are several publications that detail the many experimental and theoretical studies of compressive membrane action in concrete slabs and the various analytical models that have been proposed. The major difficulty faced by all researchers in this field has been finding a method for quantifying the amount of enhancement for all the possible variations of reinforcement percentages, span/depth ratios, edge fixity conditions, and load configurations. Johansen's classical yield-line theory makes no allowance for membrane action. Consequently the method has often been found to significantly underestimate the load-carrying capacity of restrained slabs. The COBRAS package does not attempt to derive membrane enhancement factors for any given bridge type; however, the program does provide a means by which the results of various theoretical studies can simply be included.

The method is based on the assumption that the degree of membrane strength enhancement can be represented in geometric terms by a region of the bridge deck, defined as a *membrane feature region,* in which there is a specified increase in moment capacity of the slab. This allows for variation in the amount of membrane enhancement in different areas of the deck or in zones with differing degrees of restraint (Fig. 4-4). This method is based on the approach presented by Rankin et al. (1991) for the case of restrained slabs subjected to uniformly distributed loads. The moment capacity of the slab within each membrane feature region is multiplied by the user specified membrane enhancement factor, f_m.

It is recognized that this is a rather simplistic approach to membrane action but until further research provides solutions to this problem, the above approach is one technique that allows some consideration of membrane enhancement to be included in an assessment.

4.4.6 T-Beam Effect in Beam-and-Slab Structures

The current analysis model adopts a simplified approximation technique for calculating the flexural moment capacity of a bridge deck with different cross-sectional thicknesses such as a beam-and-slab deck or slab deck with edge beams. In this approach, the moment capacity of each section of deck of different thickness is

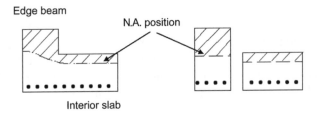

Edge beam

N.A. position

Interior slab

a) Actual N.A.
   position

b) Discrete element N.A.
   position used by COBRAS

Figure 4-5. The T-beam effect.

assessed separately. Thus a different neutral axis position will be derived for a section of slab adjacent to an edge beam element as shown in Figure 4-5.

At present, only pure rotation about the yield-line axis is considered. This has been shown experimentally to produce good agreement with tests for a variety of slab and beam-and-slab models. However, difficulties begin to arise when a yield line is found to cross a structure such as an M-beam (UK) or I-beam (United States and Australia) deck at an acute angle to the longitudinal axis of the beam, in which case it is unlikely that a long, flexural yield line will form along the beam itself. In such a situation it is far more probable that some local deviations in the yield-line pattern will occur, and some form of stepped yield line incorporating a predominantly flexural failure of the beam (or else a flexural-shear or pure shear failure) or else some form of torsional hinge might form. Modeling such behavior introduces a number of complexities that are not allowed for in the methodology developed here.

4.5 Limitations on the Analysis Method

- The primary limitations on the new generalized collapse analysis method are those inherent in the fundamental assumptions of classical plasticity theory and yield-line theory. These include the assumption that the yield-line sections will display rigid plastic behavior and that all the energy dissipated in the structure at collapse will be concentrated in localized plastic zones between undeforming, rigid plate elements.
- As an upper-bound method, there is always the possibility that other more critical failure mechanisms exist that are not considered by the program.
- The analysis method at this stage has only been implemented for flexural failure modes. Shear is not checked with this method and must be assessed separately.
- Beam-and-slab structures are currently assumed to act as independent elements in flexure and, although conservative, this could be improved by including effective flange widths into the analysis.
- Deep or heavily reinforced beams in a beam-and-slab bridge may not satisfy the ductility requirements of yield-line theory and must be considered with

caution. The yield-line method was original developed for concrete slabs although there is much evidence to also support its use with beam-and-slab structures, provided the limitations of the method are understood. The program incorporates a facility for identifying sections with potential ductility problems. Any section with high reinforcement percentages or a neutral axis location greater than user-defined limits is identified with a warning message during execution of the program.

- Provision has been made within the program for including a membrane enhancement component although the difficulty in selecting appropriate values still remains.

4.6 Validation of the COBRAS Yield-Line Analysis Program

To confirm the validity of the program, a number of forms of calibration were undertaken. Firstly, the program was checked against a number of published analytical solutions that confirmed the correct theoretical result was obtained in each case.

Secondly, a calibration study was undertaken in collaboration with the Transport Research Laboratory's (TRL) Structural Analysis Unit to compare predictions for collapse load and failure mode geometry obtained using the COBRAS program with those obtained by the TRL's nonlinear finite element program, NFES, for a number of different bridge structures under a variety of load configurations.

In this study with TRL excellent agreement between the collapse load and failure mode geometry predicted by COBRAS and NFES was obtained for all the bridges analyzed. The difference in predicted ultimate capacity was less than 4% in all the examples assessed except for four specific cases where a conservative assumption about the strength of edge-beams in the COBRAS program resulted in a maximum of 13% underestimation of the collapse load.

Thirdly, the program was used to predict the failure mode and collapse load for a number of experimental tests on concrete slabs. Although there have been numerous tests over the years to verify yield-line theory, a series of tests was conducted at Cambridge University specifically aimed at validating the theoretical predictions of collapse load obtained using the COBRAS program. To date, a total of 13 different tests have been carried out as part of an ongoing validation program. The model slabs were scaled at approximately 1/10th the size of a full-scale bridge in Scotland that had been tested to destruction by the TRL in 1992. This resulted in the model slabs being nominally 600 mm in length by 1,000 mm wide and 40 mm thick. In one set of tests the slabs were skewed at 30 degrees, and in another the slabs were widened to 1,500 mm to examine failure mechanisms contained wholly within the central region of the slab. Various reinforcement configurations were considered, with and without transverse and top steel and with varying percentages of each.

Truck loading was simulated using a two-wheel axle load that was applied at midspan in all but one of the 13 tests. In the exception a solitary point load was used. The goal was to force the model structures to fail in some form of complex fan mechanism rather than just a full-width transverse yield-line at midspan (which is often found to be critical under the uniformly distributed lane load pattern that is specified in the UK bridge assessment code (BD21/01)). Such a fan mechanism puts a greater demand on the ductility of the slab as well as on the predictive capabilities of the computer program.

4.7 Results from the Cambridge Model Tests

The results from these model tests are shown in Table 4-1, which compares the failure loads predicted using the COBRAS plastic collapse program (P_{COBRAS}) and the actual failure loads measured in the laboratory (P_{test}).

In Table 4-1 it can be seen that in all but the final two tests (A1, A2), the yield-line method was conservative in predicting the capacity of the model slabs. The mean value of the ratio of the measured failure load to the predicted load was 1.13, with standard deviation 0.13. Values ranged from 0.87 in test A2 to 1.33 in test C2, with the range being between 1.04 and 1.33 for the first 11 tests. By way of example, Figures 4-6a and 4-6d show the failure mechanism patterns obtained in two of these tests (K4 and C3), and Figures 4-6b and 4-6c show the corresponding critical yield-line pattern predicted using COBRAS overlaid on the observed soffit crack pattern. (Slab K4 was tested twice—once with a single axle load at each side of the slab.)

Examination of the specimens in tests A1 and A2 after failure suggested that a breakdown in the bond between the concrete and the smooth, shiny, 4 mm diameter bars used to reinforce the slab may have caused these two to fail at lower than the expected results. Further tests subsequently confirmed this hypothesis, which emphasizes the importance of bond in the collapse behavior of reinforced concrete structures.

An important observation in all these experimental tests in the laboratory was that substantial deformation and cracking developed well before the maximum load capacity was reached. Thus if a structure has been in service for many years with no visual evidence of distress, the assessing engineer can be reasonably confident that the structure is capable of sustaining significantly higher loads than those already experienced by the structure. Clearly this does not mean the structure is necessarily capable of sustaining the full 40 tonne load, as it may never have been subjected to loads near the maximum legal limit. However it does give some reassurance to the assessor that collapse is not imminent at the loads to which the structure has already been subjected.

Fundamental to this statement is the assumption that the critical failure mode will be flexural, and the structure is sufficiently ductile to allow such a mechanism to form. Shear failures may occur in a brittle manner and may not give warning of impending failure (although a recent test program at Cambridge on shear in beam and slab bridges has indicated that significant cracking usually precedes shear failure at loads well below the ultimate collapse load in most (but not all) cases (Ibell et al. 1997)).

As a result of the development of this yield-line analysis method and program to implement this approach, yield-line analysis is fast becoming one of the most commonly adopted alternative methods of analysis in the UK for evaluating the ultimate strength of concrete bridge decks found to be inadequate in flexure. The program

Table 4-1. Predicted Versus Actual Ultimate Loads for Model Bridge Slabs

Test No.	Hudson Tests				Kite Tests				Collins Tests			Antill Tests	
	H1	H2	H3	H4	K1	K2	K3	K4	C1	C2	C3	A1	A2
P_{test} (kN)	29.2	28.3	27.9	19.0	23.4	23.5	24.1	22.7	36.8	37.2	31.3	18.0	25.6
P_{COBRAS} (kN)	24.5	25.3	23.0	17.2	22.5	22.3	20.5	17.7	N/A	28.5	25.7	19.0	29.3
$\dfrac{P_{test}}{P_{COBRAS}}$	1.19	1.12	1.21	1.10	1.04	1.05	22.5	1.28	—	1.33	1.22	0.95	0.87

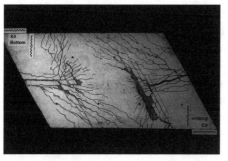

(a) Test slab K4 – actual crack pattern

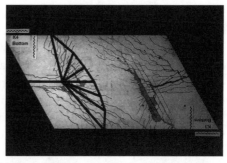

(b) Test slab K4 – predicted yield-line pattern

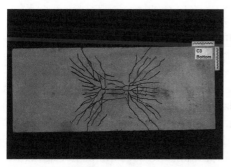

(c) Test slab C3 – actual crack pattern

(d) Test slab C3 – predicted yield-line pattern

Figure 4-6. Examples of actual and predicted failure modes for concrete bridge models.

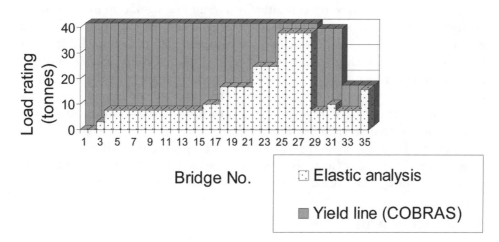

Figure 4-7. Comparison of bridge load ratings obtained using elastic and yield-line analysis.

has been employed by over 40 bridge authorities or consultants to reassess dozens of bridges that had been "failed" using conventional analysis.

The effectiveness of this approach was demonstrated in a survey of 35 bridges that had failed their original assessment and were then reassessed using this yield-line program. Twenty-eight (80%) were upgraded to the full assessment load capacity of 40 tonnes, 3 (9%) were upgraded to 38 tonnes, and 4 (11%) remained at, or were slightly upgraded to 16 tonnes. These results are shown graphically in Figure 4-7. It is evident that this approach can result in very substantial savings to bridge-owning authorities if applied in the appropriate circumstances to short and medium span concrete bridge decks.

4.8 Conclusion

There is clearly an economic imperative to refine and extend the current methods of analysis used for assessing the load-carrying capacity of existing short to medium span bridges with billions of dollars at stake. There are a number of ways in which bridge engineers could use more advanced analysis methods to more realistically predict bridge behavior. The first would be to more widely accept the definition of failure at the ultimate limit state as "collapse of part or all of a structure" rather than first yield of an individual component. Secondly, there is significant scope for much greater utilization of the lower bound theorem to manipulate element stiffness properties and hence optimize the distribution of stress resultants in a structure such that an improved estimate of load capacity is obtained. Thirdly, far wider use of inelastic and plastic methods of analysis, and in particular yield-line analysis, would result in far more realistic predictions of the load capacity of our bridges.

It would seem that the potential benefits of adopting more advanced computational tools for bridge assessment are widely recognized and acknowledged by practitioners but there has nevertheless been a reluctance to implement them.

Overall, the use of elastic analysis methods for assessing the ultimate load capacity of concrete bridges may in many situations result in a significant underestimate of strength. The development of the COBRAS yield-line program provides a very powerful alternative tool with which plastic collapse analyses of these bridges can be undertaken for a wide selection of possible failure modes and assessment load cases. As an upper-bound approach, care must be used in applying this technique; however, there is substantial theoretical and experimental evidence to support its validity for concrete bridge decks in which sufficient ductility exists to justify the assumptions inherent in yield-line theory.

There are enormous opportunities for those in the profession willing to employ more advanced methods of analysis to improve the modeling of the complex interaction between the applied live loads and the bridges that are required to support them.

Acknowledgments

The support of the Highways Agency and the Transport Research Laboratory in the development of the COBRAS analysis program described here is gratefully acknowledged. The assistance of Dr. C. T. Morley, and Mr. P. Fidler of Cambridge University, Mr. J. Wills of TRL (now retired) and Mr. B. Sadka and Mr. S. Chakrabarti of the Highways Agency are also gratefully acknowledged. Any views expressed are not necessarily those of the supporting organizations.

References

ABAQUS user's manual Vols. I and II; version 6.1. (2000). Hibbitt, Karlsson and Sorenson, Inc., Providence, R.I.

American Association of State Highway and Transportation Officials (AASHTO). (1996). *Standard specification for highway bridges,* 16th Ed., AASHTO, Washington, D.C.

American Association of State Highway and Transportation Officials (AASHTO). (1998). *LRFD bridge design specification,* 2nd Ed., AASHTO, Washington, D.C.

Armer, G. S. T. (1968). "Correspondence." *Concrete,* 2, 319–320.

BA57/90. (1990). *Design for durability of concrete structures,* Her Majesty's Stationary Office, London.

Barker, R. M., and Puckett, J. A. (1997). *Design of highway bridges,* Wiley, New York.

BD21/01. (2001). *The assessment of highway bridges and structures,* Her Majesty's Stationary Office, London.

BD44/95. (1995). *The assessment of concrete highway bridges and structures,* Her Majesty's Stationary Office, London.

Beeby, A. W. (1997). "Ductility in reinforced concrete: why is it needed and how is it achieved?" *The Struct. Engr.,* 75(18), 311–318.

Brodie, A. (1997). "Report on progress and issues arising from the assessment and strengthening program in Scotland." *5th Annual Surv. Bridge Conf.,* Nottingham, UK, The Surveyor Magazine.

Clark, L. A. (1984). "Collapse analysis of short-to-medium span concrete bridge." *Contractor's Report, CRR 528/577/124,* Transport and Road Research Laboratory, Crowthorne, UK.

COBRAS, *Concrete bridge assessment program manual, Version 1.7 Release 8.* (2002). Cambridge University Engineering Department, Cambridge, England.

Collins, M. P., and Mitchell, D. (1991). *Prestressed concrete structures,* Prentice Hall, New Jersey.

Cope, R. J., ed. (1987). *Concrete bridge engineering: performances and advances,* Elsevier Applied Science, London.

Denton, S. R., and Burgoyne, C. J. (1996). "The assessment of reinforced concrete slabs." *The Struct. Engr.,* 74(9).

DIANA. (1998). *Finite element analysis—User's manual release 7,* TNO Building and Construction Research, F. C. De Witte and P. H. Feenstra, eds., Delft, The Netherlands.

Galambos, T. V., Leon, R. T., French, C. W., Marker, M., and Dishingh, B. (1993). "Inelastic rating procedures for steel beam and girder bridges." *NCHRP Rep. 352,* Transportation Research Board, Washington, D.C.

Hambly, E. C. (1991). *Bridge deck behaviour,* 2nd Ed., E & FN Spon, London.

Ibell, T. J., Morley, C. T., and Middleton, C. R. (1997). "A plasticity approach to the assessment of shear in concrete beam-and-slab bridges." *The Struct. Engr.,* 75(19), 331–338.

Kotsovos, M. D., and Pavlovic, M. N. (1995). *Structural concrete. Finite element analysis for limit state design,* Thomas Telford, London.

McClure, S. M. (2000). *NBI Report 2000,* National Bridge Inventory Study Foundation. <www.nationalbridgeinventory.com>.

Middleton, C. R. (1997). "Concrete bridge assessment: An alternative approach." *The Struct. Engr.*, 75(23,24), 403–409.

Rankin, G., Niblock, R., Skates, A., and Long, A. (1991). "Compressive membrane action strength enhancement in uniformly loaded, laterally restrained slabs." *The Struct. Engr.*, 69(16), 287–295.

SAM-LEAP5 Bridge Software Program, Bestech Systems Limited, Horley, UK.

Shahrooz, B. M., et al. (1994). "Nonlinear finite element analysis of deteriorated RC slab bridge." *J. Struct. Engrg.*, 120(2), 422–440.

Taplin, G., and Al-Mahaidi, R. (1997). "Theoretical analysis of the reinforced concrete T-beam bridge at Baranduda." Austroads Bridging the Millennia Conf., G. J. Chirgwin, ed., Austroads, Sydney, 151–165.

West, R. (1973). "The use of a grillage analogy for the analysis of slab and pseudo-slab bridge decks." *Research Report 21*, Cement & Concrete Association, London.

Wills, J., and Crisfield, M. A. (1989). "No-tension finite-element and mechanism analyses of a half-scale beam-and-slab bridge deck." *Research Report 217*, Transport and Road Research Laboratory, Crowthorne, UK.

Wood, R. H. (1968). "The reinforcement of slabs in accordance with a predetermined field of moments." *Concrete*, 2, 69–76.

5

Geotechnical Structures

Jack Pappin

5.1 Behavior of Geotechnical Materials

This chapter covers geotechnical structures such as embankments, slopes, cuttings, caverns, and tunnels. It also includes combinations of soil and reinforced concrete or steel structures where the behavior of the geotechnical material is important to the deflections of the soil and stress in the structural elements, including piles subject to lateral loading, retaining walls, and basements.

The behavior of the geotechnical materials is an important element of the design for these situations. It is well known that soils are highly nonlinear from very small strain levels and also have quite small limiting strength. Traditionally the method of modeling these materials is to assume they have linear elastic behavior. Where failure of the soil is an important aspect, the stress states in the soil are also limited to not exceed failure. There are many computer programs that model this behavior and many are becoming available that model the full nonlinear behavior of the soil allowing for reduction of the soil shear stiffness with increasing levels of shear strain. An additional input to geotechnical design is situations where dynamic effects are important. The response of geotechnical structures and soil structure systems to seismic loading is the main case that requires this effect to be modeled.

The chapter begins by describing the main soil properties that affect the analysis of complex structures. These include failure characteristics, nonlinear stiffness properties, and the effects of cyclic loading and time. Geotechnical structures, where the behavior of the soil dominates the design, including tunnels, slopes, embankments, and reclaimed land are then discussed. This is followed by a discussion of situations where soil structure interaction effects must be considered, where both the behavior of the soil and structural elements are significant. Finally seismic loading is considered, including the effects of site response, dynamic soil structure interaction, and dynamic soil deformation on foundation elements.

5.2 Soil Properties

5.2.1 Behavior of Soils

As for computational modeling of any material, it is important to understand how the material will strain or deform under various changing stress states. This enables constitutive equations to be derived that model the various characteristics of the behavior that will significantly affect the design being analyzed. There is a wide range of design problems and different characteristics that need to be addressed for different situations. The following discussion addresses the behavior characteristics of various soils that are relevant to the design situations discussed later in the chapter. These characteristics include stress-strain nonlinearity, time dependent behavior, and response to cyclic loading.

Soil is a multiphase material comprising particles of minerals with air and or water between them. Soils range from coarse grained gravels, to sands, down to silts and clays. The stress-strain behavior of the soil particle matrix is controlled by the effective stresses that exist between the particles. The effective stresses equal the total stress acting on the soil as a whole minus the pore fluid pressure that is acting between the particles.

When a coarse grained material is loaded, the pore fluid can usually flow between the particles, and the soil reacts quickly to the applied loading in the same way as if the pore fluid had not been present (if it is at zero pressure). This is not the case when fully saturated sand is subjected to rapid loading arising from earthquakes, impact loading, or even wind loading. In these cases, unless the soil is very coarse, the pore water fluid does not have time to flow between the particles, and changes in the all round compression (or tension) stress will be directly transmitted to the pore water. This is because the pore water is essentially incompressible.

This tendency for the changes in compression stresses to be transmitted to the pore water fluid occurs even for static changes in load when clays are involved. The pore water can take a considerable time, up to several years, to be squeezed from between the clay particles and leads to long-term reduction in volume of the soil, referred to as consolidation. In the short term the clay behaves as an incompressible material but will still experience shear strains and ultimately shear failure if sufficiently large shear stresses are applied to it. The strength of the soil in these circumstances is referred to as the undrained shear strength.

5.2.2 Failure

All soils exhibit shear failure when the applied shear stresses reach a critical level. The failure strength is a function of the intergranular friction between the soil particles and the all round effective confining stresses applied to the soil matrix.

For silts and coarser materials, the failure strength can be conveniently described by an internal angle of friction ϕ'. Generally the Mohr Coulomb expression $\sigma'_1 = \sigma'_3 * \tan \phi'$ adequately defines the frictional strength. The ϕ' value varies between about 30° to 45° depending on the mineralogy, particle angularity, and the density of packing of the particles (Stroud 1988).

For long-term loading where the pore pressure has time to reach equilibrium, clay materials exhibit a similar failure behavior with ϕ' values ranging between 20° and 25°. Previous stress is very important for the short term or undrained strength of clays, however. For heavily overconsolidated clays that have been subjected to high compression stresses in the geological past, the undrained shear strength is closely related to the original preconsolidation pressure rather than to the stresses that exist at the present time. This is because the state of packing of the particles is largely controlled by the original high compressive stresses. The relationship between undrained shear strength (c_u) and overconsolidation ratio (OCR) is observed (Ladd et al. 1977) to be as follows:

$$c_u \alpha \ OCR^{0.8} \tag{5-1}$$

5.2.3 Stiffness

It is convenient to consider the stiffness behavior of soils in two parts as follows.

The *volumetric behavior* that is largely related to the change in the all round stresses or the normal effective stress $p' = (\sigma'_1 + \sigma'_2 + \sigma'_3)/3$. Both clays and sands tend to follow a similar behavior where the volumetric strain $\varepsilon_V = a * \Delta (\log p')$ where a is a constant and is much greater for primary loading than for subsequent unloading and reloading as illustrated in Figure 5-1.

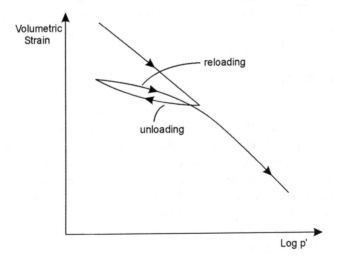

Figure 5-1. Volumetric strain against log p' for 1D compression of clay.

The *shear behavior* can be described by the shear strain γ that is generally related to the changes in applied shear stress τ. In the simplest elastic form the shear strain can be expressed as $\gamma = G * \Delta\tau$. It must be remembered that shear stress and strain are five-dimensional parameters. The shear modulus G is a function of stress level, stress history, and the state of the particle packing.

It is observed that at very small strains, the order of 10^{-5} or less, soils exhibit a linear elastic response. This small strain shear modulus G value, referred to as G_0, can be measured directly in the field and in the laboratory by measuring the shear wave velocity V_S within the soil. From physics the shear modulus $G_0 = \rho * V_S^2$ where ρ is the soil bulk density generally in the range of 1,700 to 2,000 kg/m^3. It is observed (Atkinson 2000) that the G_0 value is related to the stress level p', the state of packing of the soil, the OCR, and the soil plasticity as follows.

$$G_0 / \ A\, p'^{n} (OCR)^{m} \tag{5-2}$$

where A, n, and m are functions of soil plasticity as shown in Figure 5-2 and G_0 and p' have units of kPa. G_0 can exhibit anisotropy. In London clay, for example, where the horizontal stress is generally greater than the vertical stress due to the previous geological stress history, the shear modulus in the horizontal plane G_{0HH} is at least 50% greater than that of the shear modulus in a vertical plane G_{0VH} (Simpson 1999).

At larger shear strains significant nonlinear shear stress and shear strain behavior is observed. Figure 5-3, for example, shows how the secant shear modus G varies as a ratio of G_0 when plotted against the amplitude of shear strain imposed in simple cyclic shear tests. The associated damping, arising from the soil hysteretic behavior is also shown. It is notable how the plasticity of the soil (expressed by the Plasticity Index, PI) dramatically affects this observed nonlinearity.

For complex stress paths that can be experienced in an earthquake and other dynamic loading, the shear stiffness will be constantly changing in response to a series of random stress paths with varying amplitudes. Figure 5-4 shows how the soil is likely to react to this type of loading and how this can be represented by a series of 1D spring slider models as originally proposed by Iwan (1967).

5.2.4 Soil Dilation

The behavior described above is based on the premise that volumetric strain is the result of change in p', and that shear strain results from changes in the applied shear stress. This assumption is implicit in all elastic formulations. It is well known, however, that volumetric strain often results from changes in applied shear stress. Loose soils, for example, tend to contract when a shear stress is applied. Dense soils tend to dilate or expand. This class of behavior is referred to as dilation and is observed to be a function of the soil particle packing density, which, for clay soils, is controlled by the previous stress history or OCR (see Fig. 5-5). If the soils are saturated and undrained during this type of test, they cannot change in volume due to the pore fluid being effectively incompressible, and the resulting effective stress paths

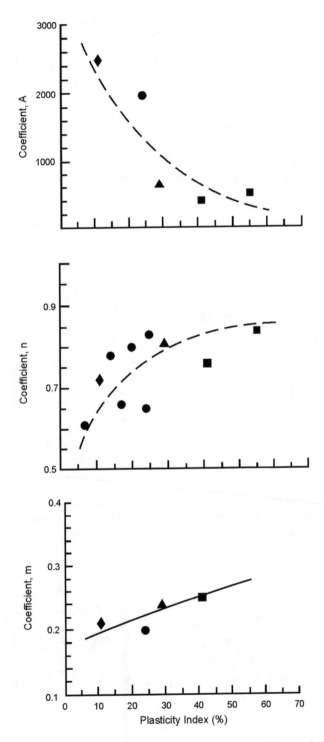

Figure 5-2. Material parameters for G_0, from Atkinson (2000), with permission of Thomas Telford.

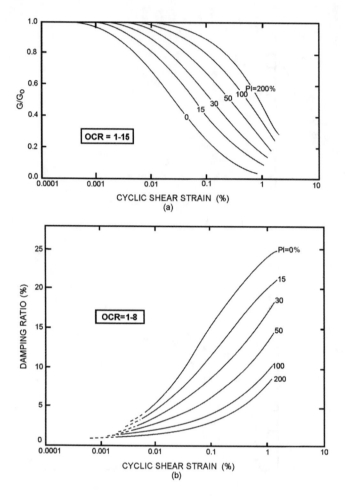

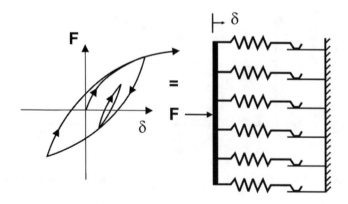

Figure 5-3. Relationships between G/G_0 and damping ratio with cyclic shear strain and soil plasticity for normally consolidated and overconsolidated clays (Vucetic and Dobry 1991), with permission from ASCE.

Figure 5-4. Iwan type soil model.

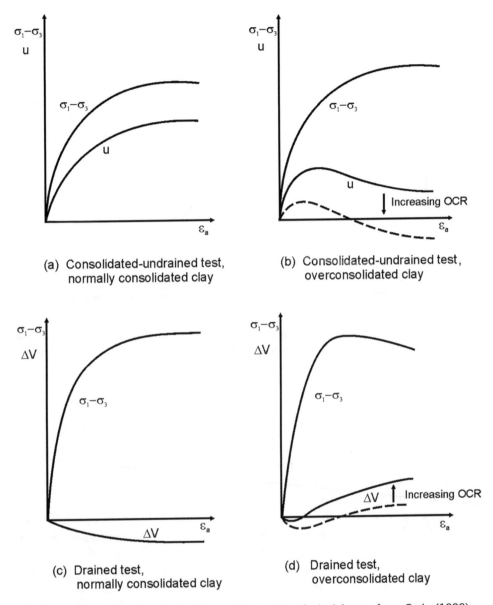

(a) Consolidated-undrained test,
normally consolidated clay

(b) Consolidated-undrained test,
overconsolidated clay

(c) Drained test,
normally consolidated clay

(d) Drained test,
overconsolidated clay

Figure 5-5. Typical results from consolidated-undrained triaxial tests, from Craig (1999), with permission from Taylor & Francis.

are shown in Figure 5-6. It is seen that as the shear stresses increase, the pore pressure within normally consolidated clay tends to increase, leading to reduced mean effective stress and low soil strengths. For heavily overconsolidated clays the pore pressure reduces increasing the mean effective stress and the soil strength. This behavior is reflected in the undrained shear strengths predicted by Equation 5-1. Cohesionless soils exhibit a similar behavior, and uniformly graded loose sands can show a reduction in strength to almost zero as illustrated in Figure 5-7.

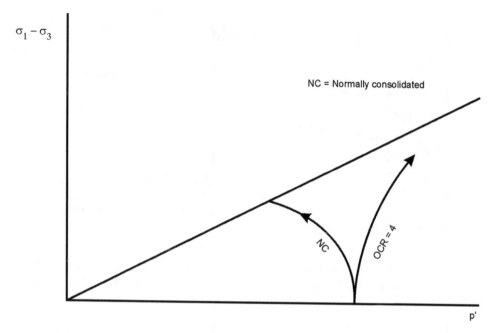

Figure 5-6. Effective stress paths observed in undrained shear tests on clays, from Lambe and Whitman (1969), with permission from John Wiley.

The soil shear strain also exhibits a response to changes in the effective mean stress p'. A classic example is the dead load test applied to loose sand where a constant shear stress is applied while the confining stress is progressively reduced. Shear strains are observed and eventually the sample experiences shear failure when the frictional strength is exceeded.

It is often considered that these effects are limited to first-time or initial loading and that subsequent unloading and reloading will follow an essentially elastic behavior. Cyclic triaxial loading tests show that this is not the case, however. Even with a dense crushed rock pavement sub-base material, it is observed that cyclic shear strains will result from stress paths only varying p', and to a lesser extent cyclic volumetric strains result from stress paths that only vary the shear stress. The diagrams comprising Figure 5-8 demonstrate these observations. Figure 5-8a shows a set of cyclic stress paths applied in a triaxial apparatus. This series of stress paths were applied at a range of p' values. The measured cyclic strains are represented by a series of contours as shown in Figures 5-8b and 5-8c for volumetric and shear strains respectively with the cyclic strains being determined by comparing the contour values at each end of the applied cyclic stress path. Horizontal cyclic stress paths are at constant shear stress, and vertical stress paths are at constant p'.

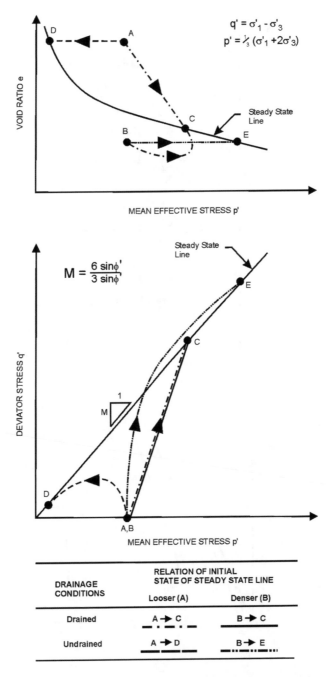

Figure 5-7. Typical steady state line, showing the effect on the soil behavior in triaxial compression, of soil state relation to the steady state line. From Sladen et al. (1985), with permission from NRC Research Press.

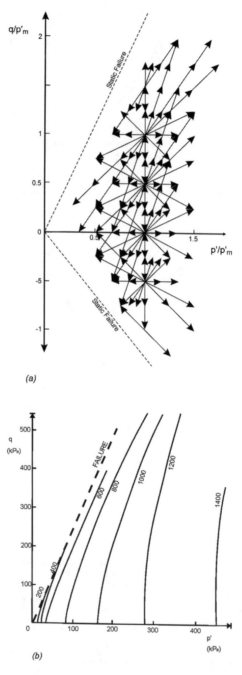

(a)

(b)

Figure 5-8. Results from cyclic local triaxial tests on samples of a dense crushed rock road sub-base material. a) Cyclic stress paths applied to sample at one value of mean normal stress p'_m ($q = \sigma_1 - \sigma_3$). b) Resilient volumetric strain contours ($\mu\varepsilon$) in $p'-q$ stress space. c) Normalized resilient shear strain contours ($\mu\varepsilon$) in $p'-q$ stress space. From Pappin and Brown (1980), with permission from Thomson Publishing Services.

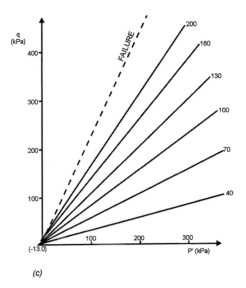

(c)

Figure 5-8. *continued.*

The Brick model, by Simpson (1992), captures this complex behavior in an ingenious way. This model is based on the premise that as a soil element is moved in strain space, a series of kinematic yield surfaces are being pulled with the soil element. Each yield surface represents part of the total stiffness response of the soil. If the strain reverses such that the current strain state is within the yield surface, an elastic response for that part of the stiffness is observed. Initially on a strain reversal the current strain state will be within all the yield surfaces, the material will behave elastically, and the soil will exhibit a relatively high stiffness. As the opposite side of the smaller yield surfaces are met, that part of the stiffness is eliminated (to behave plastically), and the stiffness reduces. This gives an identical result to the Iwan approach described above in Figure 5-4.

The model is illustrated for a 2D example in Figure 5-9. Figures 5-9a and 5-9b show the case above for a complete strain path reversal. Figure 5-9c shows the case for an undrained shear test after normal consolidation. Even though the yield surfaces are being pulled to the new strain position, they also respond elastically to the change in strain direction, and the mean effective stresses are therefore predicted to reduce even though the volume of the soil element is not changing. While only three surfaces are shown here for clarity, usually at least 10 surfaces are used in the model. Figure 5-9d shows the resulting predicted effective stress path that is seen to be similar to that in Figure 5-6 for a normally consolidated material.

This example illustrates a key feature of the BRICK model. The model also replicates failure of the soil and consolidation characteristics. A full explanation of the model and the Fortran (IBM) coding is given elsewhere (Simpson 1992).

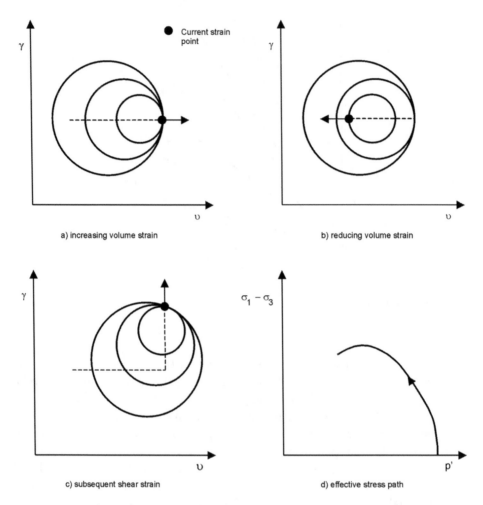

a) increasing volume strain

b) reducing volume strain

c) subsequent shear strain

d) effective stress path

Figure 5-9. Strain paths a) to c) illustrating kinematic yield surface locations following applied strain path (note positive volume strain is compression). The strain paths in b) and c) follow that in a); d) is the effective stress path resulting from the strain path c).

5.2.5 Effects of Cyclic Loading

While the behavior of soil described above covers most aspects of static or a single cycle of loading, the cumulative effects of many cycles of loading has not been discussed. It is observed, for example, that sands will usually tend to densify if subjected to repeated cycles of loading. Figure 5-10, for example, shows the observed volume changes when sand of various relative densities (expressed here by its standard penetration test (SPT) N value) is subjected to cycles of shear strain loading. As can be seen the volume strain reduces significantly for the denser materials. If the sand is fully saturated, the tendency for the sand to reduce in volume is prevented by the pore water fluid. This leads to pressure build-up in the pore water fluid and can lead to the situation where the p' value becomes almost zero, and liquefaction results with the soil having almost no shear strength. It is observed that dense sands never experience this phenomenon. A thorough review of the evaluation of liquefaction potential is presented elsewhere (Seed et al. 2003).

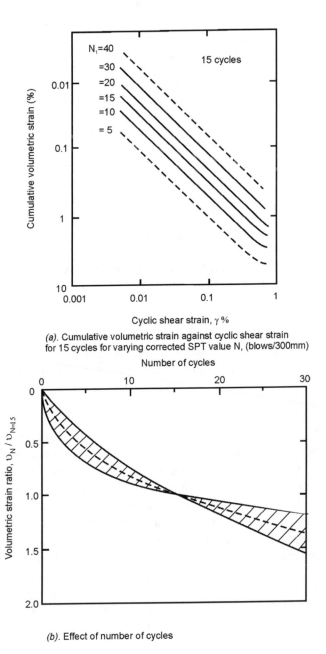

(a). Cumulative volumetric strain against cyclic shear strain for 15 cycles for varying corrected SPT value N, (blows/300mm)

(b). Effect of number of cycles

Figure 5-10. Relationships between cumulative volume strain and cyclic shear strain and number of cycles for dry sands. From Tokimatsu and Seed (1987), with permission from ASCE.

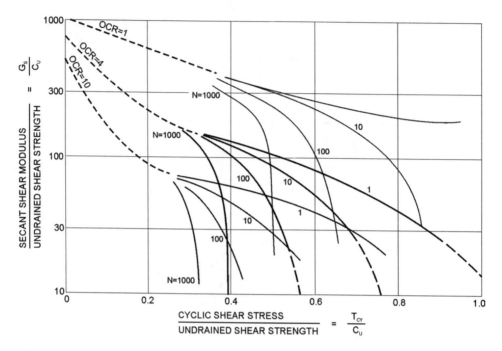

Figure 5-11. Variation of G_0/c_U with number of cycles for clays. From Andersen et al. (1982), with permission from Norwegian Geotechnical Institute.

Stiff clays show a reduced undrained strength when subjected to many cycles of loading. Figure 5-11 shows how the shear stiffness and finally the shear strength changes with number of cycles for a clay subjected to cyclic triaxial testing.

5.2.6 Effects of Time

It is observed that when soil samples are left to stand at constant stress state for a period of time, they exhibit stiffer behavior when subjected to future loading. Clays, for example, when subjected to a change in mean normal stress, continue to reduce in volume for many years even after all the excess pore water pressures have fully dissipated. This behavior is described as secondary consolidation and is illustrated by Figure 5-12. The creep is represented in the form of the reducing volume strain while the applied mean stress is constant. It is seen that the clay shows an apparent greater OCR that develops with time. This leads to an associated increased undrained shear strength and increased stiffness.

Simpson (1999) shows an example for a tunnel in London Clay where the BRICK model is modified to allow for aging and shows much better agreement for the soil surface deflection behavior when this aspect is incorporated into the prediction. He simulated this aging effect by shrinking the diameter of the kinematic yield surfaces with time and then expanding them again at the onset of new loading. This effectively moves the centers of the surfaces closer to the current strain point of the soil element as illustrated in Figure 5-13. For clarity only two surfaces are shown.

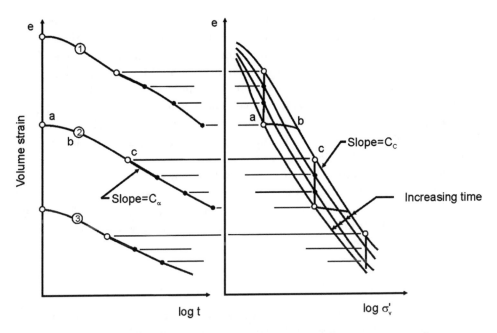

Figure 5-12. Corresponding values of C_α and C_C at any instant during secondary compression. Numbers 1, 2, and 3 between the open circles indicate effective stress increase. Closed circles indicate increasing time at constant effective stress. From Mesri and Catro (1987), with permission from ASCE.

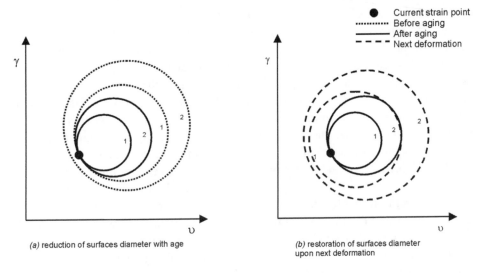

Figure 5-13. Kinematic yield surfaces of the BRICK model modified to allow for the effects of aging. From Simpson (1999), with permission from Balkema.

New rock fill and other loosely placed granular fills also exhibit a creep response in a similar fashion to normally consolidated clay. It is believed that this is due to the high stress concentration at the particle contacts, which leads to crushing of part of the particles themselves. Little research is published on this topic and the reader is directed to Pickles and Tosen (1998) and Oldecop and Alonso (2001).

In a similar manner sands show a lower tendency to liquefy as they age. Recent Holocene deposits are observed to liquefy more readily than older Quaternary deposits even though they have the same deposition history and are the same soil type (Youd and Perkins 1978).

For short term loading the rate of the loading may be important. While it has been conclusively demonstrated that sands do not have any stress rate effects (Bolton and Wilson 1989), clays are observed to experience about 6% increase in strength and stiffness arising from a static penetration test for a tenfold increase in the rate of testing (Dayal and Allen 1975).

5.3 Geotechnical Structures

The discussion in this section is limited to those structures where the behavior of the soil material dominates the design. These structures include tunnels, slopes, embankments, and reclamation.

5.3.1 Tunnels and Caverns

The design of tunnels through soil materials must achieve two primary objectives. They are to use a construction method that can ensure that the tunnels can be formed safely without risk of collapse of the tunnel and ensure that the effects to neighboring structures are kept within acceptable limits.

The tunneling process, or method employed, and the failure strength of the soils control failure of a tunnel during excavation. Generally this type of problem is addressed from experience where successful techniques are determined from previous examples of tunneling of similar diameters in similar ground conditions. For example, the need for the use of compressed air or full-face support is generally assessed by reference to past experience. In this regard the approach is essentially empirical. Unfortunately a major limitation of this type of approach is that extrapolation to a more difficult situation becomes a very uncertain process. It is much better if the mechanisms can be well modeled either physically or mathematically, and then the designer can ascertain whether failure is more or less likely.

The problem of modeling a tunnel installation is very complex. It is very difficult to represent the tunneling process as a 2D problem, and there are few 3D analysis systems available. The analysis method needs to be able to model the soil both at the face of the tunnel and the soil that is only partially supported until the following lining and grouting system is installed. It also needs to be able to model the progressive excavation process that is proceeding along the tunnel drive. Even a fully 3D analysis method, which can model the construction sequence, cannot be relied upon implicitly. It should be used both as an adjunct to the extrapolation of previous experience and to assess whether the proposed method is appreciably more risky than that used previously.

Deformation induced by the tunneling process is equally problematic. Observations from past experience show that there is a relatively narrow trough of deformation that is related to the face loss and can be expressed as a Gaussian Normal Distribution (Fig. 5-14). The face loss is found from experience to be related to soil type, groundwater condition, and excavation method. Attempting to mirror this observed behavior by 2D finite element analysis shows that a very complex soil model is required (Simpson 1999). Analysis can be used, however, to assess the likely impact of differing tunneling methods. The work by Mair (1993) examining the effect of a rigid pilot tunnel within the main tunnel drive illustrated the benefits of this type of approach as shown in Figure 5-15.

The remaining aspect of tunnel design is to determine the required strength of any permanent lining system. The purpose of the lining can be to provide permanent support to the soil, to keep water out of the tunnel, or both. While it is straightforward to design a system to resist the long-term water pressure (if this is required), the design of the lining to resist earth or rock pressure is more problematic. Again empirical methods are generally used and Duddeck and Erdmann (1982) gives good

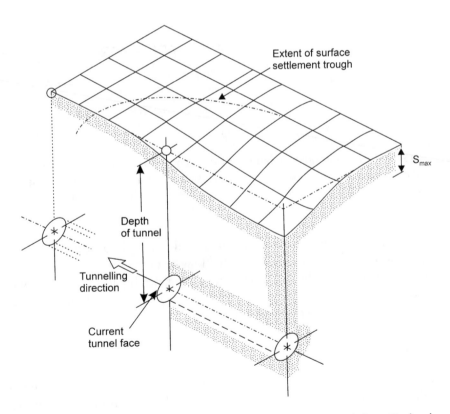

Figure 5-14. Surface settlement trough above an advancing tunnel. From Burland et al. (2001), with permission from CIRIA.

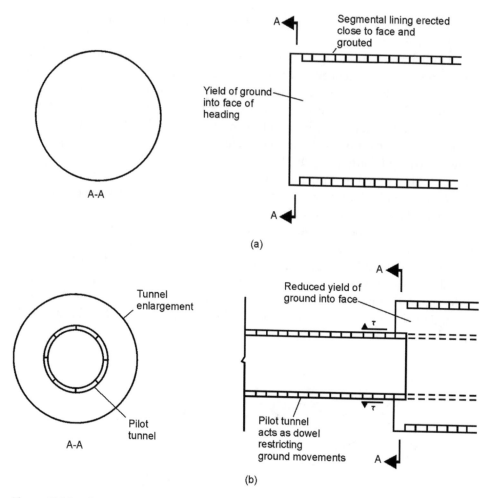

Figure 5-15. Effect of pilot tunnel in reducing ground movements. a) Principal source of ground movement for hand excavation in stiff clay without a shield. b) Pilot tunnel restricts movements. From Mair (1993), with permission from Thomas Telford.

guidance on this problem. At the completion of the tunnel construction there is no earth pressure acting the lining, but in the long term the lining must be able to resist the creep loads from the soil. These will be considerably less than the original overburden pressure and clearly the soil model will be of paramount importance in designing the lining in these circumstances.

5.3.2 Slopes

The analysis of slopes has traditionally been done using limit equilibrium methods where the strength of the soil is reduced until the soil mass forming the slope, or a significant part of it, fails by sliding down a critical slip surface. This method is well proven and successful, provided the critical slip surface or range of possible critical slip surfaces is identified and the pore water pressure regime within the slope is understood. If there is a critical geological feature present that is not identified, then no analysis method can be successful.

Designing upgrading works for slopes also generally relies on traditional limit equilibrium methods. Methods such as regrading the slopes or adding toe weighting can be successfully analyzed using this method. Methods such as soil nailing require much greater care.

Soil nails are generally designed as a passive system in that they do not experience tension, and therefore tend to stabilize the slope without the slope moving to induce that tension. Occasionally nails are designed to act in shear, but this is relatively unusual and can only really be used where blocks of rock or very stiff and strong soil are joined by the nails. The movement to induce tension may arise from creep or more probably from the slope approaching failure in the event of an extreme climatic condition. This tension will only result if the nails are oriented such that the slope movement causes them to extend. Unfortunately limit equilibrium methods are often employed that assume the nails are in tension at all times. This practice is potentially not conservative and can lead to unsafe designs. Ideally an analysis method that directly models the soil nails will be more reliable in this regard.

The 2D computer program Fast Lagrangian Analysis of Continua (FLAC) (Itasca Consulting Group 2002) has proven abilities to model nails in this manner. The spacing of the soil nails is difficult to represent, however. While a 2D analysis can consider the spacing of the nails up the slope, it cannot consider the spacing across the slope. In effect the 2D analysis models each row of nails as a continuous sheet of material. To represent the spacing along the slope is a 3D problem and should be analyzed as such. At present the horizontal spacing is set to be similar to the vertical spacing; while this appears to be reasonable, it is less conservative than the 2D analysis. It is important that the soil nails effectively restrain all parts of the soil. While this can be guaranteed by constructing a facing joining all the soil nail heads, it is rarely done in practice. In Japan it is common practice to install a grillage of reinforced concrete beams joining the nail heads, however, and in Hong Kong it is conventional practice to use square concrete nail heads about 400 mm wide with the nails spaced at between 1.5 and 2.5 m.

A particularly difficult problem to model is that of soil nails in loose materials. As stated before these materials tend to contract on shearing (Fig. 5-7), and a major concern is that the soil nails will tend to lose their resistance as the soil forming the slope begins to move as a result of extreme climate conditions. If the soil is saturated then this contraction will lead to a pore water pressure rise within the slope, which in turn will lead to lower effective confining stresses, leading to a greater chance of shear failure within the soil (Fig. 5-16a through 5-16c). To satisfactorily model this problem requires that the analysis method is either fully coupled so that the pore water and effective stresses are correctly modeled or that the strain softening and strength reduction can be modeled. Most readily available computer codes find either of these difficult to model. It is suggested that an explicit time stepping finite difference code is most likely to be able to achieve this. Cheuk et al. (2001) report on an analysis of this type of problem

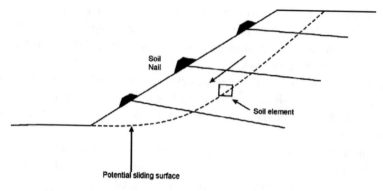

a) Geometry of potential sliding zone for a nailed slope

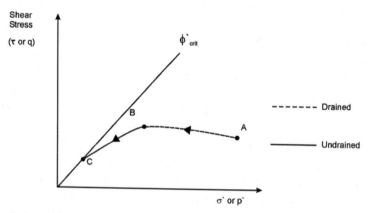

(b) Stress path in soil element as slope becomes wetter

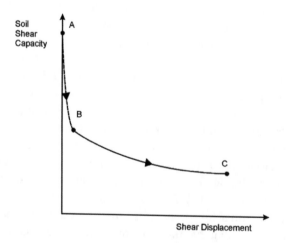

(c) Soil shear capacity versus displacement in soil element

Figure 5-16. Behavior of soil in a loose soil slope. a) Geometry of potential sliding zone for a nailed slope. b) Stress path in soil element as slope becomes wetter. c) Soil shear capacity versus displacement in soil element.

using the explicit code FLAC and show the importance of a continuous or grillage in enabling soil nails to function sensibly in loose fill.

5.3.3 Reclamation over Soft Clays

Reclamation of areas of land from the sea often involves filling over deep deposits of soft clays. As described previously the soft clays will be likely to experience large volume reductions over the years as the pore water is squeezed out from within the clays. The permeability and the compressibility of the soil control the time required for the excess pore water pressure to dissipate. Traditional consolidation theory is used to estimate this time.

Various methods can be employed to avoid ongoing large vertical settlements after the reclamation areas have been developed. These include surcharging the clay with additional soil or other materials and applying suction to the pore water above or below the clay. They rely on increasing the excess pore water pressure gradients and therefore accelerating the outflow of water from within the soil pores. An alternative method is to shorten the drainage path length by adding vertical wick drains at close spacing. The excess pore water pressure flows horizontally to the drains that allow the water to drain vertically within them. Wick drains are frequently used in combination with surcharging.

Various computational methods exist for calculating the time required for the pore water pressure to dissipate and the associated consolidation of the soil. Time stepping methods are essential, and 3D models can be used to model both the vertical drainage and the enhanced effect from the horizontal drainage resulting from the presence of closely spaced vertical wick drains.

Embankments on soft clays are a special case of the reclamation problem in that, in addition to the consolidation effects, there are significant shear stresses applied to the soft clays. If the shear stresses are too high the clays will fail leading to the phenomenon of mud waves. As noted in the first section it is known that the strength of the clay will increase as the consolidation occurs. Therefore an efficient design that enables the embankment to be constructed as quickly as possible must consider both the consolidation and the associated strength gain of the clay soils. Again a time stepping approach is required.

5.4 Soil Structure Interaction

Soil structure interaction effects are important where the soil properties and the behavior of the structural elements both have a significant effect on the overall behavior of the structural element being designed. They are generally important to the design of retaining walls for basements and for the response of piles subjected to lateral loading.

5.4.1 Flexible Retaining Walls

The design of flexible retaining walls such as steel sheet pile walls, secant bored pile walls, and diaphragm walls, is an area where soil structure interaction effects become of overriding importance. The earth pressures acting on either side of the wall depend on a wide range of soil parameters including the soil stiffness, the limiting stresses (usually referred to as active and passive pressures), and the initial horizontal stresses, as well as the flexural stiffness and yield strength of the wall itself. The sequence of construction including the installation timing, stiffness and prestress of the strutting systems, and the pore water pressure regimes at the various stages of the construction must also be designed. Figure 5-17 shows a typical construction sequence for a two-level basement wall. Conventional 2D finite element analysis that incorporates an elastic plastic soil model is generally able to predict the lateral behavior of the wall with sufficient accuracy. Predicting the settlement behind the wall requires a more complex model (Ng et al. 1998) such as the BRICK nonlinear multiple yield model. Occasionally 3D finite element analysis will be required where there is a complicated excavation sequence or a complex excavation shape.

There are many simpler analysis programs available that model the wall as a series of beam elements and the soil on either side as a series of springs that are constrained to stay within the active and passive pressure limits. These programs are easy to use and are very useful design tools. Selecting the stiffness of the soil springs is a difficult process and the end result can only be relied upon when previous experience in similar soils is available to act as a calibration on the results. A more sophisticated approximation is used in the boundary element program Flexible REtaining

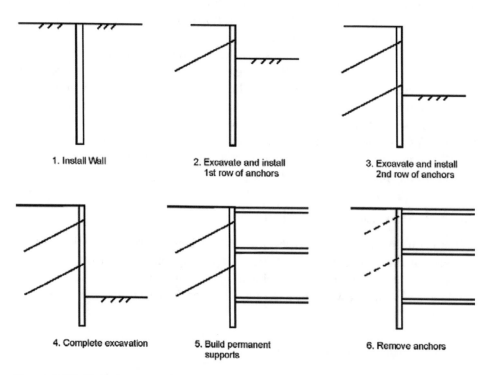

1. Install Wall

2. Excavate and install
1st row of anchors

3. Excavate and install
2nd row of anchors

4. Complete excavation

5. Build permanent
supports

6. Remove anchors

Figure 5-17. Typical construction sequence of a two-level basement excavation.

Wall analysis (FREW; Pappin et al. 1985). In this program a fully populated stiffness matrix represents the soil on each side of the wall, and the soil pressure limits allow for the effects of arching that may arise in excavations supported by struts. Again this program provides a useful design tool and is proven to be reliable in most situations. However, it is known to overpredict displacements in the case of small cantilever walls.

5.4.2 Laterally Loaded Piles

The situation of a vertical pile subjected to a lateral load is a 3D problem and requires all the parameters listed above for retaining walls to be considered. A 3D analysis program is required to correctly model this problem; however, it is found that a simple flexural beam model connected to the soil by way of a series of springs that have a maximum pressure limit generally gives sensible predictions. Figure 5-18 shows an example for a problem analyzed using the program Analysis of Laterally loaded Piles (ALP; Oasys 2001). There are many of programs available of this type. The reason these simple programs give reasonable results is that, unlike the case for flexible retaining walls, the stiffness interaction between the springs is quite limited. ALP includes the ability to impose a known soil displacement onto the pile itself. This facility is very useful when modeling the effects of tunnel construction or deep excavations on adjacent existing piles.

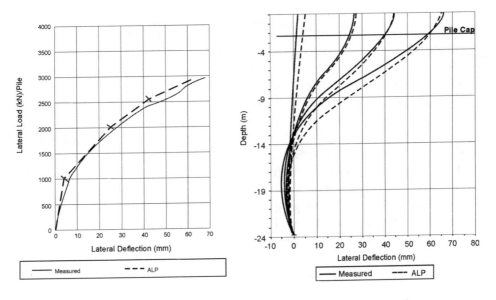

Figure 5-18. Comparison of ALP and measured pile head deflection for a pile group. From Plumbridge et al. (2000), with permission from Hong Kong Institute of Engineers.

5.5 Seismic Loading

Seismic loading is a special case of soil structure interaction in that the loading on the structure itself originates from the soil. Not only does the presence of the soil affect the shaking forces that are transmitted to the structure but the stiffness of the foundation system can also affect the way in which the structure reacts to this shaking. The effects of the soil on the shaking are referred to as site response effects, and the effects of the foundation systems are referred to as dynamic soil structure interaction. This section briefly discusses these and then addresses the design of various foundation elements such as piles and retaining walls.

5.5.1 Site Response

Considerable empirical evidence exists to estimate the amplitude, frequency, and duration of shaking that can be expected at a site due to an earthquake of known magnitude occurring at a certain distance from a site (see Kramer 1996). By combining this knowledge with the observed rate of earthquakes occurring at all the various locations around the site, a level of ground motion shaking, with a known probability of being exceeded within a predetermined time period, can be derived (Cornell 1968). These methods are most reliable for rock and hard soil sites and give a good estimate of the probability or rate at which various levels of ground motion are likely to be exceeded. It is well known, however, that local soil deposits can cause a significant change in the ground motion due to its own resonance effects. The Mexico Earthquake in 1985 was an extreme example. In this case the 30 m deep, highly plastic clays (and therefore highly elastic material as shown in Figure 5-3) were measured to amplify the peak acceleration by about a factor of 6. The natural period of the soil deposit was at around 2 seconds and the soil changed the frequency-rich input rock motion to about 15 cycles of motion that had a period of around 2 seconds. The amplification of the response spectrum at this period was about 20 times. Buildings that were sensitive to motion at this period were systematically destroyed (Booth et al. 1986).

In many cases a 1D model can be used to assess the site response effects. Various codes exist to do this including the frequency domain program SHAKE (Schnabel et al. 1972) and the nonlinear explicit time stepping program SIREN (Oasys Limited 2006). The Mexico earthquake was well modeled by these methods (Fig. 5-19). Where the bedrock is not essentially rigid, the site response analysis should allow for the presence of the soil deposit modifying the input bedrock outcrop motion. The time stepping programs can also model the effects of pore pressure rise leading to liquefaction of the underlying soil deposits.

Where the soil profile is changing dramatically with distance horizontally, a 2D or even 3D analysis needs to be used. Rassem et al. (1997) explored this problem and demonstrated that at the edges of basins 2D analyses are required; however, even at distances as close as five times the thickness of the soil deposit from the edge of the basin, a 1D analysis is adequate.

5.5.2 Dynamic Soil Structure Interaction

Structures that are effectively rigidly connected to underlying rock or very stiff soil can be analyzed relatively easily to establish their response to earthquake ground

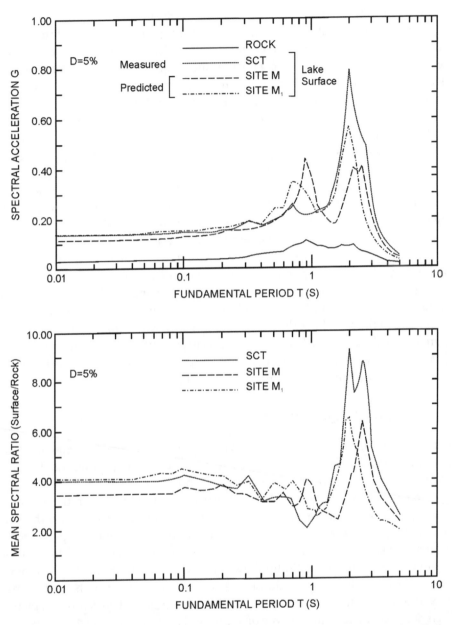

Figure 5-19. Observed and that computed by SIREN response spectra for Mexico City's "lake" zone. From Heidebrecht et al. (1990), with permission from NRC Research Press.

motion. When their foundations are relatively flexible, however, and allow the overall structure to move laterally or rotate significantly, then dynamic soil structure interaction effects need to be considered. The additional flexibility in the structural system generally leads to the structure responding at a somewhat lower frequency, leading to reduced forces but greater displacements being experienced by the structure.

There are various methods available for analyzing dynamic soil structure interaction. Some simplify the dynamic stiffness properties of the soil foundation system as an assemblage of springs and dashpots (e.g., Wolf 1997). Some computer programs model the system as being linear elastic, with a level of damping added to the system and solve the problem in the frequency domain (e.g., CLASSI and FLUSH, see Booth et al. 1988), and others use direct nonlinear time stepping explicit finite difference methods (e.g., LS-DYNA, see Hallquist 1998). With the advent of relatively cheap, high capacity computers the explicit time stepping methods are becoming more popular. As shown below they can also address a wide range of effects in the analysis.

The choice of input motion to these analyses requires careful consideration. It is generally found that several earthquake time histories should be analyzed. Some designers prefer that the time histories are actual real records measured in similar circumstances from a particular magnitude event at an appropriate distance from the site. It is the author's experience, however, that it is preferable to use artificial time histories that have response spectra similar to the smoothed design spectra. Care needs to be taken to ensure the time history is being applied correctly. If the motion is to be input at the base of a soil structure system, it is necessary to allow for the presence of the soil above the base. Programs such as SHAKE or SIREN are often required to establish a time history that is consistent with the overlying soil profile. If this is not done there can be unreasonably large ground motions predicted at the soil surface remote from the structure being modeled.

5.5.3 Design of Foundation Elements

Foundations are affected by earthquake loading in two ways, one by dynamic lateral and vertical loads applied to them by the structure they are supporting and the other by being loaded directly by the soil moving around them.

Basement walls experience additional earth pressure both from the action of the soil pushing onto them and by the structure pushing the wall into the soil. This is a 3D problem and really requires a dynamic 3D program to perform the design analysis. An example of a LS-DYNA mesh used to analyze the foundation for an ethylene tank in the Philippines is shown in Figure 5-20. The advantage of this type of analysis is that the stresses experienced by all parts of the structure and foundation can be analyzed directly. Dynamic soil structure interaction effects are also included in this type of analysis. If appropriate, yield of the various structural and foundation elements can be readily incorporated into the analysis. A potential problem with this type of analysis is reflections from the lateral boundaries. It can be seen in Figure 5-20 that the lateral boundaries were kept a significant distance away from the structure even though nonreflecting boundaries were used.

Retaining walls, while being similar to basement walls, are often sufficiently long in plan to be considered as 2D structures, and a 2D dynamic analysis will be sufficient. A useful approximation is to calculate the worst deformed shape of the soil that occurs during the earthquake using a 1D program site response analysis, such as SIREN, and

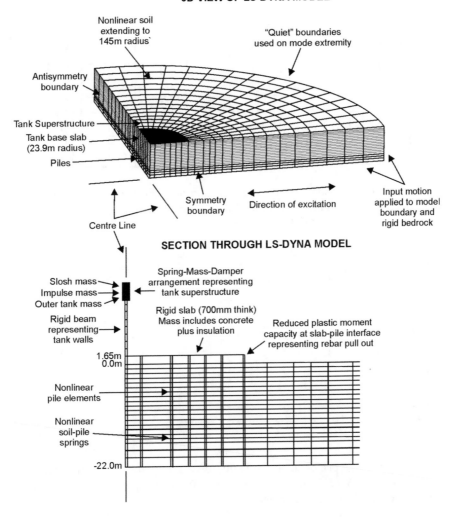

3D VIEW OF LS-DYNA MODEL

Nonlinear soil extending to 145m radius`

"Quiet" boundaries used on mode extremity

Antisymmetry boundary

Tank Superstructure

Tank base slab (23.9m radius)

Piles

Symmetry boundary   Direction of excitation

Input motion applied to model boundary and rigid bedrock

Centre Line

SECTION THROUGH LS-DYNA MODEL

Spring-Mass-Damper arrangement representing tank superstructure

Slosh mass
Impulse mass
Outer tank mass

Rigid beam representing tank walls

Rigid slab (700mm think) Mass includes concrete plus insulation

Reduced plastic moment capacity at slab-pile interface representing rebar pull out

1.65m
0.0m

Nonlinear pile elements

Nonlinear soil-pile springs

-22.0m

Figure 5-20. Finite element mesh for DSSI in LS-DYNA for an ethylene tank in the Philippines. From Lubkowski et al. (2000), with permission from New Zealand Society for Earthquake Engineering.

then applying this to a static 2D finite element analysis of the soil structure system (Fig. 5-21). To achieve this the ground displacements from the site response analysis need to be applied to each side of the model in addition to a suitable static horizontal acceleration applied to soil masses such that, in the absence of the structure, the whole soil mass will experience the desired lateral distortion. The analysis is then repeated with the underground structures in place within the finite element mesh. Dynamic loads from any superstructure also need to be applied using a combination rule to allow for the nonsynchronous motion between the soil and the superstructure. This method is essentially the same as that recommended by the Earthquake Engineering Committee (1992) for civil structures in Japan.

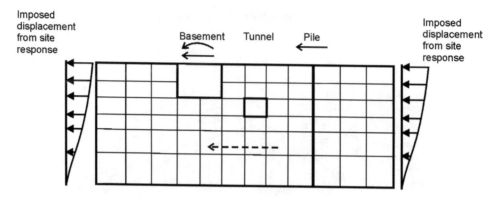

Figure 5-21. Schematic section showing seismic ground displacement profile imposed on soil (solid arrows) and imposed superstructure loads (open arrow); a horizontal body force is also imposed (dashed arrow).

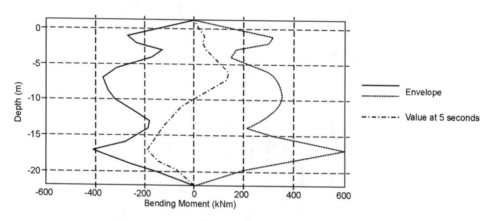

Figure 5-22. Predicted pile bending moments, lower bound soil, SSE input for the ethylene tank in Fig. 5-20. From Lubkowski et al. (2000), with permission from New Zealand Society for Earthquake Engineering.

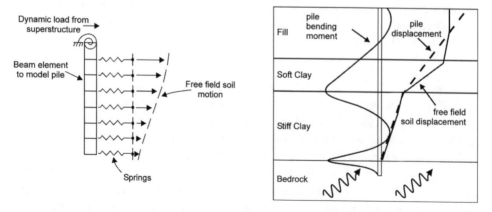

Figure 5-23. Diagrammatic model of a pile soil system and typical results. The rotational spring models the restraint from the superstructure and the horizontal springs, which are elastic/plastic, model the resistance provided by the soil to the pile moving horizontally through the soil.

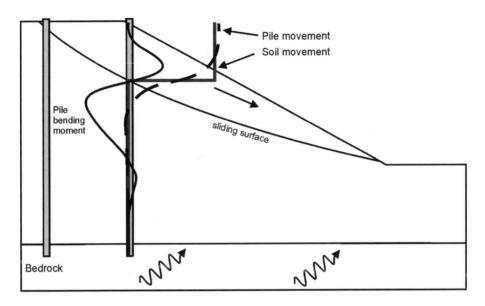

Figure 5-24. Schematic diagram of a part of a soil slope moving past a line of piles.

Piles again can be analyzed using full 3D dynamic analyses. For large pile groups where the overall lateral stiffness of the piles will have a significant stiffening effect, this is essential. Figure 5-22 shows the results for a pile within the large group under the ethylene tank in the Philippines illustrated in Figure 5-20. For single piles or cases where the piles are widely spaced and their stiffness will not have a major effect on the dynamics of the soil deposit, a simple beam element program attached to soil springs will be sufficient (Fig. 5-23). A full discussion of the design of piles in this situation is given elsewhere (Pappin et al. 1998).

This method can also be applied to the case where a pile is penetrating a soil slope that is expected to move during an earthquake as shown in Figure 5-24. It must be noted that if these slope movements are large, the pile's ability to support the superstructure may be compromised, leading to collapse of the structure. In such cases the maximum credible seismic ground motion needs to be considered when deriving the earthquake induced slope movement.

A useful summary of the considerations required when considering slope movements in earthquakes is given elsewhere (Ambraseys and Srbulov 1995). Ambraseys and Srbulov (1995) illustrate how the slope can be expected to move if the ground acceleration exceeds the static value necessary to cause slope failure, referred to as Kc. Curves are presented for expected displacement against the ratio of peak ground acceleration, Km, to Kc for a range of earthquake magnitude and distance. Figure 5-25 shows their results for a magnitude 7 earthquake. They also consider the case where the earthquake can lead to an unstable situation for a previously stable slope. This situation could arise from the soil getting weaker due to cyclic ground motion, perhaps because of pore pressure rise, or from residual soil strengths being reached because of movement during the earthquake as discussed above. While this paper is helpful for considering design scenarios, a more detailed study can be carried out by

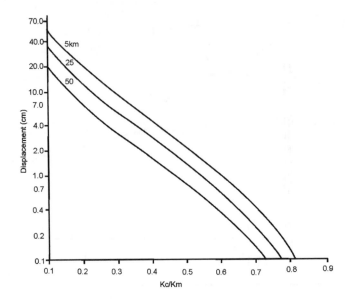

Figure 5-25. Attenuation of permanent slope displacements caused by an earthquake of surface wave magnitude 7 at source to site distances of 5, 25 and 25 km against *Kc/Km*. From Ambraseys and Srbulov (1995),with permission from Elsevier.

using nonlinear time-history analysis with a program such as LS-DYNA or similar. Clearly great care needs to be made in assessing the soil strengths and how they may change during the seismic event due to large localized shear strains or increase in pore pressure.

5.6 Conclusion

Where the behavior of the geotechnical materials is an important element of the design, the soil properties need to be understood and appropriately modeled. As discussed above, while soils at small strain levels behave in an essentially elastic manner, they become highly nonlinear with increasing strain levels and also have quite small limiting strength. This chapter presents an overview of the modeling of soil strength, small strain stiffness, and degradation of stiffness with increasing strain. It goes on to illustrate the potential importance of dilation to describe the phenomenon of soils to experience shear strains when only the mean normal stress is changed and for volume strains to be observed when only the shear stress is changed. Clearly linear elastic modeling cannot capture this behavior, and more complex modeling is required. Soil properties can also change with time. This can be because of cyclic or repeated loading leading to densification or pore pressure development in sands or due to the pore water pressure slowly changing to achieve equilibrium in clays. Clays and rock fills can also show a long-term creep behavior.

This chapter gives examples of different types of analysis that are applicable to different types of problems. Cases where the soil behavior dominates the design are discussed including:

- tunnel and cavern design where the design uses empirical methods or very sophisticated analyses,
- upgrading soil slopes using soil nails that can, with care, be designed by modeling only the soil strength, and
- building embankments or platforms over soft clays where modeling the change of the soil strength and density with time is crucial.

More complex situations such as flexible retaining walls and laterally loaded piles where soil structure interaction effects must be considered are discussed. In these cases both soil stiffness and strength must be modeled. While design methods based on linear elastic analyses with a plastic strength limit often suffice modeling the nonlinear behavior of the soil will improve the accuracy of movement predictions. In any method time dependent soil behavior must usually be explicitly considered for retaining walls in clay.

Seismic loading is generally a more complex situation where imposed forces and distortions on the structure originate from the ground. Generally nonlinear soil stiffness is important and must be modeled in addition to the soil strength. The change of the soil behavior due to cyclic loading can also be important.

While 2D analyses are often adequate, there are many situations where 3D analyses are necessary. These invariably require a high level of computing power especially for dynamic situations. Recent advances in computing power have enabled the analysis of the complex, time stepping, highly nonlinear dynamic analyses to be carried out in a practical time scale. At present the computing power required for large fully 3D problems usually demands that only a few calculations be done as confirmatory analyses. Increases in computing capability will soon enable these methods to analyze many situations such that sensitivity studies can sensibly be carried out and will enable such sophisticated tools to be used as part of the design process.

References

Ambraseys, N., and Srbulov, M. (1995). "Earthquake induced displacements in slopes." *Soil Dyn. and Earthquake Engrg.*, 14, 59–71.

Andersen, K. H., Lacasse, S., Aas, P., and Andanaes, E. (1982). "Review of foundation design principles for offshore gravity platforms." *Norwegian Geotech. Inst.*, Publication, 143.

Atkinson, J. H. (2000). "Nonlinear soil stiffness in routine design." *Geotechnique*, 50(5), 487–588.

Bolton, M. D., and Wilson, J. M. R. (1989). "An experimental and theoretical comparison between static and dynamic torsional tests." *Geotechnique*, 39(4), 585–599.

Booth, E. D., Pappin, J. W., and Evans, J. J. B. (1988). "Computer aided analysis methods for the design of earthquake resistant structures: a review." *Proc. of Inst. of Civil Engrs.*, Part 1, 84, 671–691.

Booth, E. D., Pappin, J. W., Mills, J. E., Degg, M., and Steedman, R. S. (1986). "The Mexico earthquake of 19th Sept. 1985. A field report by EEFIT." *Soc. of Earthquakes and Civ. Eng. Dyn., Inst. of Struct. Engrs.*, London.

Burland, J. B., Standing, J. R., and Jardine, F. M. (2001). "Building response to tunneling." *CIRIA*, Thomas Telford, London.

Cheuk, C. Y., Ng, C. W. W., and Sun, H. W. (2001). "Numerical analysis of soil nails in loose fill slopes." *Proc., 14th South East Asian Geotech. Conf.*, 1, A.A. Balkema, Rotterdam, 725–730.

Cornell, C. A. (1968). "Engineering seismic risk analysis." *Bull Seism. Soc. Am.*, 58, 1583–1606.

Craig, R. F. (1999). *Soil mechanics*, 6th Ed., E & FN Spon, London.

Dayal, U., and Allen, J. H. (1975). "The effect of penetration rate on the strength of remoulded clay and sand samples." *Can. Geotech. J.*, 12(3), 336–348.

Duddeck, H., and Erdmann, J. (1982). "Structural design models for tunnels." *Tunnelling*, 82, 83–91.

Earthquake Engineering Committee. (1992). "Earthquake Routine design for Civil Engineering Structure on Japan." *The Japan Soc. of Civ. Engrg.*, Tokyo.

Hallquist, J. O. (1998). "LS-DYNA Theoretical manual." Livermore Software Technology Corporation, Livermore, Calif.

Heidebrecht, A. C., Henderson, P., Naumoski, N., and Pappin, J. W. (1990). "Seismic response and design of structures located on soft clay sites." *Can. Geotech. J.*, 27, 330–341.

Iwan, W. D. (1967). "On a class of models for the yielding behaviour of continuous and composite systems." *App. Mech. Div.*, 34(E3), 612–617.

Kramer, S. L. (1996). *Geotechnical earthquake engineering*, Prentice Hall, Upper Saddle River, N. J.

Ladd, C. C., Foott, R., Ishihara, K., Schlosser, F., and Poulos, H. G. (1977). State-of-the-art Report: Stress-deformation and strength characteristics." *Proc., 9th Int. Conf. Soil Mech.*, 2, Tokyo, 421–494.

Lambe, T. W., and Whitman, R. V. (1969). *Soil mechanics*, Wiley, New York.

Lubkowski, Z. A., Pappin, J. W., and Willford, M. R. (2000). "The influence of dynamic soil structure interaction analysis on the seismic design and performance of an ethylene tank." *Proc., 12th World Conf. on Earthquake Engrg.*, Auckland.

Mair, R. J. (1993). "Development in geotechnical engineering research: Applications to tunnels and deep excavation." Unwin Memorial Lecture, *Proc. ICE Civ. Eng.*, 97(1), 27–41, Institution of Civil Engineers, UK.

Mesri, G., and Castro, A. (1987). "C_α/C_c Concept and K_0 during secondary compression." *ASCE Geot.*, 113(3), 230–247.

Ng, C. W. W., Simpson, B., Lings, M. L., and Nash, D. F. T. (1998). "Numerical analysis of a multi-propped excavation in stiff clay." *Can. Geotech. J.*, 35, 115–130.

Oasys Limited. (2001). *ALP user manual*, Oasys Limited, Newcastle, UK.

Oasys Limited. (2006). *SIREN, nonlinear 1D dynamic site response analysis*, Oasys Limited, Newcastle, UK.

Oldecop, L. A., and Alonso, E. E. (2001). "A model for rockfill compressibility." *Geotechnique*, 51(2), 127–139.

Pappin, J. W., and Brown, S. F. (1980). "Resilient stress strain behavior of a crushed rock." *Proc., Int. Symp. on Soils under Cyclic and Transient Loading*, Vol. 1, University College of Swansea, Balkema, Swansea, UK, 169–177.

Pappin, J. W., Ramsey, J., Booth, E. D., and Lubkowski, Z. A. (1998). "Seismic response of piles: Some recent design studies." *Proc., Inst. of Civ. Engrs., Geotech. Engrg.*, 131, 23–33.

Pappin, J. W., Simpson, B., Felton, P. J., and Raison, C. (1985). "Numerical analysis of flexible retaining walls." *Proc., NUMETA 85, Numerical methods in engineering: theory and applications,* University College of Swansea, Balkema, Swansea, UK, 789–802.

Pickles, A. R., and Tosen, R. (1998). "Settlement of reclaimed land for the new Hong Kong International Airport." *Proc., Inst. Civ. Engrs.,* Geotech. Engrg., 131, 191–209.

Plumbridge, G. D., Sze, J. W. C., and Tham, T. T. F. (2000). "Full scale lateral load tests on bored piles and a barrette," *Hong Kong Inst. of Engrs. Geotech. Div. 19th Ann. Seminar,* 213–220.

Rassem, M., Ghobarah, A., and Heidebrecht, A. C. (1997). "Engineering perspective for the seismic site response of alluvial valleys." *Earthquake Eng. and Struct. Dyn.,* 26, 477–493.

Schnabel, P. B., Lysmer, J., and Seed, H. B. (1972). "SHAKE; A computer program for earthquake response analysis of horizontally layered sites." *Report No. EERC, 72–12,* Earthquake Engrg. Research Center, Berkeley, Calif.

Seed, R. B., Cetin, K. O., Moss, R. E. S., Kammerer, A. M., Wu, J., Pestana, J. M., Reimer, M. F., Sancio, R. B., Bray, J. D., Kayen, R. E., and Fayis, A. (2003). "Recent advances in soil liquefaction engineering: A unified and consistent framework." *26th Annual ASCE Los Angeles Geotech. Spring Seminar,* ASCE, H.M.S. Queen Mary, Long Beach, Calif.

Simpson, B. (1992). "Retaining structures—Displacement and design." *Geotechnique,* 42(4), 541–576.

Simpson, B. (1999). "Engineering needs." *2nd Int. Symp. on the pre-failure deformation characteristics of geomaterials, Turin,* Balkema, Rotterdam.

Sladen, J. A., D'Hollander, R. D., and Krahn, J. (1985). "The liquefaction of sands, a collapse surface approach." *Can. Geo. J.,* 22, 564–578.

Stroud, M. A. (1988). "The Standard Penetration Test—Its application and interpretation." *Proc., 2nd European Symp. on Penetration Testing,* Inst. of Civ. Engrs., London.

Tokimatsu, K., and Seed, H. B. (1987). "Evaluation of settlement in sands due to earthquake shaking." *J. Geotech. Engrg.,* 113(8), 861–878.

Vucetic, M., and Dobry, R. (1991). "Effect of Plasticity on Cyclic Response." *J. Geotech. Engrg.,* 117(1), 89–109.

Wolf, J. P. (1997). "Spring-dashpot-mass models for foundation vibrations." *Earthquake Engrg. and Struct. Dyn.,* 26, 931–949.

Youd, T. L., and Perkins, D. M. (1978). "Mapping of liquefaction-induced ground failure potential." *J. Geotech. Engrg.,* 104(4), 433–446.

6

Aging Nuclear Structures

B. R. Ellingwood and D. J. Naus

6.1 Aging and Nuclear Structures

As of June 2004, 104 nuclear power reactors were licensed for commercial operation in the United States (U.S. Nuclear Regulatory Commission [NRC] 2005). These reactors produce approximately 20% of the U.S. electricity demand. The Atomic Energy Act (AEA) of 1954 limits the duration of operating licenses for most of these reactors to a maximum of 40 years. The median age of these reactors is over 20 years, with more than 60 having been in commercial operation for 20 or more years. Expiration of operating licenses for these reactors will start to occur in 2006. Therefore, continuing the service of existing nuclear power plants (NPPs) through a renewal of their operating licenses provides a timely and cost-effective solution to the problem of meeting future electricity demand. Several utilities (e.g., Baltimore Gas and Electric Company) already have selected to go this route. However, a concern as plants approach the end of their initial operating license period is that the capacity of the safety-related systems to mitigate extreme events has not deteriorated unacceptably due to either aging or environmental effects. As all but one of the construction permits for existing NPPs in the United States were issued prior to 1978, the focus of technical development and support has shifted from design to condition assessment. Here, the aim is demonstrating that structural margins of existing plants have not or will not erode during the desired service life due to aging or environmental effects. Probabilistic methods provide the quantitative tools for the assessment of uncertainty in condition assessment and are an essential ingredient of risk-informed management decisions concerning continued service of the NPP structures.

6.2 Containment Structures

A myriad of concrete- and steel-based structures are contained as part of an NPP to perform multiple functions in concert (e.g., load-carrying capacity, radiation shielding, and pressure retention). Although information provided in the balance of this chapter primarily addresses containment structures, it is also applicable to other safety-related structures, as well as civil infrastructure-related facilities. Relative to general civil engineering concrete structures, NPP concrete structures tend to be more massive and have increased steel reinforcement densities with more complex detailing.

Each boiling-water reactor (BWR) or pressurized-water reactor (PWR) unit in the United States is located within a much larger metal or concrete containment that also houses or supports the primary coolant system components. The basic laws that regulate the design (and construction) of NPPs in the United States are contained in Title 10, "Energy," of the Code of Federal Regulations (1995). The reactor containment and associated systems must provide an essentially leak-tight barrier against uncontrolled release of radioactivity to the environment and must ensure that the containment design conditions important to safety are not exceeded during postulated accident conditions. As such, the containment and associated systems are to accommodate, without exceeding the design leakage rate and with sufficient margin, the calculated pressure and temperature conditions resulting from any loss-of-coolant accident (LOCA).

6.2.1 Concrete Containments

Concrete containments are metal-lined, reinforced concrete pressure-retaining structures that in some cases may be post-tensioned. The reinforced concrete containment shell, which generally consists of a cylindrical wall with a hemispherical or ellipsoidal dome and flat base slab, provides the necessary structural support and resistance to pressure-induced forces. Leak-tightness is provided by a thin steel liner (e.g., 6 mm in thickness) that is anchored to the concrete shell by studs, structural steel shapes, or other steel products. Material systems used to fabricate concrete containments include: moderate heat of hydration and sulfate-resistant portland cement, fine and coarse aggregate and water obtained primarily from local sources, carbon steel deformed bar reinforcement having a minimum yield strength of 415 MPa, and wire or strand post-tensioning systems having capacities to 10.7 MN. Metallic liners typically are fabricated from carbon steel plate except in wet areas (e.g., fuel pool) where stainless steel or carbon steel clad with stainless steel may be used. Depending on the functional design (e.g., large dry or ice condenser), NPP concrete containments can be on the order of 40 to 50 m in diameter and 60 to 70 m high, with wall and dome thicknesses from 0.9 to 1.4 m, and base slab thicknesses from 2.7 to 4.1 m. Figure 6-1a presents a cross section for a post-tensioned reinforced concrete containment.

Current rules for construction of concrete containments are provided in Section III, Division 2 of the American Society of Mechanical Engineers (ASME) Code (ASME 1998). Code requirements for design and construction of concrete structures, other than concrete reactor vessels and containments, that form part of an NPP

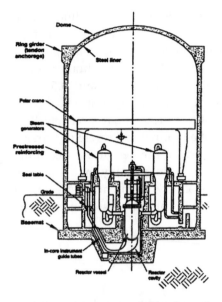

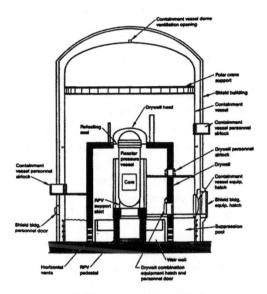

(a) Large dry post-tensioned concrete containment.

(b) Large dry metal containment in reinforced concrete shield building

Figure 6-1. NPP containment cross sections. Reprinted from U.S. Nuclear Regulatory Commission Report NUREG-1037.

and have safety-related functions have been developed (American Concrete Institute [ACI] 2001). Rules for design and construction of the metal liner that forms the pressure boundary for the reinforced concrete containments are found in Section III, Division 1, Subsection NE of the ASME Code. In the early years of NPP design in the United States, existing building codes, such as ACI Standard 318, *Building Code Requirements for Reinforced Concrete* (ACI 1999), were used in the nuclear industry as the basis for design and construction of concrete structural members. However, because the existing building codes did not cover the entire spectrum of design requirements and were not always considered adequate, the U.S. Nuclear Regulatory Commission (USNRC) developed its own criteria for design of seismic Category 1 (i.e., safety-related) structures (e.g., definitions of load combinations for both operating and accident conditions). The USNRC also has developed supplemental load combination criteria and provides information related to internal structures of steel and concrete (NRC 1981). Plants that used early ACI codes for design have been reviewed by the USNRC through the Systematic Evaluation Program to determine if there were any unresolved safety concerns (Lo et al. 1984).

Steel-lined reinforced concrete containments are designed using allowable stress design. The load combinations that usually govern design of the containment wall (shell) are:

$$\text{Normal loads:} \quad D + L + F + P_0 + T_0 + R_0 \tag{6-1}$$

$$\text{Abnormal loads:} \quad D + L + F + 1.5P_a + T_a + R_a \tag{6-2}$$

in which D and L are dead and live loads (mainly equipment, including crane), P = normal operating pressure, T_0 = normal temperature, R_0 = normal pipe reactions, P_a and T_a are the pressure and temperature arising from a design-basis LOCA, and F is due to prestressing (when applicable). Primary membrane stresses from these forces are checked against allowable stresses $0.3\ f_c'$ and $0.5\ f_y$ for Eq. 6-1 and $0.6\ f_c'$ and $0.9\ f_y$ for Eq. 6-2, in which f_c' and f_y are the specified compressive strength of concrete and yield strength of reinforcement, respectively. Base mat design is often governed by combinations of abnormal and severe or extreme environmental loads (Amin et al. 1992). Although in the design the steel liner is assumed not to contribute to the design strength, in actuality by being integral with the concrete it can account for as much as 15% of the hoop strength of the containment; this strength must be considered in condition assessment of the containment and in determining its ultimate strength.

6.2.2 Metal Containments

Prior to 1963, metal containments were designed according to rules for unfired pressure vessels that were contained in Section VIII of the ASME Code. Subsequently, metal containments were designed either as Class B vessels or as Class MC components according to rules provided in Section III of the ASME Code. The ASME Code also recognizes that service-related degradation to pressure-retaining components is possible, but rules for material selection and in-service degradation are outside its scope. Two provisions of the ASME Code are pertinent with respect to degradation. First, it is the owner's responsibility to select materials suitable for the service conditions and to increase minimum required thickness of the base metal to offset material thinning due to corrosion, erosion, mechanical abrasion, or other environmental effects. Second, criteria are provided in Section III for conduct of detailed vessel fatigue evaluations. The criteria limit the number of cycles of service pressure, temperature, and mechanical load and are based on fatigue curves provided in the ASME Code for metal containment materials.

Existing metal containments in the United States are freestanding, welded steel structures that are enclosed in a reinforced concrete reactor or shield building. The reactor and shield buildings are not part of the pressure boundary, and their primary function is to provide protection for the containment from external missiles and natural phenomena. Metal containments are typically fabricated using low-alloy or unalloyed steels. Typical metal containments range in diameter from 18 to 43 m and have shell thicknesses on the order of 25 to 51 mm, making them thin shells of revolution for the purpose of structural analysis. Figure 6-1b provides a cross section for a metal containment.

Metal containments are designed in accordance with Section 3.8.2 of the USNRC Standard Review Plan (NRC 1981) and Section III, Division 1, Subsection NE of the ASME Code. Present metal containment design procedures date from the 1960s. The load combinations that usually govern metal containment design are:

$$\text{Normal loads:} \quad D + L + P_0 + T_0 + R_0 \qquad (6\text{-}3)$$

$$\text{Abnormal loads:} \quad D + L + P_a + T_a + R_a \qquad (6\text{-}4)$$

in which the loads are as defined previously. Primary membrane stresses for these loads are checked against allowable stress $S_{mc} = 1.1\ F_u/4$, in which F_u = tensile strength. The LOCA pressure, P_a, generally is the most important load for design of a metal containment. In addition, buckling may also be important in freestanding vessels under dead load combined with lateral forces.

6.2.3 Containment Analysis and Structural Behavior

ASME Code design procedures for concrete containments (as well as metal containments) are based on elastic structural analysis using specified allowable stresses and loads, and the design procedures incorporate numerous conservatisms. In addition to the allowable stress factors, the nominal strengths and loads are specified subjectively and usually very conservatively. Also, the inelastic load-carrying capacity and ductility of steel are ignored. Numerous analyses and tests of scaled models over the past decade have confirmed that this reserve capacity is well in excess of the design basis internal pressure (Jung 1984; Walther 1992). For example, a summary of the calculated ultimate capacities, P_u, of six reinforced concrete containments designed in the 1960s to 1980s, where failure was defined as yielding of all circumferential reinforcement, lists factors of safety P_u/P_a in the range of 2.5 to 6.3 (Amin et al. 1992). Similarly studies of steel containments (Greimann et al. 1984) indicated a range of P_u/P_a from 2.2 to 5.6 based on a limiting hoop strain equal to twice the yield strain; membrane action of the containment shell was the limiting factor in all cases. Later studies (Klamerus et al. 1996) led to similar results and conclusions: median values of P_u/P_a reported were 3.0 for reinforced concrete containments and 3.4 for steel containments.

Assessments of containment safety margins through fragility modeling or other probability-based analysis requires, foremost, an estimate of the median capacity of the containment system at load levels in excess of the design basis. At such levels, the containment response as a whole is well into the inelastic range, and local strains may approach the ductility limit of the material. Steel containments can be modeled as thin shell structures, with stiffeners in both meridional and circumferential directions, and numerous transitions in shell thickness in regions where shell penetrations are required for piping and equipment access. Elastic methods of analysis or simple methods of limit analysis are inadequate for predicting the complex behaviors that occur at such load levels, as the studies above show. Any finite-element analysis used to perform the containment analyses at loads in excess of the design basis must have the capability of handling nonlinear material constitutive behavior, temperature dependence of strength and stiffness, and geometric nonlinearities due to large deformations. Such finite-element methods also are essential for fragility modeling purposes, particularly for estimating the median capacity, but also for assessing the variability in capacity due to factors known to affect containment behavior that are uncertain in nature.

A fragility assessment clearly must be tied directly to the performance requirements of the containment system, and such requirements must be couched in the context of a nonlinear structural analysis. The primary function of a containment is to confine hazardous materials in the event of an accident. Thus, its most important performance limit is loss of integrity in the pressure boundary. However, this loss of integrity can take a number of forms, with vastly different consequences, ranging from leakage involving depressurization over a period of hours to days and with the possibility of accident mitigation measures, to catastrophic rupture leading to depressurization in seconds and virtually immediate release of radionuclides. Such performance limits must be related to structural limit states involving response parameters that can be obtained from finite-element analysis and local or general structure or material failure criteria. This mapping that must occur from the performance requirement space to the structural response analysis space is exceedingly difficult and a source of significant uncertainty in the fragility assessment process.

Some notion regarding the difficulties faced by the fragility analyst in modeling the containment structural system can be gleaned from the results of scaled model tests of steel containments conducted at Sandia National Laboratory during the past decade, which have provided insights into the complexity of metallic pressure boundaries. Such tests suggest that structural failure of the containment occurs when the maximum local strains exceed the fracture ductility of the material, typically on the order of 0.25 for carbon steel. While these local strains generally occur adjacent to penetrations or transitions in shell thickness and have little impact on the global structural response of the containment, they are the points where tears initiate that lead to sudden depressurization of the containment. Lesser but still significant local strains in the vicinity of shell penetrations can cause ovalization of the penetrations and lead to failure of seals and leakage. In concrete containments with steel liners, loss of integrity is associated with liner tearing that initiates at the point where the liner studs interface with the concrete shell. Such failures can initiate when the far-field hoop strains are on the order of 0.02.

Other performance requirements in addition to integrity of the pressure boundary also play a role in a fragility assessment. Nuclear power plant structural systems are closely integrated with other safety-related mechanical and electrical systems. Excessive general shell deformations may cause malfunction of appurtenant equipment. For example, large containment shell deformations may cause interference with the polar crane bridge, piping, and adjacent structures (Amin et al. 1992). In a BWR Mark I containment, radial expansion of the containment shell may cause sufficient axial deformation in the bellows to crush the bellows and cause leakage. Such performance limits are difficult to relate to the structural responses computed from a nonlinear finite element (FE) analysis.

Thus, the structural analysis of NPP structures is exceedingly complex from the standpoint of first having to perform a nonlinear, large-deformation finite element analysis and next having to identify specific structural response quantities that can be related in a physically meaningful way to the significant performance requirements of concern. (The issue of developing appropriate load models from postulated accident scenarios using principles of thermodynamics and fluid mechanics introduces an additional level of complexity that we have not attempted to address here.) The postprocessing and interpreting of the results is particularly difficult, and it is only recently that the computational resources have become sufficient for these tasks to be performed with some confidence.

6.3 Potential Degradation Factors

Service-related degradation can affect the ability of an NPP containment to perform satisfactorily in the unlikely event of a severe accident. Degradation is considered to be any phenomenon that decreases the load-carrying capacity of a containment, limits its ability to contain a fluid medium, or reduces its service life.

Primary mechanisms that can produce premature deterioration of reinforced concrete structures include those that impact either the concrete or steel reinforcing materials (i.e., mild steel reinforcement or post-tensioning system). Degradation of concrete can be caused by adverse performance of either its cement paste matrix or aggregate materials under chemical or physical attack. Chemical attack may occur in several forms: efflorescence or leaching, sulfate attack, attack by acids and bases, salt crystallization, and alkali-aggregate reactions. Physical attack mechanisms for concrete include freeze/thaw cycling, thermal expansion/thermal cycling, abrasion/erosion/cavitation, irradiation, and fatigue or vibration. Degradation of mild steel reinforcing materials can occur as a result of corrosion, irradiation, elevated temperature, or fatigue effects. Post-tensioning systems are susceptible to the same degradation mechanisms as mild steel reinforcement plus loss of prestressing force, primarily due to tendon relaxation and concrete creep and shrinkage. Additional information related to these degradation mechanisms as well as the performance history of concrete containments is available (Naus et al. 1996).

Degradation of the containment pressure boundary (i.e., steel containments and liners of reinforced concrete containments) can be classified as either material or physical damage. Material damage occurs when the microstructure of the metal is modified, causing changes in its mechanical properties. Degradation mechanisms that can potentially cause material damage to containment steels include low temperature exposure, high temperature exposure, intergranular corrosion, dealloying corrosion, hydrogen embrittlement, and neutron irradiation. Material damage to the containment pressure boundary from any of these sources is not considered likely, however. Physical damage occurs when the geometry of a component is altered by the formation of cracks, fissures, or voids, or its dimensions change due to overload, buckling, corrosion, erosion, or formation of other types of surface flaws. Material degradation due to either general or pitting corrosion represents the greatest potential threat to the containment pressure boundary. Changes in component geometry resulting from general or pitting corrosion can affect structural capacity by reducing the net section available to resist applied loads, and pits that completely penetrate the component can compromise its leak-tight integrity. Additional information related to potential degradation mechanisms associated with these two damage classifications as well as the performance history of containment pressure boundaries is available (Oland and Naus 1996).

6.4 Time-Dependent Reliability Analysis

Evaluation of structures for continued service should provide quantitative evidence that their capacity is sufficient to withstand future demands within the proposed service period with a level of reliability sufficient for public health and safety. Structural aging may cause the integrity of structures to evolve over time (e.g., a hostile service environment may cause structural strength and stiffness to degrade). Uncertainties that complicate the evaluation of aging effects arise from a number of sources: inherent randomness in structural loads, initial strength, and degradation mechanisms; lack of in-service inspection measurements and records; limitations in available models and supporting databases for quantifying time-dependent material changes and their contribution to containment capacity; inadequacies in nondestructive evaluation; and shortcomings in existing methods to account for repair. Any evaluation of the reliability or safety margin of a containment (or other structure) during its service life must take into account these effects, plus any previous challenges to the integrity that may have occurred. Time-dependent reliability analysis methods provide the framework for dealing with uncertainties in performing condition assessments of existing and aging structures, and for determining whether in-service inspection and maintenance is required to maintain their performance at the desired level.

6.4.1 Degradation Mechanisms

Time-dependent reliability analysis and service life predictions for structures require time-dependent stochastic models of the structural strength. Strength models can be derived from: mathematical models describing the effects of the aging process resulting from service and environmental factors on concrete and steel materials and component geometry, accelerated life testing, or a combination of the two. At the current state of the art, these effects are often known qualitatively; however, quantitative models that describe material degradation processes often are empirical in nature (Clifton 1991). Service life determinations often require that these models be extrapolated outside the range of experimental data. Primary degradation mechanisms that can impact the NPP structures were described in Section 6.3. Detailed information on models for service life estimations of concrete and steel materials is available elsewhere (Naus et al. 1996; Bhattacharya and Ellingwood 1998).

6.4.2 Statistical Data on Loads and Resistance

Much of the statistical data needed to support this methodology have been collected as part of previous research to develop probability-based design requirements for

Table 6-1. Statistical Properties of NPP Loads

Load	Rate of Occurrence, $\lambda(\text{yr}^{-1})$	Duration, τ	Mean[a]	COV	PDF
Dead, D	—	—	$1.0D_n$	0.7	Normal
Live, L	0.5	0.25 yr	$0.3L_n$	0.50	Type I
Pipe, R_0	—	—	$0.85R_0$	0.30	Normal
Temperature, T_0		—	$0.85T_0$	0.16	Normal
SRV Discharge	—	1 sec	$0.8P_{SRV}$	0.14	Normal
Accid. Press., P_a	1.7×10^{-3}	20 min	$0.8P_a$	0.20	Type I
Accid. Temp., T_a	10^{-4}	20 min	$0.9T_a$	0.10	Type I
Earthquake, E	0.05	30 sec	$0.08E_{ss}$[b]	0.90	Type II[c]

[a] D_n, L_n, Q_n, P_a, and T_a are nominal loads.
[b] E_{ss} is safe-shutdown earthquake.
[c] For the PDF of annual maximum values, $F_{Eann}(x)$.v

ordinary building construction and for NPP structures (Ellingwood et al. 1982; Galambos et al. 1982; Hwang et al. 1983, 1987; Ellingwood and Mori 1992). Load and resistance models and data have been analyzed in detail (Ellingwood and Mori 1993).

6.4.2.1 Probabilistic models of loads. Structural loads that arise from rare operating or environmental events such as accidental impact, earthquakes, and tornadoes are short in duration, and such events occupy a negligible fraction of a structure's service life. Such loads can be modeled stochastically as a sequence of short-duration load pulses, occurrence of which is described by a Poisson process, with the mean (stationary) rate of occurrence, λ, random intensity, S_j, and duration, τ (Pearce and Wen 1985). If the load process is intermittent and the duration of each load pulse has an exponential distribution, the probability that the load process is nonzero at any arbitrary time is $p = \lambda\tau$. Loads due to normal facility operation or climatic variations or those that fluctuate with sufficient rapidity in time that they cannot be modeled by a sequence of discrete pulses can be analyzed as continuously parameterized stochastic processes.

A summary of generic load statistics obtained from prior research is provided in Table 6-1. When plant-specific statistics are available from either in-service monitoring programs or the site-specific hazard analyses, they should be used.

6.4.2.2 Probabilistic models of resistance. The strength, R, of a reinforced concrete component is described by (Ellingwood and Hwang 1985)

$$R = B \cdot R_m(X_1, X_2, \ldots, X_m) \tag{6-5}$$

Table 6-2. Statistical Data for Reinforced Concrete

Parameter	Mean	COV	PDF
Material strength			
f_c, psi	$960 + 0.8 fc'$	0.12	Normal
f_t psi	$6.4 \sqrt{fc}$	0.18	Normal
f_y(ASTM A 615/Gr. 60)	67 ksi	0.11	Lognormal
f_u(ASTM A 416/A 421)	270 ksi	0.04	Lognormal
Dimensions			
Overall dimensions (in.)	Nominal, h	0.4/h	Normal
Placement of reinforcement (in).	Nominal, d	0.6/d	Normal

Note: 1 in = 25.4 mm; 1 psi = 6.9 kPa; 1 ksi = 6.9 MPa.

in which $X_1, X_2, \ldots$ are basic random variables that describe yield strength of reinforcement, compressive or tensile strength of concrete, and structural component dimensions or section properties. The function $R_m(\bullet)$ describes the strength based on principles of structural mechanics. Modeling assumptions invariably must be made in deriving $R_m(\bullet)$, and the factor B describes errors introduced by modeling and scaling effects. The probability distribution of B describes bias and uncertainty that are not explained by the model $R_m(\bullet)$ when values of all variables X_i are known. The probability distribution of B can be assumed to be normal (MacGregor et al. 1983). A more accurate behavioral model leads to a decrease in the variability in B and thus in R. Probability models for R usually must be determined from the statistics of the basic variables, X_i, since it seldom is feasible to test a sufficient sample of structural components to determine the cumulative distribution function (CDF) of R directly.

6.4.2.2.1 Reinforced concrete structures. Typical statistical data on material strengths and dimensions of reinforced concrete structural elements are summarized in Table 6-2 (MacGregor et al. 1983; Ellingwood and Hwang 1985; Ellingwood and Mori 1993). The material strengths presented are based on static rates of load at designated test age. Long-term strength changes in the concrete or steel due to maturity of concrete, environmental effects, and possible corrosion of reinforcement are not reflected in these statistics. When data on the strength of concrete or steel can be obtained by in situ sampling as part of a condition assessment of a specific structure, that data should be utilized. Typical means and coefficients of variation (COV) in strength for reinforced concrete structural elements are summarized in Table 6-3. These estimates were obtained by Monte Carlo simulation using the strength and dimensional variabilities in Table 6-2. The modeling uncertainty was described by a normal distribution, with a mean $\mu_B = 1.05$ and COV $V_B = 0.06$ for flexure and

Table 6-3. Probabilistic Descriptions of Resistance of Concrete Components

Limit State	Mean[a]	COV	PDF
Beam, flexure	$1.11M_n$	0.13	Lognormal
Beam, shear	$1.22V_n$	0.18	Lognormal
Wall, shear	$135V_n$	0.18	Lognormal
Short wall, shear	$1.70V_n$	0.18	Lognormal
Slabs, flexure	$1.12M_n$	0.14	Lognormal
Short column, compression	$1.13P_n$	0.14	Lognormal
Short column, tension	$1.12P_n$	0.13	Lognormal

[a]The nominal values M_n, V_n, and P_n are the capacities that would be computed from ACI Standards 318 (ACI 1999) and 349 (ACI 1995).

compression and with $\mu_B = 1.15$ and $V_B = 0.15$ for shear. In all cases, the nominal flexural strength for concrete structural members, M_u, the nominal shear strength, V_u, and the nominal axial strength, R_u, were computed according to ACI Standards 318 (ACI 1999) and 349 (ACI 2001).

6.4.2.2 Steel structures.   Properties of steel that are required in reliability analysis of steel structures include yield strength, tensile strength, Young's modulus of elasticity, and Poisson's ratio. Results of a review of literature on this subject as part of the effort to develop load and resistance factor design procedures are summarized in Table 6-4 (Galambos and Ravindra 1978; American Institute of Steel Construction 1993). A number of ASTM designations and grades of steel are represented in these data, but they were all construction grade and are designated as "carbon" steel. A lognormal CDF fits all data in Table 6-4 reasonably well. Limited data available for steel materials used to fabricate pressure boundary materials (also available in Table 6-4) indicate that the mean strengths are somewhat more conservative with respect to nominal strengths while the coefficients of variation are smaller, indicating a higher standard of quality control (Ellingwood and Hwang 1985). Also a tendency was observed for the mean strength to decrease with increasing plate thickness. The resistance of a steel component depends on other factors besides the material strength. A simple model of resistance to a particular limit state is given by (Galambos and Ravindra 1978)

$$R = R_n MFP \tag{6-6}$$

in which R_n = nominal strength computed using material strengths, dimensions, and analytical procedures prescribed by the Code; M = material factor; F = fabrication

Table 6-4. Initial Resistance of Steel Shapes and Plates

Element	Steel	Property	Nominal (ksi)	Mean (ksi)	COV
Flanges, rolled shapes	Carbon	F_y	—	1.05 F_{yn}	0.10
Webs, rolled shapes	"	F_y	—	1.10 F_{yn}	0.11
Plates, flanges	"	F_u	—	1.10 F_{un}	0.11
Plates	"	τ_y	—	0.64 F_{yn}	0.10
Tension coupon	"	E	—	1.0 E	0.06
Tension coupon	"	V	—	0.3	0.03
$\frac{7}{16}$–$1\frac{3}{8}$ in. plate	SA 516/ Gr 60	F_y	32	47	0.05
	"	F_u	60	66	0.03
$1\frac{1}{4}$–$1\frac{3}{4}$ in. plate	SA 516/ Gr 60	F_y	36	48	0.07
	"	F_u	70	74	0.03
$\frac{1}{4}$ in. liner plate	A 285	F_y	24	37	0.04
	"	F_u	45	48	0.02

Note: 1 in = 25.4 mm; 1 ksi = 6.9 MPa

factor; and P = professional factor. M, F, and P are random variables that, as a group, model the different sources of uncertainty in resistance. With a good quality control program, the mean of F is typically close to 1.0, and its COV is typically 0.05, or less. The mean and COV of P depend on the fundamental assumptions underlying the analysis (e.g., use of simple flexural theory and neglect of strain hardening) and its rigor in modeling the behavior of interest. Since such assumptions are usually on the conservative side, the mean of P is usually greater than 1.0 while the COV typically is on the order of 0.05 to 0.10.

6.4.3 Analysis of Degrading Structures

6.4.3.1 Degradation independent of service loads. Due to the robustness of NPP structures, except for isolated instances, degradation when it occurs will primarily result from environmental factors. However, information where degradation is dependent on service loads (e.g., fatigue) is available (Ellingwood et al. 1996). The failure probability of a structural component can be evaluated as a function of (or an interval of) time if the stochastic processes defining the residual strength and the probabilistic characteristics of the loads at any time are known. The strength, $R(t)$, of the structure and applied loads, $S(t)$, are both random functions of time. Assuming that degradation is independent of load history, at any time, t, the margin of safety, $M(t)$, is

$$M(t) = R(t) - S(t). \tag{6-7}$$

Making the customary assumption that R and S are statistically independent random variables, the (instantaneous) probability of failure is

$$P_f(t) = P[M(t) < 0] = \int_0^\infty F_R(x) f_S(x) dx \tag{6-8}$$

in which $F_R(x)$ and $f_S(x)$ are the cumulative distribution function of R and probability density function (PDF) of S (Shinozuka 1983). Eq. 6-8 provides an instantaneous quantitative measure of structural reliability, provided that $P_f(t)$ can be estimated or validated (Ellingwood 1992).

For service life prediction and reliability assessment, one is more interested in the probability of satisfactory performance over some period of time, say $(0,t)$, than in the snapshot of the reliability of the structure at a particular time provided by Eq. 6-8. Indeed, it is difficult to use reliability analysis for engineering decision analysis without having some time period (e.g., an in-service maintenance interval) in mind. The probability that a structure survives during interval of time $(0,t)$ is defined by a reliability function, $L(0,t)$. If, for example, n discrete loads $S_1, S_2, \ldots, S_n$ occur at times $t_1, t_2, \ldots, t_n$ during $(0,t)$, the reliability function becomes

$$L(t) = P[R(t_1) > S_1, \ldots, R(t_n) > S_n] \tag{6-9}$$

in which $R(t_i) =$ strength at time of loading S_i.

Taking into account the randomness in the number of loads and the times at which they occur as well as initial strength, the reliability function becomes (Ellingwood and Mori 1993)

$$L(t) = \int_0^\infty \exp\left(-\lambda t \left[1 - t^{-1} \int_0^t F_S(g_i, r) dt\right]\right) f_{R_0}(r) dt \tag{6-10}$$

in which $f_{R_0} =$ PDF of the initial strength R_0 and $g_i =$ fraction of initial strength remaining at the time at which load S occurs. The probability of failure during $(0,t)$ is

$$F(t) = 1 - L(t). \tag{6-11}$$

The conditional probability of failure within time interval $(t, t + \Delta t)$, given that the component has survived up to t, is defined by the hazard function, which can be expressed as

$$h(t) = \frac{-d}{dt} \ln L(t). \tag{6-12}$$

The reliability and hazard functions are integrally related

$$L(t) = \exp\left[-\int_0^t h(x)dx \right]. \tag{6-13}$$

The hazard function is especially useful in analyzing structural failures due to aging or deterioration. For example, if the structure has survived during the interval $(0, t_1)$, it may be of interest in scheduling in-service inspections to determine the probability that it will fail before t_2. Such an assessment can be performed if $h(t)$ is known. If the time-to-failure is T_f, this probability can be expressed as

$$P\left[T_i < t_2 | T_i > t_1\right] = 1 - \exp\left(-\int_{t_1}^{t_2} h(x)dx \right). \tag{6-14}$$

In turn, the structural reliability for a succession of inspection periods is

$$L(0,t) = \prod_t L(t_{i-1}, t_i) \exp\left\{ \int_{t_1}^t h(x)dx \right\} \tag{6-15}$$

in which $t_{i-1} = 0$ when $i = 1$. Note that failures in successive intervals, (t_{i-1}, t_i), generally are not statistically independent events.

The hazard function for pure chance failures is constant with time. When structural aging occurs and strength deteriorates, $h(t)$ characteristically increases with time. Reliability assessments of nondegrading and degrading structural components can be distinguished by their hazard functions. Much of the challenge in structural reliability analysis of deteriorating structures lies in relating $h(t)$ to specific degradation mechanisms, such as corrosion. The common assumption in some time-dependent reliability studies that the failure rate increases linearly has been shown to be invalid for aging structures in nuclear plants (Ellingwood and Mori 1993). When degradation mechanisms are synergistic, $h(t)$ generally is unknown at the current state of the art. In-service inspection and maintenance impact the hazard function, causing it to change discontinuously at the time it is performed.

The reliability functions $L(t)$ and $F(t)$ are cumulative, that is, they describe the probabilities of successful (or unsuccessful) performance during the interval $(0,t)$. If $h(t)$ is very small numerically, $h(t)$ is approximately equal to $P_f(t)$ in Eq. 6-8. It should be emphasized that $F(t)$ is not equal $P_f(t)$ in Eq. 6-8; the latter is simply the instantaneous probability of failure without regard to previous (or future) structural performance. Failure to recognize the difference is a fundamental but common interpretative error.

The methods summarized above have been extended to structures subjected to combinations of structural load processes and to structural systems (Ellingwood and Mori 1992). The reliability function has a similar appearance to that of Eq. 6-10, but the outer integral on resistance increases in dimension in accordance with the number of components in the system. The system reliability may be evaluated by Monte Carlo simulation, using an adaptive importance sampling technique to enhance the efficiency of the simulation (Mori and Ellingwood 1993).

6.4.3.2. Role of inspection/repair in maintaining reliability. Forecasts of time-dependent reliability enable the analyst to determine the time period beyond which the desired reliability of the structure cannot be assured. At such time, the structure should be inspected and its condition evaluated. Intervals of inspection and repair that may be required as a condition for continued operation can be determined from the time-dependent reliability analysis. The effect of the inspection/repair is to remove larger defects from the structure and upgrade its strength, thus reducing its conditional failure rate. As the structure ages, the failure rate increases until another inspection/repair operation occurs.

6.4.3.2.1 Degradation function based on individual damage intensities. Implementation of Eqs. 6-10 through 6-15 requires that $g(t)$ be determined during the service life of the structure or component as degradation and repair take place. We begin by modeling the damage intensity by a state variable taking a value within the interval [0,1]; the values 0 and 1 indicate no damage and no residual strength, respectively. An example of this state variable describing damage would include the ratio of area of reinforcement lost due to corrosion to the original area. The following assumptions are made:

1. Initiation of damages in a component is described by a Poisson process in which the expected number of damages in time interval $(t,\ t + \Delta t;\ t > 0)$ is

$$\int_{t_1}^{t+\Delta t} v(\tau)d\tau. \tag{6-16}$$

    The damage initiation rate $v(\tau)$ is dependent on the surface area or volume of the component.
2. Damages initiate homogeneously over the surface area or volume of the component, the initial strength of the cross section is uniform, and the load effect is uniform within the component.
3. Once damage initiates at location j, it grows according to

$$X_j(t) = \begin{cases} 0 & ;\quad 0 \le t < T_{I_j} \\ C_j\left(t - T_{I_j}\right)^\alpha & ;\quad t \ge T_{I_j} \end{cases} \tag{6-17}$$

in which $X_j(t)$, $j = 1, 2, \ldots$ is the intensity of damage at time t; T_{I_j}, $j = 1, 2, \ldots$ are the random initiation times of damage; C_js are damage growth rates that are identically distributed and statistically independent random variables described by a CDF $F_C(c)$; and α is a deterministic parameter. Parameters C and α depend on the degradation mechanism. If damage grows linearly, $\alpha = 1$.

4. At most, one damage initiates at the cross section so that the degradation function $G(t)$ for a component is defined in terms of damage intensities as

$$G(t) = 1 - \max_{\text{all } j}\left\{X_j(t)\right\}. \tag{6-18}$$

Assume first that the number of damages that initiate within the interval (t_1, t), $N_I(t_1, t)$, is n. From assumption 3, the CDF of $X_j(t)$, $F_{x_j}(x; t_1, t)$, is expressed as

$$F_{X_j}(x; t_1, t) = P\left[C_j\left(t - T_{I_j}\right)^{\alpha} < x\right] \tag{6-19}$$

$$= \int_{t_1}^{t} F_C\left(\frac{x}{(t - \tau)^{\alpha}}\right) f_{T_{I_j}}(\tau) d\tau \tag{6-20}$$

where $f_{T_{I_j}}(\tau)$ is the PDF of T_{I_j}. Given $N_I(t_1, t) = n$, the rank-ordered initiation times, $T_{I_1}, T_{I_2}, \ldots, T_{I_n}$ are n order statistics of random variables $W_{I_1}, W_{I_2}, \ldots, W_{I_n}$ that are statistically independent and identically distributed with PDF expressed as (Taylor and Karlin 1984)

$$f_{W_I}(w) = \begin{cases} \dfrac{v(w)}{\displaystyle\int_{t_1}^{t} v(t)dt} & ; \quad t_1 \leq w \leq t \\[20pt] 0 & ; \quad \text{otherwise} \end{cases} \tag{6-21}$$

Therefore, the CDF of the intensity of an arbitrary damage, $X(t)$, that initiates within (t_1, t) is,

$$F_x\left(x; t_1, t\right) = \int_{t_1}^{t} F_c\left(\frac{x}{(t - \tau)^{\alpha}}\right) f_{W_1}(\tau) d\tau. \tag{6-22}$$

Note that $f_{W_i}(w)$ is an unconditional PDF. Since the C_js and W_{I_j}s are statistically independent of one another, the $X_j(t)$s are also statistically independent. Accordingly, the CDF of $X\max(t_1; t) = \max_j\{X_j(t)$ that initiated within $(t_1, t]\}$ is

$$F_{X_{max}}\left(x; t_1, t \,\middle|\, N_I\left(t_1, t\right) = n\right) = \left[F_X\left(x; t_1, t\right)\right]^{n} \tag{6-23}$$

Removing the condition that $N_I(t_1, t) = n$, we find that

$$
\begin{aligned}
F_{X_{max}}(x; t_1, t) &= \sum_{n=0}^{\infty} \left[F_X(x; t_1, t)\right]^n \cdot \frac{\left(\int_{t_1}^t v(\tau)d\tau\right)^n \cdot \exp\left(-\int_{t_1}^t v(\tau)d\tau\right)}{n!} \\
&= \exp\left[-\int_{t_1}^t v(\tau)d\tau\left\{1 - F_X(x; t_1, t)\right\}\right].
\end{aligned}
\tag{6-24}
$$

From assumption 4, the mean and variance of the degradation function are evaluated by

$$
\begin{aligned}
E[G(t)] &= E[1 - X_{max}(0;t)] \\
&= 1 - \int_0^1 [1 - F_{X_{max}}(x; 0, t)]dx
\end{aligned}
\tag{6-25}
$$

and

$$
\begin{aligned}
\mathrm{Var}[G(t)] &= \mathrm{Var}[X_{max}(0;t)] \\
&= \int_0^{-1} 2x[1 - F_{X_{max}}(x; 0, t)]dx - \{E[G(t)]\}^2.
\end{aligned}
\tag{6-26}
$$

It has been found that the variability in $G(t)$ has a secondary effect on the time dependent reliability of a component and, thus, the reliability can be evaluated considering only the mean of $G(t)$, defined as $g(t)$ in Eq. 6-10 (Mori and Ellingwood 1993).

6.4.3.2.2 Degradation function after repair. No nondestructive evaluation (NDE) method can detect a given defect with certainty. The imperfect nature of NDE methods must be described in statistical terms. The strength following repair depends on both the detectability and repair strategy adopted.

Assume that during inspection/repair the entire component is inspected, that all detected damages are repaired immediately and completely, and that the repaired parts of the component are restored to their initial strength levels. Then the effect of inspection/repair on $g(t)$ depends on the detectability function, $d(x)$, associated with the NDE method, in which x is a nondimensional measure of damage intensity consistent with Eq. 6-17. The inspection with higher $d(x)$ makes repair more likely and, accordingly, leads to higher values of the degradation function, $g(t)$. In the limit, if an inspection is perfect (i.e., $d(x) = 1$ for $x > 0$) then the component is restored to its original condition by the repair.

First assume that the detectability function, $d(x)$, is defined as

$$
d(x) = \begin{cases} 0 & ; \quad 0 \le x < x_{th} \\[2mm] 1 & ; \quad x_{th} \le x \le 1 \end{cases}
\tag{6-27}
$$

where xth is the minimum detectable value of damage. The same detection threshold values are assumed for all inspections. Following m inspections at $t_R = \{t_{R_1}, \ldots, t_{R_m}\}$, some of the damages are repaired and the CDF describing $X(t)$ and the number of damages existing at time $T > t_{R_m}$, $N(t)$, changes. The intensities of damages that initiate after t_{R_m} are independent of repair, and only the PDF of the intensities of damages initiating before t_{R_m} is updated. Let us consider damages that exist at time t and initiated within periods (A) $(0, t_{R_m}]$ or (B) $(t_{R_m}, t]$. The number of damages left unrepaired after t_{R_m}, $N(A)$, can be described by a filtered Poisson process with a parameter $p \cdot v(w)$, where

$$p = P\left[A \text{ damage is not repaired by } t_{R_m}\right] = F_X\left(x_{th}, 0, t_{R_m}\right), \qquad (6\text{-}28)$$

while the number of damages initiating within $(t_{R_m}, t]$, $N(B)$, is described by a Poisson process with a parameter $v(w)$. In other words, the number of defects existing at time t can be described by a nonstationary Poisson process within a parameter $v''(w)$ given by

$$v''(w) = \begin{cases} p \cdot v(w) & ; \quad 0 < w \le t_{R_m} \\ \\ v(w) & ; \quad t_{R_m} < w \le t \end{cases}. \qquad (6\text{-}29)$$

Therefore, the procedure to estimate the degradation function for a component before an inspection/repair can be used to estimate the function after multiple inspection/repairs, by replacing $v(w)$ by $v''(w)$, and $F_X(x; 0, t)$ by the updated CDF $F''_X(x; t_{R_m}, t)$.

By the theorem of total probability, $F''_X(x; t_{R_m}, t)$ can be expressed as

$$F''_X\left(x; t_{R_m}, t\right) = F_{X_{(A)}}\left(x; t_{R_m}, t\right) \cdot P(A) + F_{X_{(B)}}\left(x; t_{R_m}, t\right) \cdot P(B) \qquad (6\text{-}30)$$

in which $F''_X(A)(x; t_{R_m}, t)$ and $F''_X(B)(x; t_{R_m}, t)$ are the CDF of intensity of damages in group (A), $X(A)(t)$, and intensity of damages in group (B), $X(B)(t)$, respectively, and $P(A)$ and $P(B)$ are the probabilities that a defect belongs to group (A) or (B). Thus,

$$P(A) = P\left[W_1 \le t_{R_m}\right] = \int_0^{t_{R_m}} f_{W_1}(w)\,dw \quad \text{and} \qquad (6\text{-}31)$$

$$P(B) = 1 - P(A), \qquad (6\text{-}32)$$

in which $f_{W_I}(w)$ is evaluated by Eq. 6-21 replacing $v(w)$ with $v''(w)$. $F''_x(A)(x;t_{R_m},t)$ is expressed as

$$F_{X_{(A)}}\left(x; t_{R_m}, t\right) = P\left[X(t) < x \Big| X(t_{R_m}) < x_{th}\right]$$

$$= \frac{\displaystyle\int_0^{t_{R_m}} F_C\left(\min\left\{\frac{x}{(t-\tau)^\alpha}, \frac{x_{th}}{(t_{R_m}-\tau)^\alpha}\right\}\right) f_{W_I}(\tau)d\tau}{\displaystyle\int_0^{t_{R_m}} F_C\left(\frac{x_{th}}{(t_{R_m}-\tau)^\alpha}\right) f_{W_I}(\tau)d\tau}. \tag{6-33}$$

The CDF $F_{X_{max(B)}}(x;t)$ is given by

$$F_{X_{max(B)}}(x; t) = F_X\left(x; t_{R_m} \cdot t\right). \tag{6-34}$$

In general, the detectability function, $d(x)$, is not a step function but rather a non-decreasing function of damage intensity. Procedures for dealing with this more general detectability function, partial inspection/repair, and optimized inspection/repair strategies have been developed (Mori and Ellingwood 1993; Mori and Ellingwoood 1994a; Mori and Ellingwood 1994b) but are more involved mathematically.

6.4.3.3 Mathematical models of fragility. The fragility of a structural system is defined as the limit state probability or probability of "failure" conditioned on a level of demand, x (e.g., spectral acceleration, velocity, and internal pressure),

$$F_R(x) = P[LS | X = x] \tag{6-35}$$

in which $F_R(x)$ takes the form of a CDF, assuming that the limit state probability increases monotonically with increasing demand, x. The limit state depends on the performance requirements of the system considered (e.g., metal containment structural capacity is lost and rupture occurs or when deformations are large enough to interfere with the operation of attached equipment). The relation between performance limits and structural response quantities (which is where the structural analysis necessarily must stop) must be carefully established if the fragility is to be credible and useful for engineering decision analysis.

The lognormal CDF is commonly used to model fragility in the nuclear industry (Kennedy and Ravindra 1984),

$$F_R(x) = \Phi\left(\frac{\ln(x/m_R)}{\beta_c}\right) \tag{6-36}$$

in which $\Phi[\bullet]$ = standard normal probability integral, m_R median capacity (expressed in units that are dimensionally consistent with the demand parameter, x), and β_R = standard deviation (SD) in $\ln R$ (or *logarithmic standard deviation*), describing the inherent randomness in capacity of the component to withstand the demand. Parameter β_R is related to the COV, V_R, in capacity by

$$\beta_R = \sqrt{\ln\left(1 + V_R^2\right)} \tag{6-37}$$

when this COV is less than about 0.3, $\beta_R \sim V_R$.

Eq. 6-37 portrays the inherent randomness (or aleatory uncertainty) in the capacity of the component or system that it models. This source of uncertainty is inherent to the component or system; additional modeling sophistication or data collection will not change it in any significant way. Additional uncertainties in capacity arise from such things as assumptions made in the structural system analysis and limitations in the supporting statistical database. The knowledge-based (or epistemic) uncertainties are distinguished from aleatory uncertainties in that they depend on the quality of the analysis and the supporting databases; epistemic uncertainties generally can be reduced at the expense of a more comprehensive analysis. Epistemic uncertainties can be taken into account, to first order, by assuming that such uncertainties impact the estimate of the median capacity, making it a random variable rather than a constant. Accordingly, m_R in Eq. 6-36 is replaced by the random variable, M_R, assumed to be lognormal with median, m_R and logarithmic SD β_U. The fragility thus becomes a function of M_R depicted by a family of lognormal distributions defined by the parameters (m_R, β_R, β_U) and which displays the overall uncertainty in the conditional component failure probability at the value of x.

The system fragility is used in safety margin assessment to identify a level of demand at which there is a high confidence that the system will survive. It is customary in structural design to use a nominal or characteristic strength below which there is a small probability, typically 0.05, that the true strength will lie. In the presence of epistemic uncertainty, this 5% exclusion limit has a frequency distribution (analogous to a sampling distribution in classical statistics). The lower α-fractal of this frequency distribution is R_α; one might say that the 5% exclusion limit of capacity is at least R_α with confidence $(1-\alpha)$. R_α can be obtained as

$$R_\alpha = m_R \exp\left[-1.645\beta_R + k_\alpha\beta_U\right] \tag{6-38}$$

in which $k_\alpha = \Phi^{-1}(\alpha)$. In seismic margins analysis, it has been customary to set $\alpha = 0.05$ and refer to R_α as the "high-confidence, low-probability of failure" (HCLPF) value. This HCLPF can be expressed as

$$\text{HCLPF} = m_R \exp\left[-1.645\left(\beta_R + \beta_U\right)\right]. \tag{6-39}$$

The HCLPF is akin to a lower tolerance limit in a Bayesian rather than a classical statistical sense since the notion of sample size enters into determination of HCLPF only indirectly through the judgmental assessment of β_U.

An overall mean or expected fragility can be determined as

$$E\left[F_{R(x)}\right] = \int_0^\infty E\left[F_R(x)\,|\,M_R = \gamma\right]f_{M_R}(\gamma)\mathrm{d}\gamma. \tag{6-40}$$

From this, the mean fragility can be shown to be (Ellingwood 1994)

$$E\left[F_R(x)\right] = \Phi\frac{\ln(x/m_R)}{\sqrt{\beta_R^2 + \beta_U^2}}. \tag{6-41}$$

The mean fragility has a flatter slope than the median because it includes both aleatory and epistemic uncertainties.

Substituting the HCLPF from Eq. 6-39 for x into Eq. 6-41 and noting that $\sqrt{(\beta_R^2 + \beta_U^2)} \approx 0.7(\beta_R + \beta_U)$ for most structural components and systems yields

$$E\left[F_R(\mathrm{HCLPF})\right] = \Phi\left[-\frac{1.645(\beta_R + \beta_U)}{\sqrt{\beta_R^2 + \beta_U^2}}\right] = \Phi\left[-2.35\right] = 0.01. \tag{6-42}$$

In words, the HCLPF occurs at approximately the 0.01-fractal of the mean fragility. In general, the HCLPF can be estimated from

$$\mathrm{HCLPF} = m_R \exp\left[-2.33\beta_c\right] \tag{6-43}$$

in which the overall uncertainty is

$$\beta_c = \sqrt{\beta_R^2 + \beta_U^2} \tag{6-44}$$

rather than from Eq. 6-39. Some recent safety margins studies and risk analyses have used Eq. 6-43 rather than Eq. 6-39 for simplicity. However, in doing so, the sense of epistemic uncertainty in the fragility model is lost; it is important that the fragility convey a sense of quality and completeness in the analysis.

Fragilities can be constructed for a range of performance requirements and limit states shy of loss of integrity. That is, a collection of fragilities can describe progressively more severe limit states (e.g., onset of nonlinear action, onset of slow leakage, incipient instability, and loss of integrity). At any given level of demand, x, such a depiction conveys a sense of the additional margin of safety and reliability that remains at each stage. Clustered fragilities, as opposed to well-spaced fragilities, may indicate that little additional margins of safety remain once a threshold is reached. As structural degradation causes the fragility to vary with time, the fragility and HCLPF computed using degradation models provides a snapshot of the component safety margin as a function of time.

6.4.4 Illustrations of Time-Dependent Reliability Analysis

6.4.4.1 Service life prediction of reinforced concrete slab. Time-dependent reliability concepts are illustrated with a simple conceptual example of a reinforced

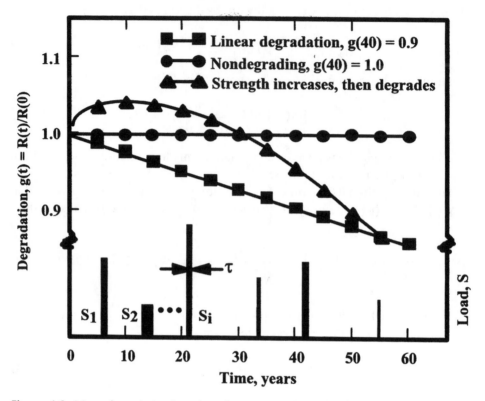

Figure 6-2. Mean degradation function of a one-way slab.

concrete slab designed using the requirements for flexure strength found in ACI Standard 318 (ACI 1999)

$$0.9R_n = 1.4D_n + 1.7L_n, \qquad (6\text{-}45)$$

in which R_n is the nominal or code resistance, and D_n and L_n are the code-specified dead and live loads, respectively. In this example three cases are considered: (1) service life prediction, (2) single inspection/repair, and (3) multiple inspection/repair. Three scenarios are considered: (1) the strength of the slab changes with time, initially increasing as the concrete matures and then decreasing due to (unspecified) environmental attack, (2) the strength degrades linearly to 90% of initial strength at 40 years, and (3) the strength remains constant with time (Fig. 6-2). In general, the behavior of resistance over time must be obtained from mathematical models describing the degradation mechanism(s) present. The statistics used in this example are contained in Tables 6-1 through 6-3.

Figure 6-3 compares the limit state probabilities, $F(t) = 1 - L(t)$ in Eq. 6-10, obtained for the three degradation models considered in Figure 6-2 for service lives $(0,t)$ ranging up to 60 years. When $R(t) = R(0)$ and no degradation of strength occurs, a result is obtained analogous to what has been done in probability-based code work to date. Neglecting strength degradation entirely in a time-dependent

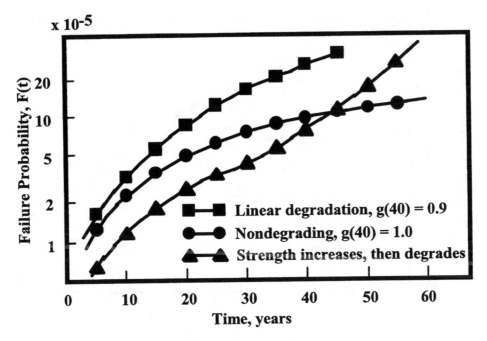

Figure 6-3. Failure probability of a one-way slab.

reliability assessment can be quite unconservative, depending on the time-dependent characteristics of strength.

6.4.4.1.1 Single Inspection/Repair. Now consider the following:

1. Every part of the structure is fully inspected and all detected damages are repaired completely.
2. The initiation of damages is described by a stationary Poisson process with a parameter $v = 5$/yr that is dependent on the surface area or volume of the structure.
3. Damage grows linearly with time as described by Eq. 6-17 with $\alpha = 1$.
4. The degradation rate, C, is lognormally distributed with mean value, $\mu_C = 0.00125$, that corresponds to $E[X(40)|T_I = 0] = 0.05$, and with a coefficient of variation, $Vc = 0.5$.

The effect on the mean degradation function of inspection/repair described by several detectability functions is illustrated in Figure 6-4. The first detectability function considered is a step detectability function in which $x_{th} = 0.03$; in the second, x_{th} is uniformly distributed [i.e., $d(x)$ is linear between x_{min} and x_{max}, where $d(x_{min}) = 0$ and $d(x_{max}) = 1$]; in the third and fourth, x_{th} is lognormally distributed with mean, $\mu_{X_{th}}$, equal to 0.03, and COV, $V_{X_{th}}$, equal to 0.3 or 0.5. It is assumed that inspection/repair is carried out at $t_{R_m} = 20$ years. The mean degradation function decreases as $V_{X_{th}}$ increases (that would result in lower reliability); however, the effect of the general shape of $d(x)$ is not significant and decreases with time elapsed since inspection. Clearly, it is important to determine x_{th}. However, the insensitivity of the mean

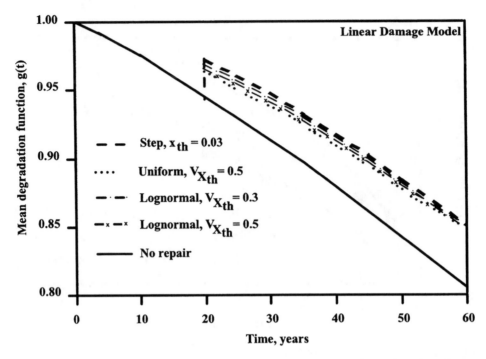

Figure 6-4. Effect of several detectability functions on mean degradation function of inspection/repair.

degradation to the shape of $d(x)$ suggests that a general detectability function might be approximated for practical purposes by a step function increasing from 0 to 1 at x_{th}. This would be advantageous for NDE technologies currently used for reinforced concrete structures because information on x_{th} may be more readily available than information on $d(x)$.

6.4.4.1.2 Multiple Inspection/Repair. The effect of multiple inspection/repair and the mean degradation function is illustrated in Figure 6-5, assuming a step detectability function and the same assumptions as used in the previous example. Inspection/repairs are carried out at 20, 30, 40, and 50 years with $x_{th} = 0.05$ when $E\left[X(40)|T_I = 0\right] = 0.05$. For comparison, the mean degradation function for a component without repair and for a component with one repair at 30 years with $x_{th} = 0.01$ is also presented in the figure. With multiple inspections/repairs, the mean degradation function can be kept within a narrow range during the service life of the structure. This suggests the existence of an optimum inspection/repair strategy in which the total expected cost, defined as the sum of the cost of inspections/repairs and expected cost (loss) due to failure, is minimized, while the failure probability of the component is kept below an established target probability during its service life. Such a strategy has been developed and is presented elsewhere (Mori and Ellingwood 1994b).

6.4.4.2 Fragility modeling of steel containment. The commercially available finite element code, ABAQUS, has the necessary features for conducting an analysis of

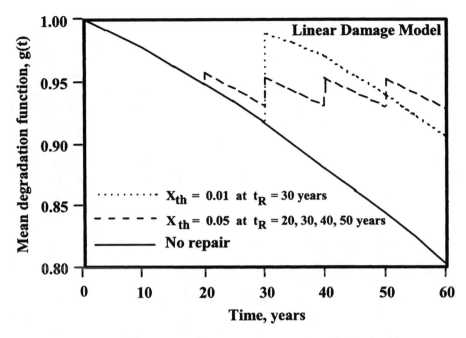

Figure 6-5. Effect of multiple inspection/repairs on mean degradation function.

a containment subjected to a pressure/temperature/time history well into the large deformation regime. The occurrence of structural limit states is identified by post-processing following completion of the finite element analysis. The limit state that will be used is (Hancock and MacKenzie 1976)

$$\varepsilon_p > \varepsilon_f \, \alpha \, \exp(-3\sigma_m/2\sigma_{vm}) \qquad (6\text{-}46)$$

in which the effective plastic strain, ε_p, computed from the finite element analysis is

$$\varepsilon_p = \sqrt{2}\left[\left(\varepsilon_1 - \varepsilon_2\right)^2 + \left(\varepsilon_2 - \varepsilon_3\right)^2 + \left(\varepsilon_3 - \varepsilon_1\right)^2\right]^{1/2}/3 \qquad (6\text{-}47)$$

in which ε_i, $i = 1,2,3$ = principal strains; ε_f = fracture ductility; and α = experimental constant. The ratio of mean stress, σ_m,

$$\sigma_m = (\sigma_1 + \sigma_2 + \sigma_3)/3 \qquad (6\text{-}48)$$

to Von Mises effective stress, σ_{vm},

$$\sigma_{vm} = \left[\left(\sigma_1 - \sigma_2\right)^2 + \left(\sigma_2 - \sigma_3\right)^2 + \left(\sigma_3 - \sigma_1\right)^2\right]^{1/2}/\sqrt{2} \qquad (6\text{-}49)$$

in which σ_i, $i = 1,2,3$, are principal stresses, reflects the influence of stress triaxiality on the limit state.

A fragility assessment requires median-centered estimates of system capacity. Most modes of structural failure of interest in fragility assessment are only credible at pressures and temperatures well beyond the design basis. Material strengths specified for design by standardizing organizations such as ASTM are usually substantially less than the actual strengths delivered in the structure. These and other sources of conservatism must be removed if a credible estimate of system capacity to withstand such demands is to be obtained. Also, the uncertainties must be taken into account.

The main sources of aleatory uncertainty affecting behavior of steel containment structures are variability in yield strength, modulus of elasticity, and fracture ductility. Variabilities in other structural parameters are negligible in comparison and can be ignored. Data collected in research to develop probability-based design codes (Galambos and Ravindra 1978; Ellingwood and Hwang 1985) and condition assessment procedures for steel structures (Ellingwood and Cherry 1999) can be used for this purpose. Table 6-4 presents initial resistance of steel plates and shapes obtained from mill tests conducted by ASTM procedures. Since the strength of carbon steel under slower rates of load is about 4 MPa less than presented in the table, the mean strengths should be reduced by this amount in any fragility assessment (Ellingwood and Cherry 1999). The fracture ductility, ε_f, in Eq. 6-46 is a random variable that for a ASME SA 516 steel plate can be described by a normal distribution with a mean of 0.28 and COV of 0.14. Furthermore, tests have shown that when the steel is corroded, the strength of the uncorroded portion remains unchanged; however, the fracture ductility decreases by approximately a factor of 2 (Cherry 1997).

Epistemic uncertainties arise from the finite element modeling of the containment for structural analysis purposes, definition of failure criteria against which the finite element results are checked, and fitting supporting experimental data with the probability models and extrapolating well beyond the realm of observation to identify design levels. Uncertainty from finite element modeling arises from selection of mesh size, element types, idealized loading and boundary conditions, execution parameters, and postprocessing. Since failures may occur as a result of local strains at a scale that the FE analysis cannot capture without costly refinement of the mesh, ε_f is multiplied by 0.5 for membrane action and 0.75 for bending action. With the further assumption that the calculated strain in the containment is an unbiased estimate of the true strain and can be determined with an accuracy of $+20\%$ of the true value with 95% confidence; the COV due to modeling is 0.10. Accordingly, the failure criterion in Eq. 6-46 is augmented with a normal random variable incorporating these parameters to account for modeling uncertainty.

The fragility is constructed by a numerical experiment that couples the nonlinear finite element analysis with Monte Carlo simulation. A stratified sampling reduction technique known as *Latin Hypercube* sampling is utilized for random sampling to reduce costs associated with the finite element analysis. Estimators obtained from this procedure can be shown to be unbiased, and their sampling error generally is close to minimum for a given sample size. In this example, the aleatory and epistemic uncertainties are considered together, as in Eq. 6-44 and the mean fragility in Eq. 6-41 is obtained. Fourteen Latin Hypercube samples were selected so that the minimum and maximum of the samples would plot at approximately the 5 and

95 percentiles using rank-median plotting positions. Parameters included in the 14 Latin Hypercube samples were assumed to be statistically independent and included elastic modulus, yield strength, ultimate strength, failure strain, ratio of corrosion penetration to section thickness, and modeling.

A PWR ice condenser plant is considered in this example. The steel containment is designed for an internal pressure of 74 kPa, and has an internal diameter of 35 m, a springline height of 35 m, and the apex of the spherical dome is at 53 m. The containment is fabricated from ASME SA 516 Gr. 60 plate, varying in thickness from 35 mm at the base mat to 12 mm at the springline. Vertical and circumferential stringers are welded to the exterior shell at spacings of approximately 1.2 m (vertical) and 3 m (circumferential). The fragility in the as-designed (uncorroded) condition establishes a benchmark for subsequent structural margin evaluations. Figure 6-6 presents the rank-ordered failure pressures for the undegraded condition plotted on lognormal probability paper for the 14 Latin Hypercube samples. The mean (and median) containment capacity is 455 kPa, and the logarithmic SD, β_c, is 0.04. The 5% exclusion limit is estimated to be at 427 kPa.

Postulating corrosion damage based on what has been observed by location, areal extent, and mean loss of shell thickness and determining the impact of such damage on strength closely models the condition assessment process as it is likely to be implemented in practice. However, in situ inspection generally would not determine the variability in loss of thickness; moreover, there are measurement uncertainties associated with such a determination. These sources of uncertainty (with respect to the mean corrosion loss) are modeled by uniform distributions. Results of

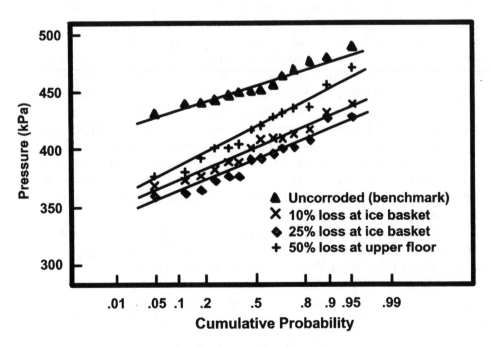

Figure 6-6. Steel containment fragility for postulated corrosion patterns.

the fragility analyses for the steel containment are illustrated in Figure 6-6 for three postulated corrosion patterns: 10% and 25% corrosion loss behind the ice basket, and 50% loss at the upper floor. Examining the condition related to 25% corrosion loss, the mean pressure at containment failure by tensile instability was 386 kPa at an effective limiting plastic strain of 0.048; the estimated 5% and 95% values of capacity are 352 kPa and 427 kPa respectively. The reduction in fragility is nonlinearly proportional to the median percentage loss in thickness, with a larger decrease occurring between the undegraded condition and the 10% loss than between the 10% and 25% losses. This reduction occurs because the fracture ductility of a corroded section decreases by a factor of two, regardless of corrosion penetration.

6.5 Commentary on Structural Reliability Methods and Perspectives on Risk

Structural reliability analysis provides the framework for analyzing uncertainties in structural loads, operational demands on a structural component or system, and structural capacity to withstand demands. This information can be utilized on several levels: to identify a review level event to be used in a subsequent structural analysis, to identify a HCLPF capacity, to determine an instantaneous probability of component or system failure, and to determine the probability of failure during any service period. A well-formulated system reliability analysis also can be used to evaluate fitness for continued service and to establish priorities for in-service inspection and repair.

In theory, component or system reliabilities can be computed from first order reliability methods (FORM) or full-distribution methods (Shinozuka 1983; Bjerager 1990). In practice, this may not be as easy as it sounds. The determination of the limit state probability of fragility requires statistical data on material strengths, dimensions and other basic random variables, modeling errors, and a verifiable structural model of behavior based on principles of mechanics to identify the limit condition (e.g., excessive inelastic deformation or instability). Moreover, modern structural analysis often is performed using finite element methods. In contrast to classical mechanics formulations, finite element analysis is algorithmic in nature, yielding structural responses (forces, displacements) at discrete points, but not a general closed form expression for the limit state function. Thus, the function describing the limit state often cannot be determined in closed form, and thus the failure domain over which the (multidimensional) joint density function must be integrated may not be well-defined.

There also is the numerical problem in evaluating the probability integral in Eq. 6-8 for realistic systems. Computation of this probability integral is numerically difficult when more than a few random variables are involved, and the domain of integration is irregular or nonlinear. Monte Carlo methods can be used to perform these computations in approximate form (Rubenstein 1981). The Monte Carlo approach has a number of practical advantages, particularly in a structural system reliability analysis (Moses 1990): complex structural behavior can be accommodated, stochastic dependency can be modeled, the possible introduction of new random variables due to inspection and repair can be dealt with, and several failure modes can be included (e.g., fracture from overload as opposed to fracture from crack growth). Perhaps one of the most useful features of Monte Carlo simulation is the way in which it facilitates visualization of the damage evolution process.

The main disadvantage of Monte Carlo methods is their lack of numerical efficiency in structural reliability analysis that involves small probabilities. If the event probabilities of interest are on the order of 10^{-N}, an unmodified random sampling procedure requires approximately 10^{N+1} samples for the failure probability estimate to be stable. The number of samples required to achieve a given sampling error, expressed by the SD (P_f), can be reduced by modifying the random sampling process. In structural reliability analysis importance sampling often has been used for this purpose (Schueller and Stix 1987; Melchers 1990; Mori and Ellingwood 1993). Importance sampling techniques can be categorized by: direct, updating, or adaptive schemes or spherical schemes/directional sampling (Engelund and Rackwitz 1993). The efficiency of these approaches depends on the number of times the structural limit state must be calculated if the SD $(P_f) < \varepsilon$; the efficiency thus is related more to the structural analysis than to the uncertainty analysis. In all methods except adaptive sampling, the positioning of the importance sampling PDF must have been achieved with a suitable algorithm, often FORM. None of the methods is optimal under all circumstances, and some experimentation is required to determine the best approach for the particular problem at hand.

Finally, there is the difficult challenge of quantifying measures of acceptable risk in terms of limit state probability or fragility benchmarks. The concept of risk involves likelihood, consequences, and context (Elms 1992). Acceptable risk guidelines are essential to distinguish alternatives that are acceptable from those that are not. The use of code calibration, which was the basis for developing the first generation probability-based limit states design codes, is not well-suited to risk management of critical facilities; where the experience base does not exist and structural components and systems tend to be one-of-a-kind, the hazards of concern are very rare, and consequences of failure are severe (Ellingwood 2001). The alternate route is to attempt to derive acceptable risks for structural components and systems from a comparison with other presumably acceptable risks, such as might be gleaned from mortality tables or other similar sources. While this approach may seem natural, such comparisons involve different contexts for risk and can be misleading. For example, thresholds of risk tend to be relatively high for those risks that are incurred voluntarily (e.g., cigarette smoking, mountain climbing). Conversely, when the hazard is unfamiliar, creates fear, or puts large numbers of people in jeopardy, acceptable risk levels are lower than can be justified on a purely actuarial basis. Moreover, thresholds of risk acceptance are higher for those risks that are viewed as being deferred. In short, people react to risks associated with different technologies in a manner that is not entirely rational to professional risk analysts (Stewart and Melchers 1997) but is consistent with their own value systems. It goes without saying that the public at large must have confidence in the process of establishing acceptable levels of risk, particularly when the technical details of the hazard or response of the facility to it are difficult to comprehend.

Work is ongoing to establish acceptable risk guidelines in specialized fields of building construction. The U.S. Department of Energy (DOE) has established target probabilistic performance goals for different categories of DOE facilities subjected to several natural phenomena hazards (DOE 1996). The U.S. Nuclear Regulatory Commission recently issued Regulatory Guide 1.174 (NRC 1998), which provides guidelines for risk-informed decision making regarding proposed changes to the licensing basis of nuclear plants. Core damage frequency (CDF) and large early release frequency (LERF) have been identified as suitable risk metrics for this purpose. The focus of these guidelines is on changes from established CDF or LERF benchmarks for the plant (in its original condition) arising from proposed changes, rather than on absolute risk, which is a reflection of the difficulty in establishing an absolute risk target. For example, when the benchmark CDF is approximately 10^{-5} per reactor-year, as it seems to be for many NPPs in the United States, proposed changes that lead to calculated increases in CDF of 10^{-6} or less per reactor-year would be considered acceptable; changes leading to increases greater than 10^{-5} per reactor-year would be unacceptable (viz. Section 2.2.4 of RG1.173, Figure 3, Region II). These probabilities provide some guidance on the levels of damage that might be considered acceptable in an aging NPP, as might be inferred from curves such as those in Figures 6-3 and 6-6, if accompanied by periodic in-service inspection. Of course, implementation of the RG 1.174 Guidelines requires, at the minimum, a Level I (or fully coupled) probabilistic safety assessment of the plant. Moreover, any such risk analysis requires a thorough examination of all uncertainties, both aleatory and epistemic, and their respective contributions to the estimated risk and ensuing decisions.

On the other hand, in a fragility analysis such as that described for the metal containment in Section 6.4.4.2, the structural performance and hazard issues are decoupled so the problem of identifying acceptably small limit state probabilities for decision analysis purposes (acceptable risks) is circumvented to a degree. The fragility analyses provide a structured framework for: identifying aging factors that are important to safety; identifying areas where the potential for degradation to impact safety is significant; focusing in-service inspection, maintenance, and repair; guiding the acquisition of additional data; and arriving at risk-informed decisions regarding suitability of an existing structure for continued service. In contrast to the typical design analysis, the fragility depicts how a structure (e.g., containment) may perform under a spectrum of possible events. A properly conducted fragility analysis must be based on rational structural mechanics and structural reliability principles and requires a supporting database to describe the various uncertainties that are likely to affect performance. The relation between performance requirements and structural limits must be clear and defensible. Similar to the risk analysis above, the starting point must be the benchmark fragility of the structure in its original as-built condition. Once the benchmark fragility is established, subsequent changes to the structure's fragility as a result of aging from operating conditions or environmental stressors can be clearly identified. One approach is to use a nonlinear finite element analysis program coupled to an efficient Monte Carlo simulation procedure to propagate uncertainties. However, simpler reliability tools might be sufficient for establishing a general relation between uncertainty and safety; an approximate analysis could be used for preliminary decision making or determining whether to invest in a more comprehensive study (Ellingwood and Cherry 1999). In support of this approximate analysis, a small number of nonlinear finite element analyses with all parameters set at their median values would be performed to anchor the fragility curve in the vicinity of its midrange (median). For application to NPP steel containments, the parameter β_c (Eq. 6-44), which defines the slope of the fragility, can be estimated from available results (e.g., Table 6-3 of Ellingwood and Cherry 1999); this value usually is between 0.05 and 0.20 for containments in the undegraded condition. With m_R and β_c determined, the mean fragility is defined by the lognormal distribution in Eq. 6-41. The 5-percentile value then could be compared to the review event. In the comparison, the rate of change in the 5-percentile values from the benchmark due to aging may be more informative than the values themselves. Decreases in this fractal of a factor of two or more over 30 years might be cause for concern. Moreover, an increase in the rate of change would indicate degradation was accelerating over time and would suggest that a comprehensive in-service examination of structural integrity should be initiated.

References

American Concrete Institute (ACI). (1999). "Building code requirements for reinforced concrete and commentary." *ACI 318R-99*, Farmington Hills, Mich.

American Concrete Institute (ACI). (2001). "Code requirements for nuclear safety related concrete structures." *ACI Standard 349–01*, Detroit, Mich.

American Institute of Steel Construction. (1993). *LRFD Specification for structural steel buildings*, Chicago.

American Society of Mechanical Engineers (ASME). (1998). *ASME boiler and pressure vessel code*, New York.

Amin, M., Eberhardt, A. C., and Erler, B. A. (1992). "Design considerations for concrete containments under severs accident loads." *Proc., 5th Workshop on Containment Integrity*, NUREG/CP-0120, U.S. Nuclear Regulatory Commission, Washington, D.C., 157–162.

Bhattacharya, B., and Ellingwood, B. R. (1998). "A damage mechanics based approach to structural deterioration and reliability." *NUREG/CR-6546*, U.S. Nuclear Regulatory Commission, Washington, D.C.

Bjerager, P. (1990). "On computation methods for structural reliability analysis." *Struct. Safety*, 9(2), 79–96.

Cherry, J. L. (1997). "Analyses of containment structures with corrosion damage," *Proc., 24th Water Reactor Safety Information Meeting*, NUREG/CP-0157, U.S. Nuclear Regulatory Commission, Washington, D.C., 333–352.

Clifton, J. R. (1991). "Predicting the remaining service life of concrete." *NISTIR 4712*, U.S. Department of Commerce, National Institute of Standards and Technology, Gaithersburg, Md.

Code of Federal Regulations. (1995). *Code of Federal Regulations, Title 10 – Energy*, Office of the Federal Register, National Archives and Records Administration, Washington, D.C.

Ellingwood B. R. (1992). "Probabilistic risk assessment." *Engrg. Safety*, McGraw-Hill, London, 89–116.

Ellingwood, B. R. (1994). "Validation of seismic probabilistic risk assessments of nuclear power plants." *NUREG/CR-0008*, U.S. Nuclear Regulatory Commission, Washington, D.C.

Ellingwood, B. R. (2001). "Acceptable risk bases for design of structures." *Progress in Struct. Engrg. and Mat.*, 3, 170–179.

Ellingwood, B. R., Bhattacharya, B., and Zheng, R. (1996). "Reliability-based condition assessment of steel containment and liners." *NUREG/CR-5442 (ORNL/TM-13244)*, U.S. Nuclear Regulatory Commission, Washington, D.C.

Ellingwood, B. R., and Cherry, J. L. (1999). "Fragility modeling of aging containment metallic pressure boundaries." *NUREG/CR-6631*, U.S. Nuclear Regulatory Commission, Washington, D.C.

Ellingwood, B. R., and Hwang, H. H. (1985). "Probabilistic descriptions of resistance of safety-related structures in nuclear plants." *Nucl. Engrg. and Design*, 88(2), 169–178.

Ellingwood, B. R., MacGregor. J. G., Galambos, T. V. and Cornell, C. A. (1982). "Probability based load criteria: Load factors and load combinations." *J. Struct. Div.*, 108(5), 978–997.

Ellingwood, B. R., and Mori, Y. (1992). "Condition assessment and reliability-based life prediction of concrete structures." *ORNL/NRC/LTR-92/4*, Martin Marietta Energy Systems, Inc., Oak Ridge National Laboratory, Oak Ridge, Tenn.

Ellingwood, B. R., and Mori, Y. (1993). "Probabilistic methods for condition assessment and life prediction of concrete structures in nuclear plants." *Nucl. Engrg. and Design*, 142, 155–166.

Elms, D. (1992). "Risk assessment." *Engrg. Safety*, D. Blockley, ed., McGraw-Hill, New York, 28–46.

Engelund, S., and Rackwitz, R. (1993). "A benchmark study on importance sampling techniques in structural reliability." *Struct. Safety*, 12(4), 255–276.

Galambos, T. V., Ellingwood, B. R., MacGregor, J. G., and Cornell, C. A. (1982). "Probability based load criteria: Assessment of current design practice." *J. Struct. Div.*, 108(5), 978–997.

Galambos, T. V., and Ravindra, M. K. (1978). "Properties of steel for use in LRFD." *J. Struct. Div.*, 104(9), 1459–1468.

Greimann, L., Fanous, F., and Bluhm, D. (1984). "Final report: Containment analysis techniques, a state-of-the-art summary." *NUREG/CR-3653*, U.S. Nuclear Regulatory Commission, Washington, D.C.

Hancock, J. W., and MacKenzie, A. C. (1976). "On the mechanisms of ductile failure in high-strength steels subjected to multi-axial stress states." *J. Mech. Phys. Solids*, 24, 147–169.

Hwang, H. H., Wang, P.C., Shooman, M., and Reich, M. (1983). "A consensus estimation study of nuclear power plant structural loads." *NUREG/CR-3315*, U.S. Nuclear Regulatory Commission, Washington, D.C.

Hwang, H. H., Ellingwood, B. R., Shinozuka, M., and Reich. M. (1987). "Probability-based design criteria for nuclear plant structures," *J. of Struct. Design*, 113(5), 925–942.

Jung, J. (1984). "Ultimate strength analyses of the watts bar, maine yankee, and bellefonte containments." *NUREG/CR-3275*, U.S. Nuclear Regulatory Commission, Washington, D.C.

Kennedy, R. P., and Ravindra, M. K. (1984). "Seismic fragilities for nuclear power plant studies." *Nucl. Eng. and Design*, 79(1), 47–68.

Klamerus, E. W., Bohn, M. P., Wesley, D. A., and Krishnaswamy, C. N. (1996). "Containment performance of prototypical reactor containments subjected to severe accident conditions." *NUREG/CR-6433*, U.S. Nuclear Regulatory Commission, Washington, D.C.

Lo, T., Nelson, T. N., Chen, P. Y., Persinko, D., and Grimes, C. (1984). "Containment integrity of SEP plants under combined loads." *Proc., ASCE Conf. Struct. Engrg. in Nucl. Facilities*, J. Ucciferro, ed., ASCE, New York.

MacGregor, J. G., Mirza, A., and Ellingwood, B. R. (1983). "Statistical analysis of resistance of reinfored and prestressed concrete members." *J. Am. Concrete Inst.*, 80(3), 167–176.

Melchers, R. E. (1990). "Search-based importance sampling." *Struct. Safety*, 9(2), 117–128.

Mori, Y., and Ellingwood, B. R. (1993). "Methodology for reliability based condition assessment-application to concrete structures." *NUREG/CR-6052*, U.S. Nuclear Regulatory Commission, Washington, D.C.

Mori, Y., and Ellingwood, B. R. (1994a). "Maintaining reliability of concrete structures I: Role of inspection/repair." *J. Struct. Engrg.*, 120(3), 824–835.

Mori, Y., and Ellingwood, B. R. (1994b). "Maintaining reliability of concrete structures II: Optimum inspection/repair." *J. Struct. Engrg.*, 120(3), 846–862.

Moses, F. (1990). "New directions and research needs in system reliability research." *Struct. Safety*, 7(2), 93–100.

Naus, D. J., Oland, C. B., and Ellingwood, B. R. (1996). "Report on aging of nuclear power plant concrete structures." *NUREG/CR-6424 (ORNL/TM-13148)*, Lockheed Martin Energy Research Corporation, Oak Ridge National Laboratory, Oak Ridge, Tenn.

Oland, C. B., and Naus, D. J. (1996). "Degradation assessment methodology for application to steel containments and liners of reinforced concrete structures in nuclear power plants." *ORNL/NRC/LTR-95/29*, Lockheed Martin Energy Research Corporation, Oak Ridge National Laboratory, Oak Ridge, Tenn.

Pearce, T. H., and Wen, Y. K. (1985). "Stochastic combinations of load effects." *J. Struct. Engrg.*, 110(7), 1613–1629.

Rubenstein, R. Y. (1981). *Simulation and the Monte Carlo method*, Wiley, New York.

Schueller, G. I., and Stix, R. (1987). "A critical appraisal of methods to determine failure probabilities." *Struct. Safety*, 4(4), 293–310.

Shinozuka, M. (1983). "Basic analysis of structural safety." *J. Struct. Engrg.*, 109(3), 721–740.

Stewart, M. G., and Melchers, R. E. (1997). *Probabilistic risk assessment of engineering systems*, Chapman and Hall, London.

Taylor, H. M., and Karlin, S. (1984). *An introduction to stochastic modeling*, Academic Press, Orlando, Fla.

U.S. Department of Energy (DOE). (1996). "Natural phenomena hazard performance categorization criteria for structures, systems and components." *DOE STD-1021–93*, Washington, D.C.

U.S. Nuclear Regulatory Commission (NRC). (1981). "Standard review plan for the review of safety analysis reports for nuclear power plants." *NUREG-0800*, Directorate of Licensing, Washington, D.C.

U.S. Nuclear Regulatory Commission (NRC). (1998). "An approach for using probabilistic risk assessment in risk-informed decisions on plant-specific changes to the licensing basis." *Regulatory Guide 1.174*, U.S. Nuclear Regulatory Commission, Washington, D.C.

U.S. Nuclear Regulatory Commission (NRC). (2005). "U.S. Nuclear Regulatory Commission information digest." *NUREG-1350*, 14, Division of Budget and Analysis, Office of the Chief Financial Officer, Washington, D.C.

Walther, H. P. (1992). "Evaluation of behavior and the radial shear strength of a reinforced concrete containment structure." *NUREG/CR-5674*, U.S. Nuclear Regulatory Commission, Washington, D.C.

Disclaimer

Research sponsored by the Office of Nuclear Regulatory Research, U.S. Nuclear Regulatory Commission, under Interagency Agreement 1886-N604-3J with the U.S. Department of Energy under Contract No. DE-AC05-00OR22725.

This report was prepared as an account of work sponsored by an agency of the U.S. Government. Neither the U.S. Government nor any agency thereof, or any of their employees, makes any warranty, expressed or implied, or assumes any legal liability or responsibility for any third party's use, or the results of such use, of any information, apparatus, product or process disclosed in this report, or represents that its use by such third party would not infringe privately owned rights.

7

Offshore Structures

Torgeir Moan

7.1 Analysis of Offshore Structures

Oil and gas are the dominant sources of energy in our society. Some of these hydrocarbons are recovered from reservoirs beneath the seabed. Various kinds of platforms are used to support exploratory drilling equipment and the chemical (production) plants required to process the hydrocarbons. Pipelines or tankers are used to transport the hydrocarbons to shore.

Structures for offshore oil and gas exploitation have been developed for more than 50 years. Initially they were piled, wooden structures in shallow, benign waters. Today a variety of structures are employed. They are supported on the seafloor or by buoyancy or a combination—in severe environments and water depths up to 2,000 m. The largest production platforms (e.g., in the North Sea), represent a capital investment of billions of U.S. dollars and significant operational costs. The continuous innovation to deal with new serviceability requirements and demanding stochastic environments, as well the inherent potential of risk of fires and explosions, has lead to an industry that has been in the forefront of development of design and analysis methodology. Figure 7-1 shows the life cycle phases of offshore structures. The focus here is on analyses for design and reassessment of design during operation.

Current practice is implemented in new offshore codes, such as those issued by the American Petroleum Institute ([API] 1993/97), International Standards Organization ([ISO] 1994, 2001), and Norwegian Technology Standards ([NORSOK] 1998, 1999), as well as by many classification societies, and is characterized by:

- design criteria formulated in terms of serviceability and safety limit states (ISO 1994)
- semiprobabilistic methods for ultimate strength design that have been calibrated by reliability or risk analysis methodology

- fatigue design checks depending upon consequences of failure (damage-tolerance) and access for inspection
- explicit accidental collapse design criteria to achieve damage-tolerance for the system
- considerations of loads that include payload; wave, current and wind loads; ice (for arctic structures); earthquake loads (for bottom supported structures); as well as accidental loads such as fires, explosions; and ship impacts
- global and local structural analysis by finite element methods (FEMs) for ultimate strength and fatigue design checks
- nonlinear analyses to demonstrate damage tolerance in view of inspection planning and progressive failure due to accidental damage

The complexity of the analysis for design is due to complex geometry; the 3D and stochastic, dynamic and nonlinear nature of environmental loads and, hence, their effects; the need to deal with very local stresses for fatigue design checks; and nonlinear ultimate strength associated with the effect of accidental loads and global collapse of damaged structures.

Moreover, reassessment of design is required during operation, for instance, because of planned change of platform function that may increase payload or because of the occurrence of damage or need to extend service life. Structural modifications to maintain an acceptable safety level for existing structures are much more expensive than during initial design, before the structure is fabricated. This applies especially to platforms permanently located offshore. For this reason other strategies than structural modifications/strengthening is used to achieve the necessary safety for existing structures. For instance, information about material and geometrical properties of the relevant structure, which is collected during fabrication, and information about structural response that is recorded during operation would normally imply reduced uncertainties in predicted resistance and load effects, and hence additional safety margins, than those envisaged at the design stage. Moreover, deterioration phenomena can be controlled by more frequent inspections and possibly repairs of smaller damages than otherwise would have been done.

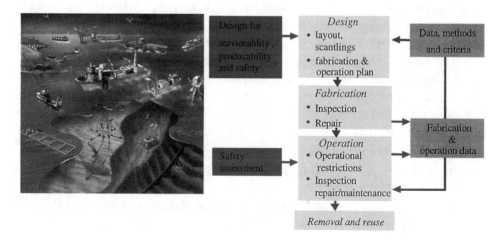

Figure 7-1. Life cycle phases of offshore structures.

In particular, analyses for demonstrating compliance with safety requirements tend to be more refined during reassessment than initial design. In this way the conservativism in simplified methods is avoided. The refinement of analysis methodology applies to all aspects of the analysis, but especially by introducing case-based risk and reliability methodology.

Section 7.2 addresses characteristic features of different types of offshore structures. Relevant design criteria are briefly described in Section 7.3. Section 7.4 deals with methods to calculate sea loads and their effects. Sections 7.5 and 7.6 describe structural analyses of space frames with fixed steel-plated and floating structures, respectively. Section 7.7 describes analysis of effects of accidental loads such as fires and explosions as well as ship impacts, while Section 7.8 deals with analysis issues relating to reassessment of design during operation.

7.2 Characteristic Features

7.2.1 General

The size and other principal features of offshore structures are primarily determined by their intended function and their environment. Different types of structures are required in the exploitation of subsea oil and gas resources.

Platforms may be used for exploratory drilling to identify producible hydrocarbons. The main mission in this case is to drill a well from the seabed to the possible hydrocarbon reservoir. Drilling operations take place by a drill string that is contained in a tube (riser) from the seafloor to the platform deck. To avoid excessive stresses in the drill string and riser, the platform needs to have limited motion. When an exploratory well has been drilled in 1 to 3 months, the platform is moved to the next location, several km away. Exploratory drilling platforms, therefore, need to be easy to relocate—to be mobile.

The most attractive mobile drilling platforms are drill ships, jack-ups and semisubmersibles, as illustrated in Figure 7-2. Drill ships are applicable in benign waters; jack-ups are limited to small water depths, while semisubmersibles are preferable in deep, harsh waters.

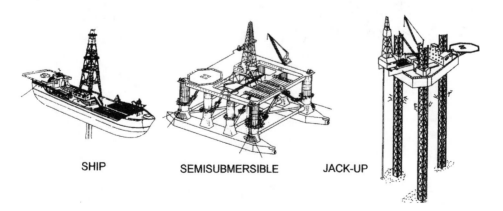

SHIP   SEMISUBMERSIBLE   JACK-UP

Figure 7-2. Drilling platforms.

Production platforms serve as a base for drilling and completion of the production wells, the hydrocarbon production itself, and for transporting hydrocarbons to shore. Hence, they need load carrying capacity and space to accommodate separators, pumps, piping, as well as to support risers and other parts of the chemical process plant used to produce hydrocarbons say for 20 years or more. Figure 7-3 shows a production facility on a floating platform. Storage capacity may be an additional requirement, depending upon the transport infrastructure used. In harsh environments limited motion is desirable to reduce the stress level in steel risers and maintain efficiency of the separation process. If flexible risers are applied, however, the motion constraint would be relaxed.

In small water depths the functional requirements are fulfilled at the lowest costs by using structures supported on the seafloor (Fig. 7-4a). However, in increasing water depths installation of fixed platforms may not be feasible or very expensive. Moreover, the sea loads on a platform tower structure primarily act at the water surface level, implying that load effects will increase very fast with water depth and make fixed platforms less economically feasible with depth. In greater water depths buoyant structures shown in Figure 7-4b are more attractive.

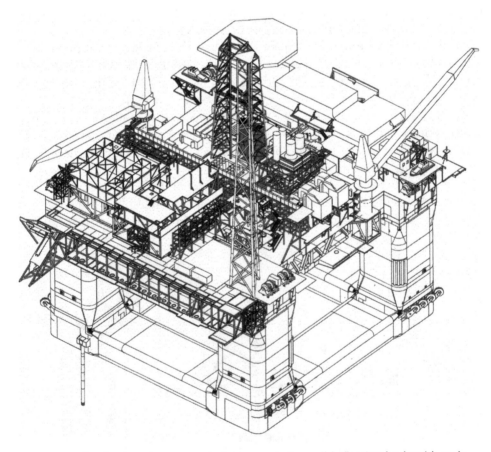

Figure 7-3. Production plant supported on a semisubmersible floating body with main dimensions 85 by 85 m. Figure courtesy of Aker Technology.

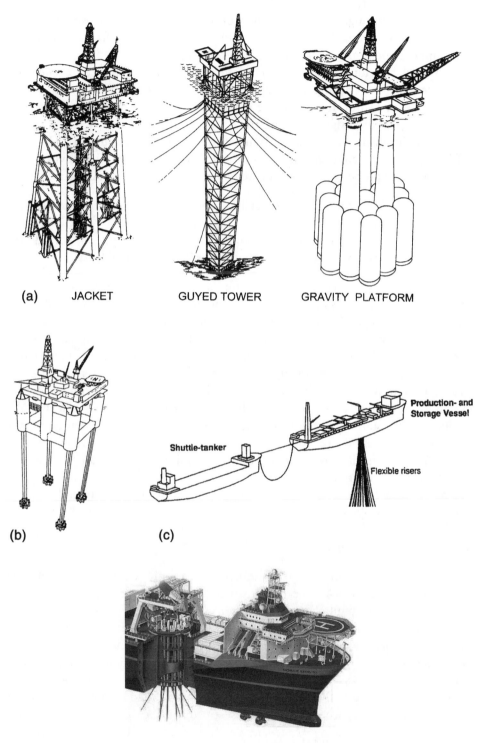

(a)   JACKET   GUYED TOWER   GRAVITY PLATFORM

(b)   (c)

Figure 7-4. Production platforms. a) Platforms supported on the seafloor. b) Buoyant structures, semisubmersible platform with tension-leg mooring. c) FPSO and shuttle tanker at offshore site. d) Turret arrangement for FPSO bow with turret.

Figures 7-2 through 7-4 show a variety of structural layouts. The overall layout is to a large extent governed by functional requirements such as payload capacity (e.g., riser support, and limited motion) and water depth.

The deck structure of a production platform is a complex and compact chemical process plant, consisting of three to seven stories. The significant latent energy associated with hydrocarbons flowing through or stored in production platforms obviously represents a significant hazard.

The environmental and hydrocarbon hazards make safety an important issue for offshore platforms. Safety requirements are specified to avoid fatalities, environmental damage, and property damage and are related to the following failure modes:

- overall rigid body instability (capsizing) under permanent, variable, and environmental loads
- failure of (parts of) the structure foundation or mooring systems, considering permanent, variable, and environmental as well as accidental loads

Safety criteria have been issued by various kinds of regulatory bodies—including national authorities, classification societies, and bodies such as API and NORSOK—set up by the industry. Currently a significant effort is underway by ISO to establish codes for offshore structures that could serve as harmonized international codes.

Stability requirements for floating platforms affect the layout and the internal structure—subdivision in compartments. Moreover, criteria to prevent progressive structural failure after fatigue failure or accidental damage would have implications on overall layout of all types of platforms. Otherwise, structural strength criteria affect the scantlings of the stiffened flat and cylindrical panels that constitute offshore structures.

The location far offshore makes evacuation and rescue difficult, but on the other hand accidents on offshore plants affect the general public less than accidents on similar facilities on land.

In the subsequent sections the characteristic features of some typical offshore structures are described.

7.2.2 Template Platforms

The most common type of offshore platform is the fixed, pile-supported steel template platform, often called jacket. Jackets consist of a plate girder or truss deck structure, supported by a welded tubular steel space frame that is piled to the seafloor. The legs are battered and the area/volume exposed to waves and current are reduced toward the surface to increase the overall moment capacity and reduce the loads, respectively. Jackets are installed by transporting the space frame as a self-floating body or on a barge to the offshore site, where they are lifted or ballasted in upright position on the seafloor and piled to the seafloor. Finally the deck is lifted in one or more pieces and joined to form a space frame. The largest structure is the 67,000 tons Bullwinckle platform that was placed in 492 m water depths in the Gulf of Mexico and designed to resist 23 m high waves, winds of 76 m/s, and current of 1.2 m/s.

The design of jackets is primarily determined by requirements associated with the permanent operational conditions but may be influenced by temporary conditions during transport, launching, and offshore installation.

In deep water and soft soils, the natural period of jackets may exceed 4 to 5 s and lead to dynamic amplification.

The design of the joints between the circular tubular members in the truss is challenging because they exhibit complex shell behavior and may suffer ultimate or fatigue failure.

7.2.3 Compliant Towers

An alternative to piling the steel space frame to the seafloor would be to provide some kind of hinge at the seafloor and support the tower by mooring lines (e.g., guyed tower, Fig. 7-4a) or buoyancy provided by tanks located below the water surface to minimize wave forces, yet high enough to provide the uprighting moment when the platform heels. With a natural period of say 25 to 40 s above wave periods, such a platform will move with waves, and wave loads will essentially be balanced by inertia forces set up by the "rigid body" motion.

7.2.4 Jack-Ups

Jack-up platforms are mobile drilling platforms with a floatable deck and axially moveable legs. Jack-ups are moved between sites with raised legs. On location the legs are lowered to the seabed, and the deck is lifted above the sea surface by hydraulic or electric jacks. During operation the platform behaves like a fixed platform. The unfavorable motion and stability characteristics of the jack-up during transit phase with elevated legs above the deck, make this phase critical from a stability and strength point of view and make it necessary to have relatively long periods with calm weather to carry out transit operations. Due to the frame type structure during permanent operation, the response induced by environmental loads, partly because of dynamic effects, increases rapidly with water depth. Jack-ups have been used up to a water depth of about 130 m in benign waters, and about 70 m in harsh waters. To limit the penetration of the legs into soft soils, spud cans are used at the base of the legs to provide additional seafloor bearing areas.

7.2.5 Gravity Platforms

Gravity platforms are used as production platforms and consist commonly of a steel deck, a concrete framework and caisson as well as steel skirt foundation. Gravity platforms have been made of reinforced concrete for water depths up to 300 m in harsh North Sea environments. The risers are contained in the caisson and shafts, and are hence protected against the action of severe wave loads and ship impacts. The caisson provides storage space. Gravity platforms are built in upright position, towed out to the offshore site, and installed by ballasting. The largest body moved on earth is a concrete gravity platform (Gullfaks C in the North Sea) with a tow-out weight of 1.5 million tons.

Since the deck and equipment are in -place on the substructure already during tow-out, and skirt penetration takes place directly during installation, the offshore installation time is reduced to a minimum.

The concrete structure consists of cylindrical, conical, and spherical shells. Thus, the caisson consists of 12 to 24 cylinders with spherical end caps (domes). In particular the 3D intersections between shells are challenging from analysis and design point of view.

Gravity steel platforms have also been designed, but only for relatively shallow water. They are trusswork platforms with a relatively wide base and are equipped with large buoyancy tanks at the base.

7.2.6 Semisubmersibles

Semisubmersible platforms may be used as exploratory drilling or production platforms. They consist of submerged horizontal pontoons, columns, and a deck. Both steel and reinforced concrete have been used as hull material. Pontoons and columns provide buoyancy while columns stabilize the platform. The fact that a large part of the hull is submerged and the small waterplane area of such platforms results in small wave excitation forces and small motions. While typically two pontoons are used on drilling platforms (Fig. 7-2), a ring pontoon is used for production platforms (Fig. 7-3, 7-4b). This is because exploratory drilling platforms need to be mobile and hence exhibit limited hydrodynamic resistance when they are moved from location to location. To reduce the effect of wave forces (and, hence, steel weight) in the deck of such semisubmersible drilling platforms (Fig. 7-2), they are equipped with a bracing system that is located above the transit water line.

However, the stability and draught of semisubmersibles are sensitive to payload and position. Owing to this fact they do not normally have any oil storage capacity.

Many current floating production systems are based on converted drilling rigs. However, a larger payload capacity is required for production than for drilling. Converted units can, hence, only produce from small reservoirs. Purpose-built semisubmersibles are required in order to produce from large reservoirs such as the North Sea (Fig. 7-3).

Pontoons and columns are subdivided into compartments to store ballast water and to limit the buoyancy loss in case of flooding (e.g., due to impact on the platform).

The deck is a grillage of girders or trusses. It may have to be designed to ensure survival of the platform in case of accidental damage that causes buoyancy loss in pontoons or columns.

While requirements of payload, buoyancy, and limited motions determine the overall size and shape of pontoons and columns, a possible bracing system, internal structure, and scantlings are determined by structural strength requirements.

Semisubmersibles are kept in position by a mooring or thruster system. Such systems are designed to resist the "steady" wind, current, and slow-drift wave forces and not the first order wave loads, which are balanced by inertia forces. Mooring forces are hence an order or two of magnitude smaller than the first order wave forces acting on the platform.

7.2.7 Tension-Leg Platforms

A tension-leg platform (TLP) hull made of steel or concrete is similar to that of semisubmersibles with ring pontoons (Fig. 7-4b) and is used as a permanent production platform. However, the hull is designed with excessive buoyancy to provide a vertical-tension mooring system (tethers). As a consequence, the platform will move as an inverted pendulum. The tension should be so large that tethers do not experience compression (slack) in waves and would typically be 10% to 25% of the total buoyancy. By the use of pretensioned tethers, the heave and pitch motions are almost eliminated, and conventional steel risers can be applied. Usually there are three to four tethers in each of the four platform corners. A tether is a tubular made of 12- to 15-meter-long sections connected by welds or threaded joints. The tether foundation is by gravity skirts or piles.

TLPs are fabricated in upright position, towed to the site, and installed.

7.2.8 Ship-Shape Structures

Ships may be used for exploratory drilling or production. Such vessels can easily be designed as combined floating production, storage, and offloading (FPSO) units. However, due to the large volume displaced close to the water line, ships have poor heave (and pitch) response characteristics and can only be used for drilling operations using steel risers in areas with long periods of calm weather. Production, however, can take place in heavy weather by using flexible risers that can sustain the motions. Figure 7-4c shows an FPSO.

The hull structure of an FPSO is similar to that of an oil tanker. For this reason many FPSOs have been converted tankers, especially for benign waters. However, the trend is to use purpose-built FPSOs. Such vessels are designed more like barges with larger block coefficient than tankers.

Ship-shape facilities are positioned by thrusters or catenary mooring systems. In harsh environments the mooring system needs to be combined with a turret arrangement (Fig. 7-4d). The turret is made possible by a cylindrical opening in the forepart of the hull. In this opening a cylindrical structure is supported in such a way that the vessel can rotate relative to this cylinder. The cylindrical structure is made geostationary by a catenary mooring system anchored to the seafloor. The turret arrangement and production equipment on deck make up the main differences in the hull design compared to conventional tankers. In addition there are differences in operations that have implications on still-water and wave-induced loads. For instance, FPSOs are located offshore all the time while tankers partly are at sea and partly in ports and may be operated to avoid severe sea conditions.

7.3 Limit State Requirements

7.3.1 General

Adequate performance of offshore structures is ensured by designing them to comply with serviceability and safety requirements for a service life of 20 to 40 years. Serviceability criteria are introduced to make the structure provide the function required and are specified by the owner.

Safety requirements are imposed to avoid ultimate consequences such as fatalities, environmental damage, or property damage. Depending upon the regulatory regime, separate acceptance criteria for these consequences are established. Property

Table 7-1. Safety Criteria

Limit states	Description	Remarks
Ultimate (ULS)	— Overall "rigid body" stability — Ultimate strength of structure, mooring, or possible foundation	Different for bottom–supported, buoyant, —. Component design check.
Fatigue (FLS)	— Failure of joint	Component design check depending on residual system strength after fatigue failure.
Accidental collapse (ALS)	— Ultimate capacity[a] of damaged structure (due to fabrication defects or accidental loads) or operational error	System design check.

[a]Capacity to resist "rigid body" instability or total structural failure.

damage is measured in economic terms. But fatalities and pollution obviously have economic implications. In particular, the increasing concern about the environment could cause even small damages to have severe economic implications. While fatalities caused by structural failures would be related to global failure (i.e., capsizing or total failure of deck support), smaller damages may result in pollution or property damage that is expensive to repair (e.g., for an underwater structure).

Safety criteria for offshore structures may be classified as shown in Table 7-1 and are briefly outlined in the following sections.

7.3.2 Ultimate and Fatigue Limit State Criteria

Ultimate limit state (ULS) criteria for overall stability of bottom-supported structures are based on overturning forces due to wave, current, wind, and stabilizing forces due to permanent and variable payloads. Stability of floating structures is considered in terms of overturning moment by wind only and uprighting moment due to hydrostatics of the inclined body (Clauss et al. 1991).

ULS and fatigue limit state (FLS) criteria for structural components have been developed for the relevant failure modes dependent upon geometry and load conditions. Permanent and variable payloads, fluid pressure loads as well as environmental loads are considered. Environmental loads due to waves, current, wind, and possibly ice and earthquakes are considered. Load and resistance factor design is commonly used for ULS.

Figure 7-5 illustrates the characteristic features of design check relating to sea loads. Load effects are determined by recognizing the stochastic character of the waves and using hydrodynamics to calculate wave forces and structural mechanics to obtain load effects in an appropriate format for the design check.

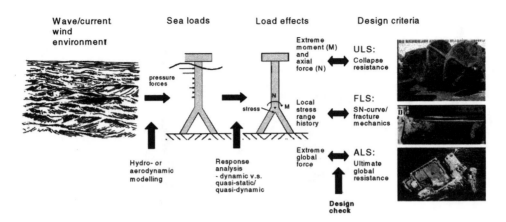

Figure 7-5. Analysis for design of offshore structures.

Extreme loads with an annual exceedence probability of 10^{-2} are normally required for ULS check, while for FLS check the local stress range history is needed. Estimation of relevant sea load effects requires consideration of a variety of sea states, elaborate multidegrees-of-freedom hydrodynamic and structural models, and are time consuming. Ultimate strength formulations used in design are traditionally based on strength of material formulations and substantiated by extensive test results. Stiffened flat panels and cylindrical shells are the main components in offshore structures. However, design based on direct ultimate strength analysis, by using FEMs and by accounting for nonlinear geometric and material effects, is emerging.

Fatigue is an important consideration for structures in areas with more or less continuous storm loading (such as the North Sea) and especially for dynamically sensitive structures. Fatigue strength is commonly described by SN-curves that have been obtained by laboratory experiments. Fracture mechanics analysis of fatigue strength have been adopted to assess more accurately the different stages of crack growth including calculation of residual fatigue life beyond through-thickness crack, which is normally defined as fatigue failure. Such detailed information about crack propagation is also required to plan inspections and repair.

Inspection, maintenance, and repair are important measures for maintaining safety in connection with fatigue crack growth as described in Section 7.3.4.

7.3.3 Accidental Collapse Limit State Criteria

Accidental collapse limit state (ALS) requirements are motivated by the design philosophy that small damages, which inevitably occur, should not cause disproportionate consequences. The criterion, which was introduced for offshore structures in Norway in 1984, is illustrated in Figure 7-6.

The initial damage should correspond to events that are exceeded with an annual probability of 10^{-4}, such as fire explosions, ship impacts, or fabrication defects causing abnormal fatigue crack growth as identified by risk analyses. The (local)

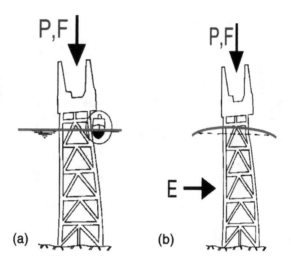

Figure 7-6. Design check for accidental limit state. a) Estimation of damage under permanent (P) and variable (F) deck loads. b) Survival check of damaged platform under P, F, and environmental loads (E).

damage, permanent deformations, or rupture of components needs to be estimated by accounting for nonlinear effects.

The structure is required to survive the various damage conditions without global failure. Compliance with this requirement can in some cases be demonstrated by removing the damaged parts and then accomplishing a conventional ULS design check based on a global linear analysis and component design checks using truly ultimate strength formulations. However, such methods may be very conservative, and more accurate nonlinear analysis methods should be applied, as described in Section 7.5.

Global nonlinear analyses are beneficial to estimate residual strength after fatigue failure in connection with inspection planning and survival analysis in ALS checks.

7.3.4 Inspection, Monitoring, and Repair

Inspection and monitoring (IM) and maintenance and repair (MR), if necessary, are important measures for maintaining safety, especially with respect to fatigue, wear, and other deterioration phenomena. But their effect on reliability depends upon the quality of inspection, for example in terms of detectability vs. size of the damage, as well as time for repair. Hence, an inspection and repair measure can contribute to safety only when there is a certain damage tolerance. This implies that there is an interrelation between design criteria (fatigue life, damage tolerance) and the inspection and repair criteria. To some extent these issues are reflected in fatigue design criteria in offshore codes (NORSOK, ISO) where the fatigue design check depends upon the consequences of failure and access for inspection and allowable cumulative damage, D_{all}, varies between 0.1 and 1.0 (NORSOK N-001). The consequence measure is based on whether the structure fails in a condition with an expected variable load and 100-year sea loads after fatigue failure of the relevant joint, and is thus linked to the ALS. The treatment of both the consequence and inspection issue, however, could be improved by taking D_{all} to be dependent on a more precise measure of reserve strength and to be an explicit measure of the effect of inspection. This issue has been pursued in connection with reassessment of existing structures, as described in Section 7.8.

The quality of visual or nondestructive inspection (NDI) methods depends very much upon the conditions during inspection. Large-volume offshore structures are accessible from the inside, while, for example, jackets and tethers in TLPs, which are made of relatively slender tubulars, are not. Underwater inspections (e.g., of jackets) are carried out under difficult conditions due to the presence of marine growth, bad visibility, wave motions, and so forth. Before an underwater NDI can be carried out, a time-consuming cleaning of the welded joints is needed (MTD 1992). For this reason and the high costs of diver operations, underwater inspections become expensive, yet with a quality deficient to that of dry inspections.

7.3.5 Risk and Reliability Methods in Offshore Engineering

Since about 1970 the offshore industry has been concerned with a rationalization of safety measures through application of risk and reliability methods. Initial efforts were focused upon establishing risk-based storm load criteria (e.g., Marshall 1969) but also on application of structural reliability to calibrate ULS code requirements (Fjeld 1977; NPD 1977). Increased applications of such methods follow the general development of structural reliability methods (Melchers 1987). The significant effort by API (Lloyd and Karsan 1988) and Canadian Standards Association (Jordan and Maes 1991) to develop a reliability-based load and resistance factor design code for fixed platforms is noted. In certain situations when a new design falls outside the scope of existing codes, reliability analysis has been applied ad hoc to establish design criteria. This was the case when the first offshore production ship was designed some years ago. It then became evident that application of the ship rules for trading vessels and existing offshore codes differed significantly, implying a difference in steel weight of the order of 20% to 30%. Moan (1988) conducted a study to establish ultimate strength criteria for ship type offshore production installations that complied with the inherent safety level in the existing NPD code for offshore structures. An evaluation of previous efforts on calibration of offshore codes was provided by Moan (1995) in conjunction with the ISO effort to harmonize codes for offshore structures (ISO 1994).

Code calibrations described above all refer to ultimate strength criteria. Because fatigue in offshore structures is mainly caused by one type of loading, there is limited merit of calibrating fatigue criteria in the same sense as ULS criteria relating to multiple loads. However, to achieve consistent design and inspection criteria, fatigue design criteria should be calibrated to reflect the consequences of failure and inspection plan. Moan, Hovde, and Blanker (1993a) show the allowable cumulative damage (D) in design can be relaxed when inspection is carried out.

While this calibration is done on a generic basis, it is important that information obtained (e.g., by inspections during operation) is used, and the inspection plan is updated accordingly. For this reason probabilistic fracture mechanics analysis of crack growth needs to be carried out for each individual structure, while ultimate strength criteria have been based on generic code calibration by code committees.

In particular it is noted that inspection planning for template, space-frame structures is based on a simplified systems reliability approach, considering one failure mode at a time (Vårdal et al. 1997). The basis is to require that the system failure probability, $P_{FSYS(i)}$ associated with a fatigue failure of member (i) complies with a target level $P_{FSYS(T)}$, such that

$$P_{FSYS(i)} = P_{FSYS/FF(i)}P_{FF(i)} \leq P_{FSYS(T)}. \tag{7-1}$$

The background for this approach is described in Section 7.8.

In the late 1970s research on Probabilistic Risk Assessment (PRA) in the offshore industry (Moan and Holand 1981) was based on experiences in the nuclear and aeronautical industries.

The first guidelines on PRA were issued by NPD (1981). The most serious offshore accident to date in terms of fatalities, the Piper Alpha Disaster (Cullen Report 1990), was the direct reason for introducing PRA, or quantitative risk analysis (QRA), in the UK in 1992, which certainly was a milestone.

Significant developments of reliability and risk methodology, including Bayesian updating techniques, took place in the 1980s. Methods for dealing with the effect of inspection on reliability were developed, with due account of previous research in the aeronautical field. Some of the research relating to system reliability and Bayesian updating were directly motivated by needs in the offshore industry, as discussed subsequently.

7.4 Sea Loads and Load Effects

7.4.1 General

All offshore structures are subjected to permanent loads, payloads, and hydrostatic pressures as well as wave, current, and wind loads. Seismic loads may be important for bottom-supported platforms in some offshore sites. In arctic or subarctic regions, ice loads may be important. However, the main challenge is associated with sea loads, which are commonly the dominant load. Moreover due to their time variation, they can cause different types of dynamic effects and both ultimate and fatigue failure. This section is therefore confined to sea loads and their effects.

7.4.2 Environmental Condition

Data about the time and space variation of wave height, period, and direction as well as current speed and direct are required. An adequate description of the long-term variation of sea states requires data for several decades. Such data partly form the basis for formulating extreme loads and partly describe the frequently occurring loads that cause fatigue.

Table 7-2. Environmental Conditions at Offshore Sites Worldwide

100-year value	North Sea	Gulf of Mexico	Campos Basin (Brazil)	West Africa (Nigeria)
Wave height (m)	25–32	20–25	10–15	6–8
Wind speed (m/s) (10 min. mean at 10 m elevation)	40–42	45	30–35	30–35
Surface current speed (m/s)	1.0–2.0	0.5–1.5	2.0–3.0	1.0–1.5

Typical sea environments in different geographical regions are indicated by the 100-year wave height, surface current velocity, and wind speed, as shown in Table 7-2.

7.4.3 Wave and Current Forces

7.4.3.1 Kinematics. Surface waves on the sea involves fluid particle velocities, v_w, accelerations, a_w, and dynamic pressure variations.

A regular wave in deep water may be described by the linear (Airy) theory and by the wave height, H and length, λ. The wave length, λ, is expressed by the wave period, T, or radian frequency $\omega = 2\pi/T$, and

$$\lambda = \frac{g}{2\pi}T^2 = \frac{2\pi g}{\omega^2}. \tag{7-2}$$

The wave number is $k = 2\pi/\lambda$.

In regular waves the undulating water surface elevation, η, is given by:

$$\eta = \frac{H}{2}\sin(\omega t - kx) \tag{7-3}$$

where t is time, and x is a horizontal coordinate pointing in the direction of wave propagation with the origin at still water level under a wave crest at time zero. Taking z as a vertical coordinate pointing upwards with origin at the mean water level, the pressure p at any point under the water surface is given by:

$$p = \rho g \left(-z + \eta e^{(kz)}\right) \tag{7-4}$$

where ρ is the density of water. Horizontal velocity v_x and acceleration a_x are given by:

$$v_x = \omega \frac{H}{2} e^{kz} \sin(\omega t - kx), \text{ and} \tag{7-5}$$

$$a_x = \omega^2 \frac{H}{2} e^{kz} \cos(\omega t - kx). \tag{7-6}$$

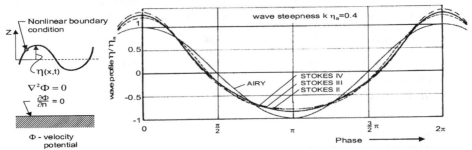

Figure 7-7. Ocean wave theory.

In shallow water where the depth, d, is less than half a wavelength, these expressions are modified as described by Sarpkaya and Isaacson (1981).

For waves whose height exceeds 1/50 of the wavelength, the water surface is no longer sinusoidal; the wave crests are steeper, and the troughs are flatter. Higher order wave theories, Stokes 3rd, Stokes 5th, or stream function, give more accurate solutions and a better fit to the boundary conditions, as illustrated in Figure 7-7. The theoretical limiting steepness H/λ for deepwater waves is 1/7. Also, shallow water changes the wave shape (Sarpkaya and Isaacson 1981).

A real sea state can be described by superposition of regular waves with different frequencies (wavelengths), phase angles, and directions.

Current represents a particle velocity, v_c.

Buoyancy forces on a body in the water are a result of the vertical pressure gradient acting on its surface. When a body such as a cylindrical member is subjected to a horizontal pressure gradient, lateral forces, analogous to buoyancy, result; furthermore, since the body partially blocks the flow (lateral acceleration) of the surrounding water, an "added mass" effect creates an additional force in phase with the pressure gradient and water particle acceleration. A turbulent wake behind the body creates a drag force that is proportional to velocity squared.

7.4.3.2 Wave forces. Wave forces, F per unit length of a slender vertical rigid component with diameter, D. and no motion, may be expressed by the empirical Morison's equation as

$$F = \tfrac{1}{2}C_D \cdot \rho D v |v| + (C_A + 1)\rho \frac{\pi}{4}D^2 \cdot a \qquad (7\text{-}7)$$

where $v = v_w + v_c$, $a = a_w$, C_D and $C_A + 1 = C_M$ are empirical drag and inertia coefficients respectively, and ρ is the density of water. If the structural members are slender, the structure does not affect the kinematics and v_w and a_w obtained from the incident wave can be used.

Generally accepted design values for use with regular storm waves acting on template type offshore structures range from 0.7 to 1.2 for C_D and 1.2 to 2.0 for

$C_A + 1 = C_M$. Various kinematic models for random waves, and the corresponding force coefficients, are discussed by Sarpkaya and Isaacson (1981).

For inclined and horizontal members, the total wave particle velocities and accelerations are first computed, and then the vector components normal to the member axis are used in Morison's equation.

The inertia term is due to two effects: pressure effects due to undisturbed incident waves (Froude-Krylov force) and pressure effects due to the relative acceleration between wave particles and structure (added mass).

For structures, which undergo significant dynamic motion, a modified form of Morison's equation is used. Relative velocity (water particle velocity minus structure velocity) is used in the drag force term. The inertial wave force term is unchanged, but added mass is included with the structure's inertial mass (using C_A for the added mass coefficient).

If the structure diameter is large compared to the wave length, say $D/\lambda > 1/3$, inertia forces dominate. Moreover, in this case the incident waves are diffracted; the structure sets up an additional wave field.

If a large volume structure is floating, the waves will induce motion of the structure. These motions in turn set up a wave pattern. In this case the relative velocity and acceleration cause hydrodynamic damping and added mass terms. Numerical methods are generally required to deal with the complex wave patterns and the hydrodynamic pressure that cause the structural loads in such cases.

7.4.3.3 Nonlinear effects. Sea waves have a period in the range of 2 to 20 s. Linear wave forces and motions have the same period. Nonlinearities in wave forces may cause steady state loads with a period that is a fraction 1/2, 1/3, ... or a multiple 2, 3, ... of the wave period. The nonlinearity in the drag force means that the force will be a nonlinear function of the wave height, H. The total force may be written as

$$Q = c \cdot H^{\alpha} \,. \tag{7-8}$$

For submerged horizontal members, $\alpha = 2.0$, while for surface piercing members $\alpha > 2.0$. This is because the wetted area subjected to wave force increases with wave height.

Nonlinearities associated with large volume structures also lead to higher order force components. The higher order wave force components are normally an order of magnitude less than the first order (linear) wave forces. However, if the period of the higher order component coincides with a natural period, the effect of such forces can be large. Commonly load components with a frequency higher than the wave frequency will affect the response in fixed platforms while it is possible that low frequency load components are most important for floating structures.

7.4.4 Load Effects

It is not yet feasible in a single load effect analysis to account for the stochastic features of sea waves and current, interaction between incident waves and structure, possible nonlinear loads, dynamic behavior, and the structural geometry to the detail required for fatigue analysis. Load effects are, therefore, determined in a hierarchy of analyses.

Extreme values for conventional ULS design check typically corresponding to an annual probability of exceeding of 10^{-2}, are obtained by

- stochastic analysis considering all relevant sea states in a long-term period of, say, 20 years and typically using crude structural models to identify the load pattern (forces or nominal stress) that can be used to determine the extreme response
- deterministic analysis, using a hierarchy of refined models of the structure and using the simplified load pattern

For tower type structures with static behavior the load pattern is typically based on an equivalent regular wave, that is, an appropriately determined wave height and period.

Structural ULS design of platforms is commonly based on a regular (design) wave approach. The design wave corresponding to a 100-year return period is established by using the 100-year design wave height and varying the wave period within a reasonable range. For floating platforms, usually the wave period is the most critical wave parameter. Then the most critical wave period (length) is first identified on the basis of response to regular waves, and the wave height corresponding to that length is established based on stochastic analysis. When dynamic effects are of concern, a model that recognizes the stochastic features of waves is necessary. The seaway is then modeled as a Gaussian process, comprising regular waves with different frequency, phase lag, and direction (Gran 1992). For linear systems the stochastic response is most efficiently and accurately obtained in the frequency domain. For a single response variable, X, the response is concisely described by the response spectrum, $S_X(\omega)$, which can be obtained by

$$S_X(\omega) = |H_X(\omega)|^2 S_\eta(\omega) \qquad (7\text{-}9)$$

where $H_x(\omega)$ and $S_\eta(\omega)$ are amplitude transfer function from wave to response and wave spectrum, respectively. The distribution of individual response peaks, and their maximums are known to be a Rayleigh distribution and Gumbel distribution, respectively, for narrow-band Gaussian response, and the distribution parameters are readily obtained from $S_X(\omega)$. Also for wide-band response there is a theoretical basis that makes the analysis efficient. For nonlinear problems a very limited theoretical basis is available. Theoretical methods apply only to very special cases involving static response of individual cylinders to waves. Time-domain simulations are in general necessary to obtain the response in complex offshore structures with dynamic behavior. It is generally important to ensure that the use of refined stochastic mechanics models are consistent with current design practice. This means, for instance, that a stochastic analysis approach should be consistent with the design wave approach for structures with quasi-static behavior. Moreover, dynamic effects should preferably be considered by their additional forces as compared to the quasi-static ones. These two issues are illustrated for a three-legged jack-up platform in Section 7.5.2.

The stochastic long-term analysis becomes especially challenging when non-linearities are important. It is then necessary to apply time-domain simulation, as discussed by Moan (2001). Since proper account of frequency dependent mass and damping in general requires an integral-differential equation (convolution of mass with time and acceleration), such analyses become very time consuming. The computer efforts involved make it necessary to improve the efficiency of the long-term analysis. For instance, long-term extreme response values can be effectively calculated by identifying the (few) sea states contributing to the maximum response (Moan 2001). Fatigue analysis requires the total time history of stresses. Since moderate sea states contribute most to fatigue, linear analyses often suffice. Frequency domain analyses of wave response in the various sea states can then be efficiently carried out with coarse finite element models of the structure. The structural model used in such analyses at best provides nominal stresses. More refined analyses, as described in Sections 7.5.3 and 7.6.2, need to be carried out to obtain the local stresses for fatigue design checks.

In some cases fatigue load effects are influenced by nonlinear effects and time-domain simulation combined with rain flow counting of stress ranges is necessary. Systematic studies carried out for offshore structures suggest that the long-term response of the response variables, x can be well described by a two-parameter Weibull distribution:

$$F_X(x) = 1 - \exp\left\{-\left\{\frac{x}{\theta}\right\}^h\right\}$$

(7-10)

where

$$\theta = x_0 / \left(\ell n N_0\right)^{1/h} ; P\left[X \geq x_0\right] = 1/N_0.$$

(7-11a-b)

Guidance on the magnitude of the shape parameter, h, for distributions of the stress range is available (Marshall and Luyties 1982; Odland 1982). The scale parameter, θ is directly related to the extreme response value (x_0) required for ULS design checks and can be estimated in connection with determination of extreme response values. In this way, fatigue loading, at least for initial design and screening to identify the importance of fatigue, can be readily accomplished.

7.5 Structural Design Analysis of Space-Frame Structures

7.5.1 Global Static Analysis

The first step in design analysis of space-frame structures (jackets, compliant towers, and jack-ups) subjected to gravity deck loads as well as wave and current loads is a global analysis comprising space-frame structure and piles or other relevant foundation and appropriate models of the loads.

For platforms with natural periods 3 s or less, a static analysis can be used based on design waves propagating in different directions.

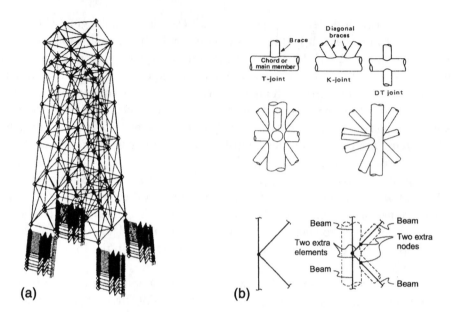

Figure 7-8. Integrated soil-pile-structure model. a) Global model. b) Tubular joints and flexibility model of K-joint.

While conventional ultimate member and joint design checks are based on internal forces calculated by means of linear elastic space-frame models, local analysis of tubular joints is necessary to estimate stresses for fatigue design check, as described in Section 7.5.3.

Structural members are slender, and beam models yield an accurate model. To account for the joint behavior the member length may be modeled by its length face to face with connecting members and rigid joints or center to center and account of joint flexibility. Joints are modeled by shell elements. Systematic studies with such models have resulted in parametric formula for the joint flexibility. A comparison of various formulations shows good agreement between them (Hellan 1995).

The pile foundation can be represented by springs or by considering the piles as a steel structure and representing the interaction between the pile and soil by springs. Soils exhibit nonlinear stress-strain behavior, even at low load levels, that needs to be accounted for. Response analyses are carried out as integrated structure-pile-soil analyses. A typical model is shown in Figure 7-8a.

7.5.2 Global Dynamic Analysis

If the fundamental natural period exceeds 3 s, a dynamic model of the structure and foundation system is necessary. Figure 7-9 shows the typical range of natural periods for offshore structures.

It is crucial that the natural period lies outside, say, the wave period range of 6 to 25 s. Higher order load components corresponding to fractions of the wave period may excite flexible modes in jackets, jack-ups, articulated towers, and even

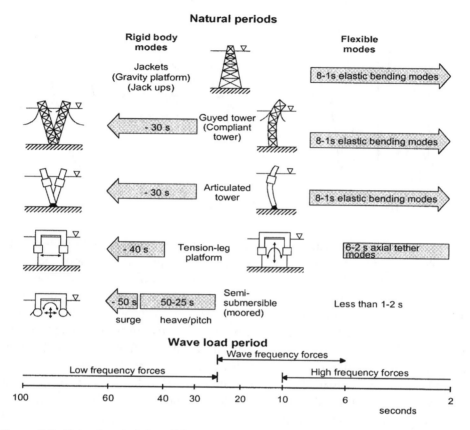

Figure 7-9. Natural periods for offshore structures and typical period range for wave loads.

heave motion of TLPs. Higher order load components with a period that is a multiple of the wave period may be of importance to rigid body motions of floating structures. In such situations, as well as for impact loads and other transient loads, proper dynamic analysis is necessary.

Since the natural periods for structural modes of vibration are below the wave load periods, dynamic effects would be largest for wave conditions with a small wave period. This means that dynamic amplification becomes important for wave conditions relevant for FLS before ULS.

The dynamic model is established by using a relative motion formulation for hydrodynamic loads. The drag term then implies damping, while there is an added hydrodynamic mass associated with the structural acceleration. Nonlinear pile-soil behavior is modeled based on the cyclic (hysteretic) behavior. Structural damping for elastic behavior and radiation damping in the soil should be explicitly incorporated.

If the dynamic effects are most important for fatigue the nonlinearity associated with drag forces and soil behavior is limited, and it can be linearized at the load level contributing stresses most relevant for fatigue damage.

Dynamic behavior of tower type platforms may be illustrated by considering two towers subjected to a regular wave with period T (Fig. 7-10). One tower is a jacket

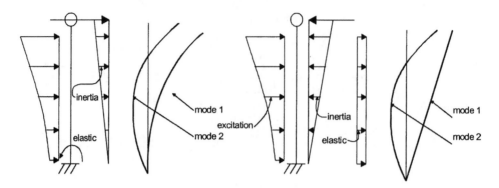

Figure 7-10. Dynamic behavior of tower type platforms.

structure with flexible structure and soil and a natural period below of the wave period. The second tower is a compliant one, with pin-joint support at the seafloor and buoyancy restoring forces along the tower. The dominant reaction forces for the two cases are elastic (restoring) and inertia force, respectively. More importantly, dynamic equilibrium requires the restoring force to balance the wave load and inertia force induced in the first system, while wave forces are balanced inertia forces, see Figure 7-10 and Moan (2001). This fact also would imply that the bending moment and shear forces in a compliant tower is an order of magnitude less than in a "fixed" tower.

When structural dynamic effects are important, a stochastic dynamic approach is conveniently used to calibrate a dynamic amplification factor (DAF). Since the dynamic effects are due to inertia forces, they can alternatively be accommodated first by applying inertia forces in a quasi-static analysis. This is illustrated for the jack-up platform in Figure 7-11, which has a first natural period of 5.7 s at the extreme load level (Karunakaran et al. 1994). A relevant model for the jack-up may then be a simple stick (beam) model to represent the mass, stiffness, and damping properties. However, it is important to determine the loads by properly including the phase lag of the load on different components. For this reason it is convenient to include elements in the model that are only used to introduce loads properly. The first step is to determine the DAF, as the ratio between the dynamic extreme response and the quasi-static extreme response. It is important to determine these responses for representative sea states and to perform time domain simulations such that statistical uncertainties do not affect the results too much (i.e., by using the same wave sample) (Karunakaran et al. 1993). The DAF is consistently determined as the ratio of the expected maxima obtained by time-domain analyses with a dynamic and static model, respectively. Also, it is important that the stochastic quasi-static model is calibrated to be consistent with the regular wave approach used, as measured by the factor R_{QS} in Table 7-3. The dynamic effects can most conveniently be accounted for by applying inertia loads—mass times accelerations—on the deck and tower structure masses. Since the masses are given, the acceleration field is tuned such that the DAF for the base shear and overturning moment are fairly accurately represented for the extreme wave condition. This approach is applied for jack-ups, jackets, and gravity platforms where dynamic response is dominated by one flexible model of vibration.

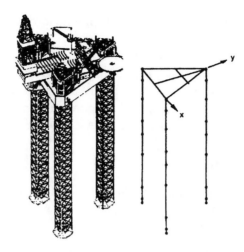

Figure 7-11. Jack-up platform and model.

Table 7-3. Extreme Load Effects in Three-Legged North Sea Jack-Up in a Sea State with H_{m0} = 14.8 m and T_P = 16 s and Design Wave H = 27 m and T = 14.5 s (Karunakaran et al. 1994)

	Load Effect[a]					
	Base Shear		Overtuning Moment		Deck Displacement	
	R_{QS}	DAF	R_{QS}	DAF	R_{QS}	DAF
C_D = 1.0 in time-domain analysis: Gaussian waves	1.07	1.14	1.04	1.29	1.04	1.25
Non-Gaussian waves	1.26	1.13	1.33	1.24	1.33	1.20
C_D = 0.7 in time-domain analysis: Gaussian waves	0.76	1.14	0.74	1.29	0.74	1.25
Non-Gaussian waves	0.92	1.13	0.96	1.24	0.95	1.20

[a]For each load effect two characteristics are given:

(1) The ratio, R_{QS} of the expected maximum load effect obtained by stochastic analysis and the load effect obtained by design wave approach with no dynamics accounted for.

(2) Dynamic amplification factor (DAF) obtained as the ratio of the expected maximum load effect obtained in stochastic analysis based on a dynamic and quasi-static model, respectively.

The behavior of compliant towers is more complex since dynamic contributions stem from two modes with natural periods on either side of the dominant wave excitation period. This means that inertia forces in the first mode balance excitation forces while the inertia forces in the second (flexible) mode add to the excitation. However, it has been demonstrated (Vugts et al. 1997) that the dynamic effects also can be accommodated by applying inertia forces in a quasi-static model for this case.

Obviously, the method that is outlined above is expected to yield accurate estimates when the dynamic response is dominated by a single mode, the response is narrow-banded, and the dynamic response is associated with wave period well-separated from those that cause quasi-static response. This approach is, for instance, adopted in design approaches for jack-ups (SNAME 1994).

7.5.3 Fatigue Analysis

Fatigue crack growth is primarily a local phenomenon and requires that local stresses are calculated in a sequence of analyses as illustrated in Figure 7-12a. The analysis, hence, includes the following elements:

1. Long-term wave climate is the starting point fatigue analysis. This is the aggregate of all sea states occurring yearly (or for longer periods of time). Obtaining this data often requires a major effort, with significant lead times.
2. Global-scale space-frame analysis is performed to obtain structural response in terms of nominal cyclic member stresses for each sea state of interest.
3. Geometric stress concentrations at all potential hot spot locations within the tubular connections must be considered since fatigue failure initiates as a local phenomenon.
4. Accumulated stress cycles are then counted and applied against suitable fatigue criteria (e.g., Miner's rule) to complete the analysis of fatigue damage.

In view of the scatter and uncertainty in fatigue, the choice of target calculated fatigue life requires careful evaluation of the economic and risk factors involved. Typically, the target life, $T_{fatigue} = T_{service}/D_{all}$ is a multiple of the required service life, $T_{service}$.

The global analysis of nominal stress histories, considering all relevant sea states is carried out according to the principles outlined in Sections 7.5.1 through 7.5.2.

The geometric stress concentration is determined by strain gauge measurements or by finite element analysis based on shell or solid elements.

Traditionally fatigue design criteria for welded civil and marine metal structures were based on nominal stress definition of load effects and strength (SN-curves). In this case SN-curves for various structural details need to be established. However, in situations where a variety of joint types (Fig. 7-8b) and geometrical and load parameters affect the fatigue performance, it is more convenient to adopt the so-called hot spot method. This method has been used for welded tubular joints in offshore structure for the last 10 to 20 years and is also described in recent codes (e.g., API 1993/1997, ISO, 19902, NORSOK N-004). Hot spot stress places many different connection geometries on a common basis. The microscopic notch effects, metallurgical degradation, and incipient cracks at the toe of the weld are empirically built into the SN-curve. A single SN-curve is then applied, and the definition of stress needs to be consistent with the SN-curve. Stresses have to be calculated with due account of weld geometry. Moreover, the mesh needs to be fine to accommodate the

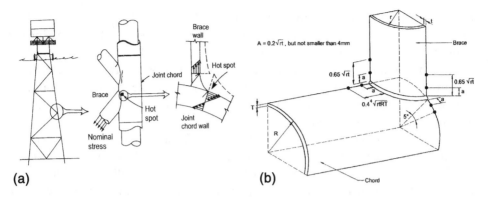

Figure 7-12. Fatigue analysis procedure. a) Hierarchy of analyses. b) Hot spot stress extrapolation.

stress gradient within the zone of shell bending, characterized by the elastic length parameter $\alpha \sqrt{rt}$ where r and t are the radius and thickness of the relevant tubular member, respectively. Due to the notch in the transition between tubular members in the joint, stresses are not directly calculated but extrapolated from points located as given in Figure 7-12b. Based on systematic studies of unstiffened tubular joints with different configurations and relative scantlings, parametric formulae for the stress concentration factor have been developed, (e.g., NORSOK N-004 and forthcoming ISO 19902).

A simple expression for the cumulative damage can be obtained by assuming that the SN-curve is defined by $NS^m = K$, and the number, $n(s)$, of stress ranges is given by the Weibull distribution in Eq. 7-11 with $x = s$ (stress range). The cumulative damage D in a period T with N_T cycles is then

$$D = \sum_i \frac{n_i}{N_i} = \sum \frac{n(s_i)}{N(s_i)} = \frac{N_T}{K} \left[\frac{s_0^h}{\ell n N_0} \right]^{m/h} \Gamma(m/h + 1) \qquad (7\text{-}12)$$

where $\Gamma(\ )$ is the Gamma function.

The Weibull distribution (Eq. 7-10) is readily obtained for linear(ized) problems based on frequency domain analyses for each sea state by assuming Rayleigh distribution and combining the stress ranges for each sea state by their probability of occurrence.

For nonlinear systems the stress ranges need to be determined by appropriate counting of cycles, for example by the rain flow method. Eq. 7-10 is convenient as a basis for an early screening of fatigue proneness, using a simple (conservative) estimate of the extreme response s_0 and assuming the shape parameter, h, of the Weibull distribution. The parameter h depends upon the environmental conditions, relative magnitude of drag and inertia forces, and possible dynamic amplification. For a quasi-static response in an extratropical climate like the North Sea h may be around 1.0 while h may be as low as 0.4 to 0.6 for Gulf of Mexico platforms subjected to infrequent hurricanes. For structures with predominantly drag forces, h will be smaller than for predominantly inertia forces. Dynamic effects may start to affect load effects relevant for fatigue when the natural period exceeds 2.0 s. Increasing the natural period from 2 s to 4 s may increase h from 0.7 to 1.1 and from 0.9 to 1.3 for Gulf of Mexico and North Sea structures, respectively. The implication is a factor of the order of 10 on fatigue damage. It can be shown that the main contribution to fatigue damage is caused by wave-induced stresses that are of the order of 10% to 30% of the 100 year values of the stress.

More detailed fatigue analysis would require consideration of 70 to 400 directional sea states occurring over the life of the structure, with stresses and cumulative fatigue damage being examined at thousands of potentially critical locations throughout the structure. Commonly frequency domain analysis methods are applied toward this aim. Moderate nonlinearities may be handled with "quasi-transfer" functions based on time-domain analysis. In some situations direct time-domain analyses combined with rain flow counting of cycles is necessary. In the Gulf of Mexico, such efforts are undertaken only for deepwater structures where dynamics amplify the effects of everyday small waves (Marshall and Kinra 1980). For monumental North Sea structures, bigger consequences of failure and a more severe wave climate also create fatigue problems that must be analyzed. More detailed analysis at the local or microscopic level, to obtain weld toe notch effects or fracture mechanics stress intensity factors, are usually done only for research and code development purposes or for special design and inspection planning situations.

7.5.4 Nonlinear System Capacity Assessment for Ultimate or Accidental Collapse Limit States

Current ultimate strength code checks of marine structures are commonly based on load effects (member and joint forces) that are obtained by a linear global analysis while resistances of the members and joints are obtained by experiments or theory that accounts for plasticity and large deflections. This methodology, hence, focuses on the first failure of a structural component and not the overall collapse of the structure, which is of main concern in view of the failure consequences. The advent of computer technology and FEM have made it possible to develop analysis tools that include second order geometrical and plasticity effects to account for possible redistribution of the forces and subsequent component failures until system collapse. By using such methods a more realistic measure of the overall strength of structures is achieved.

Initially such methods were developed for seismic analysis (Marshall et al. 1977) and for calculating the residual strength of systems with damage (e.g., according to the accidental limit state checks) (Moan et al. 1985). More recently, such methods have also been applied for reassessment of aging structures to determine the ultimate capacity of the intact system as well as the global strength after fatigue-induced fracture of members in connection with inspection planning.

Recently, Skallerud and Amdahl (2002) prepared a state-of-the-art review of methods for nonlinear analysis of space-frame offshore structures.

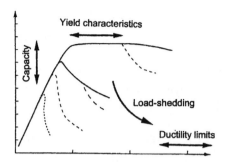

Figure 7-13. Nonlinear behavior characteristics of structural components.

7.5.4.1 Members. Models that have been used to idealize the structure range from phenomenological models to general-purpose finite element models. Techniques that have been developed for the analysis of space frames are addressed in this section (see Hellan et al. 1994, 1995; Søreide and Amdahl 1994; Nichols et al. 1997). Cost-effective solutions are obtained by using large deformation theory for beam elements and special displacement functions (e.g., Livesley or "stability" functions) and concentrating the material nonlinearities in yield hinges at predefined locations or at maximum stress. Yield hinge models are developed at different refinement, from yield hinges with zero extension along the element to models that account for the extension of the yield hinge with elastic-perfectly plastic or gradual plastic behavior of the cross section, strain hardening, and the Bauschinger effect. The joint behavior may be modeled by a plastic potential with interaction between the axial force, in plane bending, and out of plane bending. Formulations have also been published that account for brace to brace interaction by adding beam elements between the brace ends (Fig. 7-8b). Typical nonlinear component behavior is indicated in Figure 7-13.

7.5.4.2 Local failure criteria. It is important that the nonlinear analysis method can predict failure in accordance with recognized failure criteria (Ultiguide 1999). Hence, besides modeling of stiffness and capacity, yield characteristics, postultimate behavior, and ductility limits (if applicable) should be represented, including local failure modes such as local denting, local buckling, joint overload, and joint fracture. Local buckling may account for a reduction in section capacity to the level prescribed by the design (ISO 19902) equations. The capacity of dented members can be expressed by the reduced cross-sectional properties. Another alternative is to account for local buckling in the postcollapse range and growth of the buckle as member deforms (Taby and Moan 1985). General shell FEM formulations may be able to capture local buckling through the shell modeling, but initial geometric imperfections are generally needed to initiate the local buckle. Obviously a global analysis based on a model that accommodates local buckling modes by shell elements will be very time consuming.

The presence of hydrostatic pressure leads to more brittle postultimate behavior of the member. Local buckling will occur for lower D/t ratios than members in air and the local buckle will develop more rapidly once it is formed.

Tensile fracture is defined as deconnecting of the member due to excessive tensile straining. This will be influenced by the presence of cracks, welds, residual stresses, and stress or strain concentrations.

Most normalized steels used in offshore platforms show rather good ductility in tensile coupon tests with ultimate strain values larger than 15% to 20%. Geometric notches and crack-type defects are introduced in the members when they are welded together. Recent investigations (Ultiguide 1999) suggest a limit of 5% nominal tension strain averaged over a length of maximum 20 times the thickness. This applies to welded connections without cracks, provided that fabrication is performed such that overmatch welds are achieved. Nominal strain is defined as the strain derived from a beam element model without including any stress concentrations factor. For structures with cracks of larger size, the strains in the cracked region should not exceed the critical strain for the crack as defined by a fracture mechanics assessment.

Since marine structures are subjected to cyclic loading, cyclic degradation of members after buckling should be accounted for (see Skallerud and Amdahl 2002).

7.5.4.3 Tubular joints. Under extreme loads, the nonlinear deformations of the joint and failure characteristics can influence the disposition of forces and the overall structural response.

Failure of tubular joints generally involves some combination of the following local and global modes:

- Local plastic deformation (yield) of the chord around the brace intersection
- Cracking in the chord at the weld toe (and propagation to severance)
- Local buckling in compression areas of the chord
- Ovalization of the chord cross-section
- Beam shear failure across a gap K joint chord
- Beam bending of the chord especially for T/Y action

The specific response depends upon the type of joint (T/Y, X, K; simple, stiffened, grouted; etc.), the loading (axial—tension/compression; bending—in-plane bending/out of plane bending, etc.), and the joint geometry parameters (β, γ, etc.).

Figure 7-14 compares typical load-deformation responses for axially loaded joints as seen in isolated component tests performed with idealized boundary conditions.

The representation of joint stiffness and failure are made by phenomenological approximations of the behavior shown in Figure 7-14 using ultimate capacity according to codes or by nonlinear shell finite elements. Figure 7-8b illustrates the introduction of separate joint elements or springs to represent joint behavior.

Analytical and experimental results indicate that the collapse load of cracked tubular joints can be predicted by multiplying the capacity of the uncracked joint by an area reduction factor, F_{AR}. The reduction factor is given in PD 6493 (1991) as

$$F_{AR} = \left(1 - \frac{A_C}{A}\right) \cdot \left(\frac{1}{Q_\beta}\right)^{m_q} \tag{7-13}$$

where A_C and A are the cracked and total cross-sectional area, respectively. Q_β is the tubular joint geometry modifier, and m_q is the power allocated to Q_β.

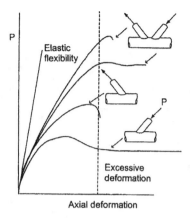

Figure 7-14. Load-deformation characteristics of tubular Y- and K-joints.

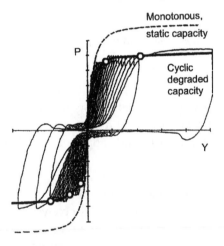

Figure 7-15. Cyclic degradation of soil.

7.5.4.4 Pile-soil behavior. Fixed platform analyses are carried out by modeling the pile-soil behavior, by equivalent linear or nonlinear concentrated springs or distributed springs along the piles, or by continuum (finite element) model that represent stiffness and foundation capacity appropriately using the material properties in the different soil layers. The effects of global seabed scour and local scour in granular soils and the partial loss of pile-soil contact in cohesive soils should be accounted for. When modeling the individual piles in a pile group, nonlinear soil P-y and T-z curves have to be adjusted to account for pile group effects (Horsnell et al. 1996; Lacasse and Nadim 1996).

Cyclic loads cause deterioration of the lateral bearing capacity as indicated in Figure 7-15. The soil capacity and the nonlinear P-δ characteristics given in most codes represent the fully degraded properties of the soil. The capacity under monotonic, static loading can be significantly higher, as shown in Figure 7-15.

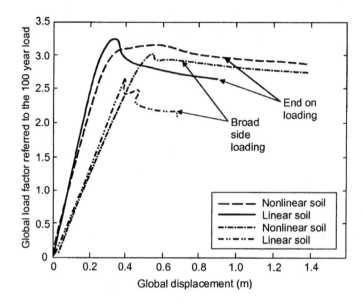

Figure 7-16. Load-displacement relationship for jacket in Figure 7-8a as a function of pile-soil modeling.

High loading rates (compared to laboratory test loading) can also increase the capacity up to 40% for lateral loading and 50% for axial loading for wave loading. It is noted that this strength increase only applies to the dynamic (variable) part of the soil reaction.

7.5.4.5 Global analysis. The space-frame model for ultimate capacity analysis could be simplified compared to models used for conventional component ULS and FLS design checks. The focus should be on the primary framework, while the stiffness and strength of secondary components such as conductors, risers, boat landings, and other appurtenances can be omitted. However, wave loads on such components should obviously be included. In particular, the deck structure can be simplified.

As demonstrated by Moan et al. (1997), the choice of pile-soil model can affect the load distribution in the structures and, hence, the failure mode and corresponding ultimate strength. The most important issue is, of course, that a pure, linear pile-soil model would not represent a possible soil failure and hence overestimate the system strength if the pile-soil is the critical part of the system. For the jacket in Figure 7-16 with plugged piles, the pile foundation is not critical. Yet the difference in jacket failure mode when using a linear instead of a nonlinear model results in an ultimate load that is about 15% smaller for the former case. See Table 7-4.

Determination of the global ultimate capacity by monotonically increasing wave loading has become a well-established approach, as reported elsewhere (API RP2A 1993/1997).

The dynamic behavior of fixed platforms under load levels that ensure linear elastic behavior is stiffness dominated, and inertia forces amplify the response as

Table 7-4. Ultimate Capacity of an Eight-Legged North Sea Jacket with Plugged Piles (Fig. 7-8a) Using a Nonlinear Pile-Soil Model (Moan et al. 1997)

Limit State	Static Load Capacity Factor (100-year load)		Cyclic Dynamic Load Capacity Factor (static capacity)	
	End-on Loading	Broad-side Loading	End-on Loading	Broad-side Loading
First member failure	1.94	1.79	—	—
Ultimate limit	2.89	2.73	1.12	0.96

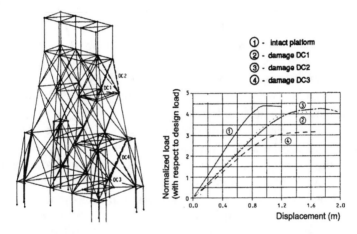

Figure 7-17. Global ultimate strength of jacket with damage.

discussed above. However, as the ultimate strength is approached, the stiffness decreases and the system becomes inertia dominated. In this situation the external forces are partly balanced by inertia forces, and the ultimate strength of the system increases (Stewart 1992; Emami and Moan 1995; Moan 2001). However, this increase in the ultimate capacity requires a ductile static load-displacement characteristic.

The nonlinear system analysis predicts a more realistic estimate of the ultimate strength of the system than a conventional design approach, based on linear global analysis and ultimate member strength. This is particularly evident for systems with initial damage, for example due to accidental damage or fatigue failure of a component. An illustration of the damage tolerance of an eight-legged jacket is shown in Figure 7-17.

7.6 Structural Analysis of Plate and Shell Structures

7.6.1 Global Stress Analysis for ULS Check

The hull of floating structures and gravity platforms with large storage capacity are large volume structures. The structural analysis of such hulls is made considering:

- functional load conditions during operation (i.e., hydrostatic pressures and various ballast conditions),
- wave loads, and
- installation loads. For instance, the deck simply will be supported, when it is installed (mated).

It is noted that the load effects due to functional loads present before deck mating need to be calculated by a different model than load effects introduced after deck installation. When the deck is installed it is initially pinned at all column supports, and the loads in the deck are transmitted into columns according to this feature of the deck. The installation is completed by welding the deck to the columns, ensuring moment transfer. The variable loads later introduced on the structure will be transferred accordingly.

In principle waves cause loads with the wave frequency and at lower frequencies (slow-drift forces). However, the latter normally yield small accelerations and can be neglected when analyzing the hull forces.

Based on a stochastic long-term analysis using a crude frame or shell element model of the hull, a regular design wave approach can be defined, that is, the height and period of the regular wave that gives the same section forces in representative members as the more accurate approach can be established. Commonly, different design waves would have to be used for various response variables and different wave directions. Afterwards a detailed structural analysis is carried out with approximately 15 to 25 deterministic regular wave conditions identified. Besides the wave excitation forces, inertia reaction forces associated with the motions are then properly taken into account.

The global model to determine stresses for ULS check could be based on modeling the stiffened panels by

- shell elements only,
- shell elements for shell plating and girders and beam elements for stiffeners,
- layered, orthotropic shell (sandwich) elements, possibly using substructuring, superelement, or submodel techniques.

Figure 7-18 shows a global shell model of the platform in Figure 7-3. The mesh refinement is chosen so as, for instance, to accommodate so-called beam behavior by account of shear deformations and effective flange effects, and the rigid connections between pontoons and columns.

To estimate inertia forces properly, masses and their centers of gravity need to be specified. The masses associated with equipment are accounted for by mass points supported by simple framework structures.

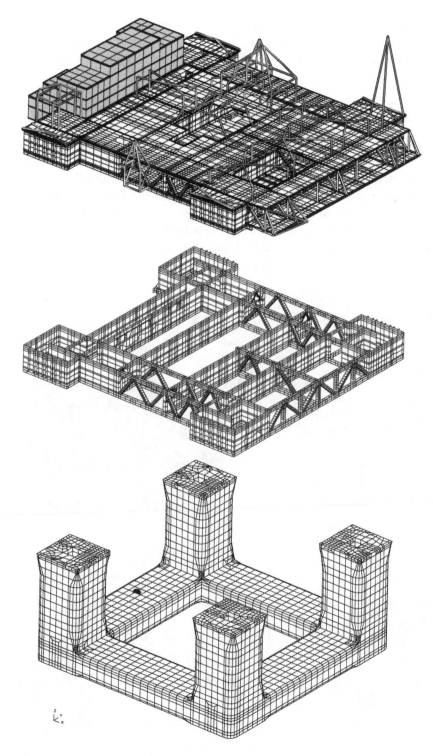

Figure 7-18. Global finite element model of the semisubmersible in Figure 7-3. Figure courtesy of Aker Technology.

In the modeling of the deck plates, only a stressed skin behavior (i.e., shear stiffness) is modeled. This is because the plates are thin and will buckle under the normal stresses and don't contribute stiffness nor load carrying capacity for normal stresses.

7.6.2 Stress Analysis for Fatigue

Obviously, the global analysis is also the starting point for the fatigue analysis. In this case deck plates are modeled. A screening of possible joints that need a refined fatigue design check is based on the method indicated in Section 7.5.3.

The final fatigue analysis is based on a linear stochastic response analyses in the frequency domain, by means of transfer functions and wave spectra. Moreover, the local stresses need to be properly estimated. This can in principle be achieved by using substructures in the global analysis or zooming/submodeling. The reason is to

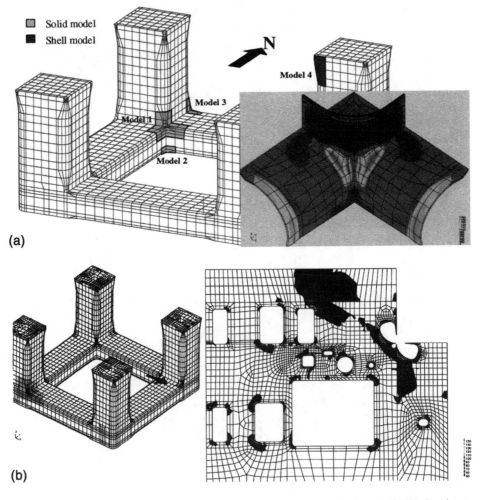

Figure 7-19. Local models of the platform in Figure 7-3. Figure courtesy of Aker Technology.

improve the description of the geometry in critical areas as well as to obtain a better resolution of stresses. Figure 7-19a indicates selected areas where castings are used to improve the fatigue endurances. These are example areas that are modeled by more refined meshes as submodels. Model one is shown in Figure 7-19a. The boundary conditions for these submodels are introduced by imposing (interpolated) displacements obtained in the global analysis along the submodel boundaries. Another submodel is shown in Figure 7-19b. This figure shows the local model and stress contour plot for the internal longitudinal bulkhead in the column-pontoon connection. In particular it is noted that the stress singularity in the reentrant corner is eliminated by introducing cutouts in the bulkhead. Yet, the stress level is high in this area, and the possible buckling of the free edges of the cutout needs to be investigated.

7.7 Analysis of Accidental Load Effects

7.7.1 General

As mentioned above, the ALS check is a survival check of the structural system that is damaged due to accidental actions or abnormal strength. Accidental actions are caused by human errors or technical faults and include fires and explosions, ship impacts, dropped objects, unintended distribution of variable deck loads and ballast, and change of intended pressure difference. Figure 7-20 shows accident rates for mobile (drilling) and fixed (production) platforms according to initiating event of the accident. Heavy weather would normally affect capsizing and foundering as well as structural damage. In most cases some kind of human error or omission by designers, fabricators, or operators of the given installation was a major contributor to the accident. Most notable in this connection is of course accidental loads such as ship impacts, fires, and explosions that should not occur but do so because of operational errors and omissions.

In some cases, accidents have been caused by deficient general engineering practice—at the time of design as compared to current practice. If, for instance, design against accidental loads had been introduced at an earlier stage in the industry at large, the risk level would have been reduced (Moan 2000b). The same applies to methods to deal with loads on jackets and static strength of tubular joints as well as fatigue life prediction. For instance, observations of cracks in North Sea jackets reveal a distinct difference of crack occurrence in platforms installed before and after the year 1978 (Vårdal and Moan 1997).

The accidental actions and abnormal conditions of structural strength are supposed to be determined by risk analysis (Vinnem 1999; Moan 2000b), accounting for relevant factors that affect the accidental loads. In particular, risk reduction can be achieved by reducing the probability of initiating event such as leakage and ignition (that can cause fire or explosion) or ship impact, or by reducing the consequences of hazards. Passive or active measures can be used to control the magnitude of the accidental event and, thereby, its consequences. For instance, the fire action is limited by sprinkler/inert gas system or by fire walls. Fenders can be used to reduce the damage due to collisions.

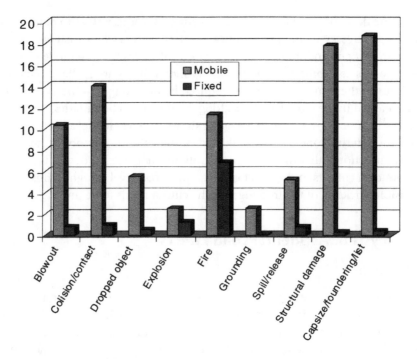

Figure 7-20. Accident rates for mobile and fixed offshore structures based on initiating event. Mobile platforms especially include semisubmersibles, drill-ships, and jack-ups while jackets are the dominant fixed platforms. (Based on data collected from WOAD, 1996).

ALS checks apply to all relevant failure modes as indicated in Table 7-5. Account of accidental loads in conjunction with the design of the structure and equipment as well as safety systems is a crucial safety measure to prevent escalation of accidents. Typical situations where direct design may affect the layout and scantlings are indicted in Table 7-6.

Different subsystems, like:

- load-carrying structure and mooring system,
- process equipment, and
- evacuation and escape system

are designed according to criteria given for the particular subsystems. For instance, safety criteria for structural design are given in terms of ULS, FLS, and ALS with specific target levels, and, hence, imply a certain residual risk level.

ALS is carried out by checking the system strength after the effect of accidental actions with annual exceedance probability of 10^{-4} (as determined by risk analysis).

A complete identification of a Design Accidental Event should also include an estimate of the probability of occurrence. For each physical phenomenon (fire, explosions, collisions, etc.) there is normally a continuous spectrum of accidental events. A finite number of events has to be selected by judgment. These events represent different action intensities at different probabilities. The characteristic

Table 7-5. Examples of Accidental Actions for Relevant Failure Modes of Platforms

Structural Concept	Failure Mode	Relevant Accidental Action
Fixed platform	Structural failure	All
Floating platform	Structural failure	All
	Instability	• Collision, dropped object, unintended pressure, unintended ballast that initiates flooding
	Mooring system strength	• Collision on platform • (Abnormal strength)
Tension leg platform	Structural failure	All
	Mooring system – slack – strength	• Accidental actions that initiate flooding • Collision on platform • Dropped object on tether • (Abnormal strength)

Table 7-6. Design Implications of Accidental Loads

Load	Structure	Equipment	Passive Protection System
Fire	Columns /deck (if not protected)	Exposed equip. (if not protected)	Fire barriers
Explosion	Topside (if not protected)	Exposed equip. (if not protected)	Blast / Fire barriers
Ship impact	Waterline structure (subdivision) (if not protected)	Possibly exposed risers (if not protected)	Possible fender systems
Dropped object	Deck Buoyancy elements	Equipment on deck, risers, and subsea (if not protected)	Impact protection

Direct design of:

• structure (to avoid progressive structural failure or flooding)
• equipment and protective barrier (to avoid damage and escalation of accident)

accidental load on different components of a given installation could be determined as follows:

• establish exceedance diagram for the load on each component,
• allocate a certain portion of the reference exceedance probability (10^{-4}) to each component,
• determine the characteristic load for each component from the relevant load exceedance diagram and reference probability.

If the accidental load is described by several parameters (e.g., heat flux and duration for a fire, pressure peak and duration for an explosion) design values may be obtained from the joint probability distribution by contour curves (NORSOK N-003 1999). However, in view of the uncertainties associated with the probabilistic analysis, a more pragmatic approach would normally suffice. Yet, significant analysis efforts are involved in identifying the relevant design scenarios for the different types of accidental loads.

For each design accident scenario the damage imposed on the offshore installation needs to be estimated, followed by an analysis of the residual ultimate strength of the damaged structure in order to demonstrate survival of the installation. To estimate damage (permanent deformation, rupture, etc.) of parts of the structure, nonlinear material and geometrical structural behavior need to be accounted for. While in general nonlinear FEMs need to be applied, simplified methods, such as those based on plastic mechanisms, are developed and calibrated using more refined methods to limit the computational effort required.

In the following sections the estimation of damage due to accidental loads will be exemplified.

7.7.2 Fires and Explosions

The dominant fire and explosion events are associated with hydrocarbon leakage from things like flanges, valves, equipment seals, and nozzles. Commonly the effects of 40 to 60 scenarios need to be analyzed. This means that location and magnitude of relevant hydrocarbon leaks, likelihood of ignition, as well as combustion and temperature development (in a fire) and pressure-time development (for an explosion) need to be estimated, followed by a structural assessment of the potential damage.

The fire thermal flux may be calculated on the basis of the type of hydrocarbons, release rate, combustion, time and location of ignition, ventilation, and structural geometry using simplified conservative semiempirical formulae or analytical/numerical models of the combustion process. The heat flux may be determined by empirical, phenomenological, or numerical method (SCI 1993; BEFETS 1998). Typical thermal loading in hydrocarbon fire scenarios may be 200 to 300 kW/m^2 for a 15 min to 2 hour period. The structural effect is primarily due to the reduced strength with increasing temperature.

In case of explosion scenarios, the analysis of leaks is followed by gas dispersion and possible formation of gas clouds, ignition, combustion, and development of overpressure. Tools such as FLACS, PROEXP, or AutoReGas are available for this effort (Moan 2000b; Czujko 2001). Typical overpressures for topsides of North Sea platforms are in the range of 0.2 to 0.6 barge, with a duration of 0.1 to 0.5 s.

The damage due to an explosion should be determined with due account of the nonlinear and dynamic character of the load effects. Simple, conservative single degree of freedom models may be applied (NORSOK N-004 1998). In particular cases where simplified methods have not been calibrated, nonlinear time-domain analyses based on numerical methods like FEM should be applied. A recent overview of such methods may be found in Czujko (2001). For instance, the behavior of the

Figure 7-21. Layout of deck structure (topside) of a six-legged North Sea jacket (Czujko 2001), with permission from FORCE Technology, Norway.

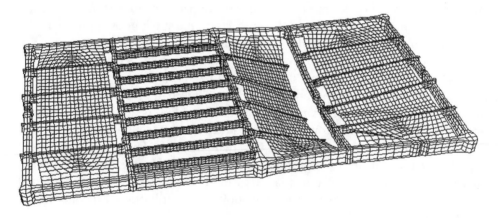

Figure 7-22. Failure mode of lower deck in topside structure in Figure 7-21 (Czujko 2001), with permission from FORCE Technology, Norway.

topside structure of the six-legged North Sea jacket shown in Figure 7-21 under blast loading has been studied. Figure 7-22 shows the failure mode of the stiffened lower deck. The analysis was carried out by LS-DYNA using relevant strain-based rupture criteria (Czujko 2001). As indicated in this figure the final failure is a rupture.

Fire and explosion events that result from the same scenario of released combustibles and ignition should be assumed to occur at the same time, in other words, to be fully dependent. The fire and blast analyses should be performed by taking into account the effects of one on the other. The damage done to the fire protection by an explosion preceding the fire should be considered.

7.7.3 Ship Impacts

Ship collision loads are characterized by a kinetic energy, described by the mass of the ship including hydrodynamic added mass and the speed of the ship at the instant of impact. If the collision is noncentral (the contact force does not go through the center of gravity of the platform (installation) and the ship), a part of the kinetic energy may remain as kinetic energy after the impact. The remainder of the kinetic energy has to be dissipated as strain energy in the installation and, possibly, in the vessel. Generally this involves large plastic strains and significant structural damage to either the installation or the ship or both.

All ship traffic in the relevant area of the offshore installation should be mapped and possible future changes in vessel operational pattern should be accounted for. The velocity can be determined based on the assumption of a drifting ship, or on the assumption of erroneous operation of the ship. Ship traffic may therefore for this purpose be divided into categories: trading vessels and other ships external to the offshore activity, offshore tankers, and supply or other service vessels. Merchant vessels are often found to represent the greatest platform collision hazard.

The most probable impact locations and impact geometry should be established based on the dimensions and geometry of the offshore structure and vessel and should account for tidal changes, operational sea state and motions of the vessel, and structure that has free modes of behavior. Impact scenarios should be established representing bow, stern, and side impacts on the structure as appropriate.

While historical data provide information about supply vessel impacts, risk analysis models are necessary to predict other types of impacts, such as those involving trading vessels (NORSOK N-003 1999; Moan 2000b).

The collision problem comprises both internal mechanics related large, inelastic deformations at the point of contact as well as global hull bending of struck vessel and interaction with the surrounding fluid (added mass, viscous forces, etc.). A fully integrated analysis is fairly demanding. It is, therefore, often found to be convenient to split the problem into two uncoupled analyses, namely, the external collision mechanics dealing with global inertia forces and hydrodynamic effects and internal mechanics dealing with the energy dissipation and distribution of damage in the two structures (NORSOK N-004 1998).

Now, the main purpose of the analysis of ship impacts is to determine the damage in the installation. The first step toward this aim is to estimate the amount of collision (kinetic) energy to be dissipated as strain energy. This is achieved by assuming that the dominant forces during collision are inertia forces (including hydrodynamic added mass) and the collision force. Viscous forces are neglected. Assuming a central impact and applying the principle of conservation of momentum and conservation of energy the amount of kinetic energy to be dissipated as strain energy, E_s, is given by the formula (see NORSOK N-004, Appendix A, 1998):

$$E_s = \frac{1}{2}(m_s + a_s)v_s^2 \frac{\left(1 - \dfrac{v_i}{v_s}\right)^2}{1 + \dfrac{m_s + a_s}{m_i + a_i}} \tag{7-14}$$

where m_s = ship mass; a_s = ship added mass; v_s = impact speed; m_i = mass of installation; a_i = added mass of installation; and v_i = velocity of installation. The strong dependence of the mass ratio is noticed.

The next step will be to estimate how the energy is shared among the installation and the ship. Methods for assessing the damage caused by ship impacts are thoroughly described by Amdahl (lecture notes, 1999). Here a brief outline is presented.

The structural response of the ramming ship and installation can formally be represented as load-deformation relationships as illustrated in Figure 7-23. The strain energy dissipated by the ship and installation equals the total area under the load-deformation curves. The total energy dissipation may be expressed by:

$$E_s = E_{s,s} + E_{s,i} = \int_0^{w_{s,\max}} R_s \, dw_s + \int_0^{w_{i,\max}} R_i \, dw_i \qquad (7\text{-}15)$$

where R_i and R_s are the resistance of installation and ship, respectively; and dw_i and dw_s are the deformation of installation and ship. As the load level is not known a priori, an incremental procedure is generally needed. It is customary to establish the load-deformation relationships for the ship and the installation independently of each other assuming the other object infinitely rigid. Figure 7-23b shows approximate indentation resistance for ships. This approach may imply severe limitations because both structures will inevitably dissipate some energy regardless of their relative strength. Care should therefore be exercised in selecting or calculating the load-deformation curves to ensure that they are representative for the true, interactive nature of the contact between the two structures.

Based on the relative energy absorption capabilities of the installation and ship, the design of the installation different design principles may be distinguished; namely: strength design, ductility design, or shared energy design. As indicated in Figure 7-24, the distribution depends upon the relative strength of the two structures. *Strength design* implies that the installation is strong enough to resist the collision force with minor deformation so that the ship is forced to deform and dissipate the major part of the energy. *Ductility design* implies that the installation undergoes

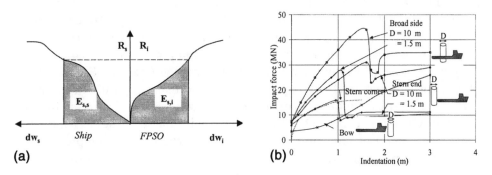

Figure 7-23. Energy absorption based on force-indentation relationship. a) Impact force-indentation. b) Impact force-indentation for supply vessel impact on rigid cylindrical column.

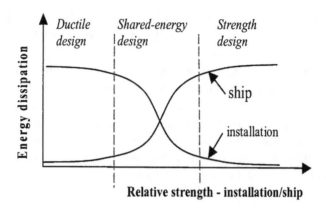

Figure 7-24. Ship impact design principle based on relative energy sharing between ship and installation.

large, plastic deformations and dissipates the major part of the collision energy. *Shared energy design* implies that both the installation and ship contribute significantly to the energy dissipation.

From the calculation point of view strength design or ductility design is favorable. In this case the response of the "soft" structure can be calculated on the basis of simple considerations of the geometry of the "rigid" structure. For instance, Figure 7-23b can be used when the ship is soft while the platform is rigid to carry out a strength design of the platform in the case of supply vessel impact. In shared energy design both the magnitude and distribution of the collision force depends upon the deformation of both structures. The analysis has to be carried out incrementally on the basis of the current deformation field, contact area, and force distribution over the contact area. It is the current weaker structure that is forced to deform most, whereas the damage of the other may remain virtually unchanged during an incremental step. The relative strength of the two structures may vary both over the contact area as well as over time.

Recent advances in computers and algorithms have made nonlinear finite element analysis (NLFE) a viable tool for assessing collisions. There are generally two methodologies available: implicit analysis and explicit analysis. Implicit methodologies require solution of equation systems. This places demands on the equation solver and the computer capacity, especially in terms of memory resources. Explicit systems do not require equation solving. Equilibrium is solved at element level. However, to maintain stability, very small time steps are needed. Careful choice of element type and mesh are required. It is found that a particularly fine mesh is required in order to obtain accurate results for components deforming by axial crushing. A major challenge in NLFE analysis is prediction of ductile crack initiation and propagation. This problem is not yet solved. Crack initiation and propagation should be based on fracture mechanics analysis, using the J-integral or crack tip opening displacement method rather than simple strain considerations. A difficulty in this connection is that the strain depends upon the element mesh. The simplest approach to the problem is to remove elements once the critical strain is attained.

This is fairly easily done in an explicit code because there is no need to assemble and invert the effective system stiffness matrix. However, deleting elements disregards the fact the large stresses can be maintained parallel to the cracks. An improved modeling is to introduce a double set of nodes such that the elements are allowed to separate once the critical stress is attained. A drawback with a double set of nodes is that the potential location of cracks needs to be defined prior to analysis.

While the main concern about ship impacts on fixed platforms is the reduction of structural strength and possible progressive structural failure, the main effect for buoyant structures is damage that can lead to flooding and, hence, loss of buoyancy. The measure of damage in this connection is the maximum indentation implying loss of watertightness. However, in case of large damage, reduction of structural strength is also of concern for floating structures.

For semisubmersible steel platforms the indentation in the column is a useful measure of collision damage. A collision involving an energy of 14 MJ would normally imply an indentation of 0.4 to 1.2 m, while 75 to 100 MJ may be required to cause an indentation equal to the column radius (Moan and Amdahl 1989). The effect is highly dependent upon the location of impact contact area relative to decks

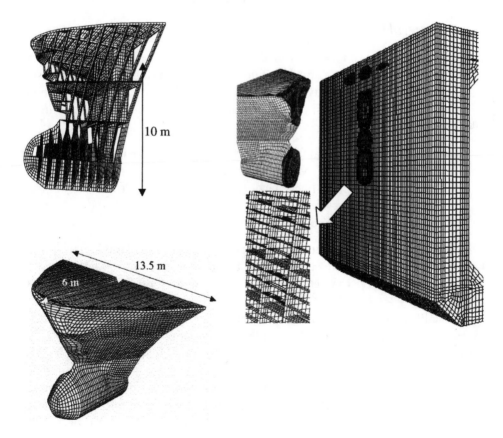

Figure 7-25. Impact of supply vessel against FPSO ship side (Moan and Amdahl 2001).

and bulkheads in the column. Moan and Amdahl (2001) considered a bow impact by supply vessels or merchant vessels with a displacement of 2,000 to 5,000 tons on a typical FPSO side structure with displacement of about 100,000 tons. For this case the limiting energy for rupture of the outer side is approximately 10 MJ (700 mm bow displacement). The energy at 1,500 mm bow intendation is approximately 43 MJ while the limiting energy for rupture of inner side is about 230 MJ, at 3,700 mm bow displacement. This corresponds to a critical speed of 18.5 knots for a 5,000 ton displacement vessel. It should be recalled that rupture of plate material is very uncertain so that the quoted energy level should be assessed carefully. If the FPSO side structure is assumed sufficiently rigid so that all energy is absorbed by the bow, the maximum collision force (peak force) is 30 MN. But most of time the force oscillates between 15 and 25 MN (Fig. 7-25). Integrated collision analysis of both structures shows that the actual FPSO side selected undergoes very small deformations (< 200 mm), yielding contact forces very close to those obtained for rigid side. This shows that it is practically feasible to design the side of FPSO sufficiently strong to impose most of the energy dissipation to the bow of the ramming ship for this vessel size category.

7.8 Requalification of Offshore Structures

7.8.1 General

Requalification means the process of demonstrating fitness for further use of an installation during operation and, if required, proposing modification to the installation to maintain serviceability and acceptable safety for people, the environment, as well as national and company assets. Requalification may be necessary to undertake, for example, due to:

- occurrence of overload damage or deterioration, or other incidents that may imply doubt about the safety of installations,
- desired change of use and hence payloads,
- desired extension of service life, or
- reduction of operational costs.

Requalification is a relevant issue for all kinds of industrial plants and infrastructure and is hence treated in the general ISO Standard for structural design (ISO 2394, draft for approval, 1998) as well as in the ISO Standard for offshore structures (ISO 19900 1999).

Requalifications thus have been undertaken after overload damage due to hurricanes in the Gulf of Mexico (Dunlap and Ibbs 1994), subsidence of North Sea jackets (Broughton 1997), as well as frequently after accidental loads such as explosions, fires, ship impacts, or in conjunction with increased payload. Requalification is also necessary in connection with fatigue damage that accumulates during transport and installation as well as operation. Reassessment of design is required in planning inspection, maintenance, and repair in view of crack growth and especially in connection with service life extension.

If the safety level is found to be too low, mitigation may be achieved by implementing one or more of the following measures:

- replacement of the platform with a new one;
- removing unnecessary conductors, equipment, and marine growth to reduce both functional and environmental loads;
- strengthening the jacket by grinding of cracks, grouting, extra clamped braces, or the pile by shirt piles or plugs;
- reducing potential failure consequences by demanning; or by evacuation and shut-down in the case of severe wave conditions; and
- implementing monitoring or inspection and repair strategy when there is a gradual reduction of safety level, perhaps due to subsidence and crack propagation, and a limited remaining operation life.

Basically, requalification involves the same assessments as carried out during initial design, except that it is carried out with more specific information obtained through monitoring and inspection during fabrication and operation. However, depending upon the inherent excessive damage tolerance ensured by initial design, the measures that have to be implemented to improve the strength of an existing structure may be much more expensive than to design and build a new structure. This fact commonly justifies more advanced analyses of loads, responses, and resistances as well as use of reliability analysis and risk-based approaches than in the initial design (Moan 2000a).

7.8.2 Requalification of Offshore Platforms in View of Fatigue

Fatigue is an important consideration for structures in areas with continuous storm loading such as the North Sea and especially for dynamically sensitive platforms. While the first generation of offshore structures in the North Sea in the early 1970s was not designed against fatigue failure, fatigue design is an important consideration today. Yet, state-of-the-art design checks for fatigue imply highly scattered estimates. The approaches differ especially in methods used to determine wave kinematics in the splash zone, hydrodynamic loads, deterministic vs. stochastic pile-soil flexibility, and stress concentration factors (SCFs) as a function of geometry and load pattern. Fatigue life estimates vary by one to two orders of magnitude repair (Vårdal and Moan 1997).

Fatigue cracks gradually develop from fabrication defects, possibly until they result in rupture. However, the gradual development of cracks allows control by an inspection/repair strategy. A Design, Inspection, Maintenance and Repair (DIMR) strategy involves

- design for certain robustness against progressive failure,
- identification of areas for surveillance, inspection, and repair,
- damages of concern (e.g., size of cracks to look for),
- frequency of inspection or monitoring,
- inspection or monitoring method (and its quality), and
- repair strategy (e.g., size of crack to be repaired, repair method, time required for repair).

It is important that the inspection program (methods, frequency, repair strategy) established at the design stage is sequentially updated based on the additional information that becomes available during fabrication and use.

The most refined semiprobabilistic fatigue design practice is to let the allowable damage depend upon consequences of member failure and access for inspection (e.g., NPD 1992). Inspection requirements are not very specific. Since typically about 20% of the joints are inspected during the service life, it is important to prioritize inspections. Initially, North Sea inspections were executed more or less arbitrarily. Later, when fatigue analyses become common, fatigue life estimates were used to prioritize NDIs to detect cracks. More recently, failure consequences are also considered, for example, those expressed by the residual strength after fatigue failure of the relevant member. The underestimation of fatigue damage has been experienced with current prediction methods for long fatigue lives for nonredundant structures. Further refinement of the inspection planning has been made by introducing probabilistic methods (Moan 2000a).

Inspired by the NPD ALS criterion for ultimate strength a simplified system reliability approach is applied to ensure acceptable safety in inspection planning. In general the systems failure probability when the effect of inspection is considered may be written as

$$
p_{FSYS} = P[FSYS|I] = P\left[\bigcup_{i=1}^{N}\bigcap_{j=1}^{n_i}\left(g_{i_j}^{(\cdots)}(\ )\leq 0\right)\bigg|\bigcap_{i,k,\ell}I_{\ell_k}^i\right] \tag{7-16}
$$

where $I:\bigcap_{i,k,\ell}\left(I_{\ell k}^i\right)$ denotes a set of inspection events (outcomes).

For a multicomponent system and an inspection event tree with multiple branches, the computational efforts to calculate Eq. 7-16 are significant (Moan 1994).

The calculation of the system failure probability may be drastically simplified by assuming :

$$
\begin{aligned}
p_{FSYS|up} = P[FSYS|I] &\approx P[FSYS(U)] \\
&+ \sum_{j=1}^{n}P[F_j|I]\cdot P[FSYS(U)|F_j]+\dots
\end{aligned} \tag{7-17}
$$

This formulation implies that ultimate failure is based on modeling the system as a component. This is a reasonable model in view of the large correlation that exists between load effects in different components (especially in tower type platforms), and the fact that the uncertainty in load (effects) dominates the failure probability.

Moreover, initial fatigue failure (and system failure) events are (conservatively) considered to be independent. Moreover, the effect of inspection is considered to affect only the initial fatigue failure probability. The formulation for updated failure probability in Eq. 7-17 is applicable when the inspection event I aims at detecting cracks before the failure of individual members, in other words, before they have caused rupture of the member. The possible effect of inspection on ultimate component failure is then not accounted for. A further simplification may be to update the failure probability of each joint based on the inspection result for that joint. This is conservative if no cracks are detected, but nonconservative if cracks are detected.

Finally it is noted that the failure probability may be referred to a year or service life. In particular, the so-called hazard rate may be applied.

Another inspection strategy would be to apply visual inspection to detect member failure and repair after the winter season in which the member failed. In this case the Eq. 7-17 will have to be modified as follows: the individual fatigue failures of components (F_j) would not depend on the inspection event. Rather, such an inspection and repair strategy will have implication on the time period for which the failure probability $P[FSYS(U)|F_j]$ should be calculated.

Now, revert to Eq. 7-17. Based on this equation the following simplified requirement with respect to crack growth control can be established:

$$p_{FSYS|up_j} = P\left[F_j|I\right] \times P\left[FSYS(U)|F_j\right] \leq p_{FSYS|} \qquad (7\text{-}18)$$

where $p_{FSYS|T}$ is the target level for system failure, as referred to a single system failure mode initiated by fatigue failure of one joint. This approach is applied in inspection planning with respect to crack control (Aker 1997). $P[F_j|I]$ and $P[FSYS(U)|F_j]$ are then taken to be the fatigue failure probability in the service life and the annual conditional failure probability given fatigue failure of the relevant joint, respectively. Obviously, the target level $p_{FSYS|T}$ must be defined in a manner that reflects this fact. The fatigue failure probability is determined by using a probabilistic fracture mechanics approach, while the latter probability is calculated using ultimate strength analysis methods indicated in Section 7.5. It is recognized that the application of the criterion in Eq. 7-18 may imply a true system failure probability that is several times the target value $p_{FSYS|T}$.

7.8.3 Requalification of a Semisubmersible Production Platform

A particular issue is concerned with documenting adequate safety against fatigue failure for an existing structure when the service life is to be extended beyond the initially planned value. A main issue is how much fatigue life remains and how much it may be extended by inspection. Another issue is whether the acceptance level of safety for an existing structure should be the same as for new designs. The Veslefrikk B (VFB) case will be used to exemplify the treatment of these issues.

The VFB platform is a semisubmersible production unit (Fig. 7-26a). The platform was initially constructed in 1985 and converted to a production unit and installed on the Veslefrikk field in 1989. The platform was dry docked for modification in 1999, partly to increase payload and partly for local reconstruction of the corner areas of column-pontoon connections. Additional payload capacity was achieved by two additional columns and sponsons on the existing columns were required for the purpose of stability (Fig. 7-26a).

The corner areas of column-pontoon connections facing the center of the structure needed improved local design with an increased fatigue capacity. This was

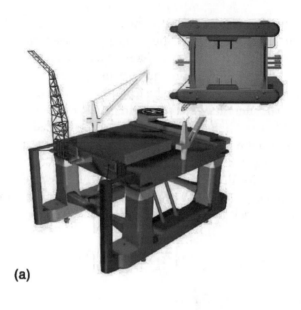

(a)

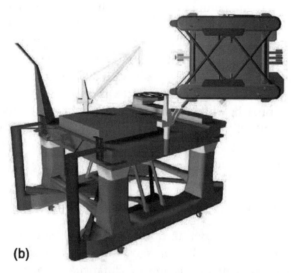

(b)

Figure 7-26. Alternative structural layouts of Veslefrikk B semisubmersible production platform. a) Platform with additional buoyancy (sponsons). b) Platform with sponsons and additional horizontal bracing; figures courtesy of Aker Technology.

achieved by introduction of casted components. A detailed FEM analysis revealed also a high fatigue potential for the two horizontal braces, see Figure 7-27.

Traditional deterministic SN-curve estimates identified the most critical areas of the braces to have a 5- to 10-year fatigue life. Upon 10 years in-service, a requirement of an additional 15 to 20 years in operation motivated consideration of different structural modification alternatives and other measures to improve the fatigue endurance. One alternative to achieve acceptable extension of the service life was

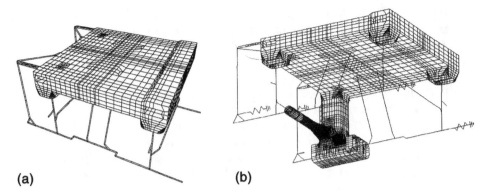

(a)   (b)

Figure 7-27. Hybrid beam-shell element models of the Veslefrikk B semisubmersible production platform. a) Model for redundancy analysis. b) Model for fatigue analysis of brace-column.

to add horizontal braces as shown in Figure 7-26b. This design modification would reduce the stress level as well as the consequence of a brace failure. However, the solution selected was to document adequate safety with respect to fatigue failure by utilizing the extensive inspection history of "No Crack Growth Detection," application of weld improvement methods (grinding and hammer peening), as well as more frequent in-service inspections (Vårdal et al. 2000; Johannesen et al. 2000). To do so, a probabilistic fracture mechanics approach was applied. Such a method also allowed for treatment of the interaction between design, inspection, maintenance, and repair requirements. By considering the quality of the inspection in terms of the probability of crack detection (POD) and by using Bayes theorem, it was possible to better predict the future crack growth potential. The information obtained by an extensive Eddy Current/Magnetic Particle Inspection (EC/MPI) upon 10 years in service for the VFB structure was also utilized. The approach chosen, involving detailed analyses and follow-up during operation, was estimated to imply a significant cost saving over the service life (Vårdal et al. 2000).

7.9 Conclusion

A diversity of structural forms, design and reassessment criteria, as well as load and structural analysis methods for offshore structures have been briefly presented. It is hoped that this information can stimulate further interest in offshore structural engineering and exchange of information between this field and related fields of structural engineering.

References

Aker. (1997). "PIA theory manual." *Doc. No. 580235-N-Q20–004, Rev. 2.1*, Aker Maritime, Bergen, Norway.

Almar-Næss, A., ed. (1985). *Fatigue handbook for offshore steel structures*, Tapir, Trondheim, Norway.

American Petroleum Institute (API). (1993/1997). "Recommended practice for planning, designing and constructing fixed offshore platforms." *API RP2A-WSD*, and "Assessment of existing platform." Supplement 1, *Sect., 17.0,* API, Dallas, Tex.

Barltrop, N. D. P., and Adams, A. J. (1991). *Dynamics of fixed marine structures,* Butterworth's, London.

BEFETS (1998). "Blast and fire engineering for topside systems, Phase 2." *SCIPublication No. 253,* Steel Construction Institute, Ascot, UK.

Broughton, P. (1997). "Engineering challenges of Ekofisk, the first North Sea oil field." *Proc., 8th BOSS Conf.,* Vol. I, Pergamon Press, New York, 41–70.

Clauss, G., Lehmann, E., and Østergaard, C. (1991). *Offshore structures,* Vol. 1, Springer, Berlin.

Cullen Report. (1990). *Public inquiry into the Piper Alpha disaster,"* Her Majesty's Stationary Office (HMSO), London, Cmnd 1310.

Czujko, J., ed. (2001). "Design of offshore facilities to resist gas explosion hazard." *Engrg. Handbook,* Woodhead Publishing, Cambridge, UK.

Department of Energy. (1986). *Background to new fatigue design guidance for steel welded joints in offshore structures,* HMSO, London.

Dunlap, W. A., and Ibbs, C. W., eds. (1994). *Proc., Int. Workshop on Assessment and Requalification of Offshore Production Struct.,* Offshore Technology Research Center, Texas A&M University.

Emami Azadi, M. (1998). "Analysis of static and dynamic pile-soil-jacket behaviour." PhD thesis, *MTA-report 1998:121,* Norwegian University of Science and Technology (NTNU), Trondheim.

Faltinsen, O. M. (1990). *Sea loads on ships and offshore structures,* Cambridge University Press, New York.

Farnes, K. A., and Moan, T. (1994). "Extreme dynamic, non-linear response of fixed platforms using a complete long-term approach." *J. Appl. Ocean Research,* 15, 317–326.

Farnes, K. A., Skjåstad, O., and Hoen, C. (1994). "Time domain analysis of transient resonant response of a monotower platform." Massachusetts Institute of Technology, Boston, Mass., 7th Behavior of Offshore Structures (BOSS) Conference.

Fjeld, S. (1977). "Reliability of offshore structures." *Proc., 9th Offshore Tech. Conf.,* 4, OTC Offshore Technology Conference, Richardson, Tex., 459–472.

Gran, S. (1992). *A course in ocean engineering,* Elsevier, Amsterdam.

Heideman, J. C., and Weaver, T. O. (1992). "Static wave force procedure for platform design." *Proc., Civil Engng. in the Oceans,* V, College Station, Tex., 496–517.

Hellan, Ø. (1995). "Non-linear pushover and cyclic analyses in ultimate limit state design and reassessment of tubular steel offshore structures." *MTA report (PhD),* 108, Norwegian Institute of Technology, Trondheim.

Hellan, Ø., Drange, S. O., and Moan, T. (1994). "Use of non-linear pushover analyses in ultimate limit state design and integrity assessment of jacket structures." *Proc., 7th BOSS Conf.,* 3, Pergamon, New York, 323–345.

Hellan, Ø., Skallerud, B., Amdahl, J., and Moan, T. (1991). "Reassessment of offshore steel structures: shakedown and cyclic non-linear FEM analysis." *Proc., 1st Int. Offshore Polar Conf.,* Int. Soc. of Offshore and Polar Engineers (ISOPE), Edinburgh, 34–42.

Horsnell, M. R., and Toolan, F. E. (1996). "Risk of foundation failure of offshore jacket piles." Paper No. 7997, *Proc., 28th Offshore Tech. Conf.,* OTC Offshore Technology Conference, Richardson, Tex., 381–392.

International Standards Organisation (ISO) (1994). "Petroleum and Natural Gas Industries—Offshore Structures—Part 1: General Requirements." *ISO 19900,* London.

International Standards Organisation (ISO) (1999). "Control and mitigation of fires and explosions on offshore production. Installations—Requirements and guidelines." *ISO 12702,* London.

International Standards Organisation (ISO) (2001). "Petroleum and Natural Gas Industries—Offshore Structures—Part 2: Fixed Steel Structures." *ISO 19900, Draft,* London.

Johannesen, M. J., Moan, T., Vårdal, O. T. (2000). "Application of probabilistic fracture mechanics analysis for reassessment of fatigue life of a floating production unit—Theory and validation," *Paper No. 00–2079, Proc., 19th Offshore Mechanics and Arctic Engineering (OMAE) Conf.* New Orleans, ASME, New York.

Jordaan, I. J., and Maes, M. A. (1991). "Rational for load specifications and load factors in the new CSA code for fixed offshore structures." *J. Civil Engng.* 18(3), 454–464.

Karunakaran, D., Spidsøe, N., Gudmestad, O., and Moan, T. (1993). "Stochastic dynamic time-domain analysis of drag-dominated offshore platforms." *Proc., EURODYN '93,* A.A. Balkema, Rotterdam.

Karunakaran, D. N., Spidsøe, N., and Haver, S. (1994). "Nonlinear dynamic response of jack-up platforms due to non-gaussian waves." *Proc. OMAE Conf.,* ASME, New York.

Lacasse, S., and Nadim, F. (1996). "Model Uncertainty in Pile Axial Capacity Calculations." *Proc., Offshore Tech. Conf.* Paper No. 7996, OTC Offshore Technology Conference, Richardson, Tex.

Lloyd, J. R., and Karsan, D. I. (1988). "Development of a reliability-based alternative to API RP2A." *Proc., 20th Offshore Tech. Conf.,* Vol. 4, OTC 5882, OTC Offshore Technology Conference, Richardson, Tex., 593–600.

MacCamy, R. C., and Fuchs, R. A. (1954). "Wave forces on piles: A diffraction theory." *Memo no. 69,* U.S. Army Corps of Engineers, Beach Errosion Board, Washington, D.C.

Marshall, P. W. (1969). "Risk evaluations for offshore structures." *American Society of Civil Engineers St. Div.* 95 (12).

Marshall, P. W. (1992). *Design of tubular connection—Basis and use of AWS code prov.,* Elsevier, Amsterdam.

Marshall, P. W., Gates, W. E., and Anagnostopoulos, S. (1977). "Inelastic dynamic analysis of tubular offshore structures." *Proc., 9th Offshore Tech. Conf.,* OTC 2908, OTC Offshore Technology Conference, Richardson, Tex.

Marshall, P. W., and Luyties, W. H. (1982). "Allowable stresses for fatigue design." *Proc., BOSS '82,* McGraw-Hill, New York.

Maus, L. D., Finn, L. D., and Danaczko, M. A. (1996). "Exxon study shows Compliant Piled Tower cost benefits." *Ocean Industry,* 20–25.

Melchers, R. E. (1987). *Structural reliability analysis and prediction,* Ellis Horwood, Chichester, UK.

Moan, T. (1988). "The inherent safety of structures designed according to the NPD regulations." *Report No. F88043,* Div. Structural Engineering, The Foundation for Scientific and Industrial Research (SINTEF), Trondheim, Norway.

Moan, T. (1995). "Safety level across different types of structural forms and material—Implicit in codes for offshore structures." *SINTEF Report STF70 A95210,* Prepared for ISO/TC250/SC7, SINTEF, Trondheim, Norway.

Moan, T. (2000a). "Recent research and development relating to platform requalification." *J. OMAE,* 122, 20–32.

Moan, T. (2000b). "Accidental actions. Background to NORSOK N-003." Department of Marine Technology, Norwegian University of Science and Technology, Trondheim.

Moan, T. (2001). "Chapter 5: Wave loading." *Dynamic loading and design of structures,* A. J. Kappos ed., Spon Press, London.

Moan, T., and Amdahl, J. (1989). "Catastrophic failure modes of marine structures." *Struct. Failure*, T. Wierzbicki and N. Jones, eds., Wiley, New York.

Moan, T., and Amdahl, J. (2001). "Risk analysis of FPSOs, with emphasis on collision risk." *Report RD 2001–12*, American Bureau of Shipping, Houston, Tex.

Moan, T., Emami Azadi, M., and Hellan, Ø. (1997). "Nonlinear dynamic vs. static analysis of jacket systems for ultimate limit state check." *Proc., Int. Conf. on Advances in Marine Structures*, Defence Evaluation and Research Agency (DERA), Dunfirmline, UK.

Moan, T., and Holand, I. (1981). "Risk assessment of offshore structures experiences and principles." *Proc., 3rd International Conference on Structural Safety and Reliability (ICOSSAR)*, Elsevier, Amsterdam, 803–820.

Moan, T., Hovde, G. O., and Blanker, A. M. (1993a). "Reliability-based fatigue design criteria for offshore structures considering the effect of inspection and repair." *Proc., 25th Offshore Tech Conf.*, 2, OTC 7189, OTC Offshore Technology Conference, Richardson, Tex., 591–599.

Moan, T., Karsan, D., and Wilson, T. (1993b). "Analytical risk assessment and risk control of floating platforms subjected to ship collisions and dropped objects." *Proc., 25th Offshore Tech. Conf.*, 1, OTC 7123, OTC Offshore Technology Conference, Richardson, Tex., 407–418.

Moses, F. (1985). "Implementation of a reliability-based API -RP2 A format." *Report API-PRAC 83–22*, American Petroleum Institute (API), Dallas, Tex.

Moses, F. (1987). "Load and resistance factor design-recalibration LFRD." *Report API PRAC 87–22*, API, Dallas, Tex.

Nadim, F., Dahlberg, R. (1996). "Numerical modelling of cyclic pile capacity in clay." *Proc., Offshore Tech. Conf.*, 1, paper 7994, OTC Offshore Technology Conference, Richardson, Tex., 347–356.

Nichols, N. W., Birkenshaw, M., Bolt, H. M. (1997). "Systems strength measures of offshore structures." *Proc., 8th BOSS Conf.*, 3, Pergamon, 343–357.

Norwegian Technology Standards. (1998). "Steel structures." *NORSOK N-004*, Oslo.

Norwegian Technology Standards. (1998). "Structural design." *NORSOK N-001*, Oslo.

Norwegian Technology Standards. (1999). "Actions and action effects." *NORSOK N-003*, Oslo.

Norwegian Petroleum Directorate (NPD). (1977). "Regulation for the design of fixed structures on the Norwegian continental shelf." Stavanger.

NPD. (1981). "'Guidelines for safety evaluation of platform conceptual design." Stavanger.

NPD. (1991). "Regulations relating to implementation and use of risk analysis in the petroleum activities." Stavanger.

Odland, J. (1982). "Response and strength analysis of jack-up platforms." *Norwegian Maritime Research*, 10(4).

PD6493. (1991). *Guidance on methods for assessing the acceptability of flaws in fusion welded structures*, 2nd Ed., British Standard Institution, London.

Sarpkaya, T., and Isaacson, M. (1981). *Mechanics of wave forces on offshore structures*, Van Nostrand Reinholdt, New York.

Skallerud, B., and Amdahl, J. (2002). *Nonlinear analysis of offshore structures*, Research Studies Press, Baldock, Hertfordshire, UK.

SNAME. (1994). "Site specific assessment of mobile jack-up units." *Technical & Research Bulletin 5–5A of the Soc. of Naval Architects and Marine Engrs.*, New York.

Søreide, T. H., and Amdahl, J. (1994). "USFOS—A Computer Program for Ultimate Strength Analysis of Framed Offshore Structures." *Theory manual, Report STF71 A86049*, SINTEF Structural Engineering, Trondheim.

Stell Construction Institute (SCI). (1993). "Interim guidance notes for the design and protection of topside structures against explosion and fire." *Document SCI-P-112/503,* London.

Stewart, G., Moan, T., Amdahl, J., and Eide, O. (1993). "Nonlinear reassessment of jacket structures under extreme storm cyclic loading, Part I. Philosophy and acceptance criteria." *Proc., 12th OMAE,* ASME, New York, 492–502.

Taby, J., and Moan, T. (1985). "Collapse behaviour and residual strength of damaged tubular members." *Proc., 4th BOSS Conf.,* Delft University of Technology, Delft.

Ultiguide. (1999). *Best practice guidelines for use of nonlinear analysis methods in documentation of ultimate limit states for jacket type offshore structures.* DNV-SINTEF-BOMEL, Oslo, Trondheim, London.

Vårdal, O. T., and Moan, T. (1997). "Predicted versus observed fatigue crack growth: Validation of probabilistic fracture mechanics analysis of fatigue in North Sea jackets." *Proc., 16th OMAE, Paper No. 1334,* Yokohama, Japan, ASME, New York.

Vårdal, O. T. , Moan, T., and Bjørheim, L. G. (2000). "Applications of probabilistic fracture mechanics analysis for reassessment of fatigue life of a floating production unit-philosophy and target levels," Paper No. 00–2078, *Proc. 19th OMAE Conference,* New Orleans, La.

Veritec/SINTEF. (1988). "Handbook of accidental loads." Det Norske Veritas, Oslo.

Vinnem, J. E. (1999). *Offshore risk assessment,* Kluwer Academic Publishers, Doordrecht.

Vugts, J. H., Dob, S. L., Harland, L. A. (1997). "Strength design of compliant towers including dynamic effects using an equivalent quasi-static design wave procedure." *Proc. BOSS '97,* Delft University of Technology, Delft.

Wheeler, J. D. (1970). "Method for calculating forces produced by irregular waves." *J. Petroleum Engrg.,* 359–367.

WOAD. (1996). "Worldwide offshore accident databank." Det Norske Veritas, Oslo.

Wolf, J. P. (1994). *Foundation vibration analysis using simple physical models,* PTR Prentice Hall, Englewood Cliffs, N.J.

8

Automotive Structures

Tim Keer and Richard Sturt

8.1 Automobiles as Structures

An automobile is a mass-produced structure. The production volumes of the typical vehicle (as many as several hundred thousand units per year) are significantly greater than the production volumes of most of the other types of structure discussed in this book. The economic incentives to optimize the design to reduce cost and to improve performance (e.g., fuel economy) are strong. Optimization has typically been performed with extensive testing of full-scale physical prototypes during the design and development process.

During recent years, analysis has become a cheaper, quicker, and more powerful alternative to prototype testing. The objective of most analysis should not just be to calculate a vehicle's performance but to develop an understanding of the vehicle's structural behavior so that the performance can be improved. The role of analysis in the vehicle development process is described in this chapter, as are some of the techniques used to ensure that the analysis is accurate and can be used to improve the design.

8.2 Structural Performance of the Automobile Body

An automobile's structural performance is determined by the performance of its body, usually referred to as the "body-in-white." The body-in-white is a steel (or aluminum or composite) shell structure composed of a series of stressed panels fixed together. It supports all the other vehicle components, such as closures (doors, hood, deck lid), trim (seats, instrument panel), power train (engine, transmission), and chassis (suspension, wheels). During the first half of the twentieth century, most vehicles used *body-on-frame* construction, where the vehicle's stiffness and strength

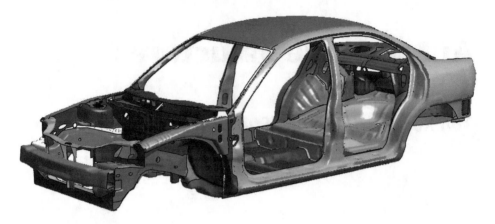

Figure 8-1. An automotive body-in-white.

came from a ladder frame to which the (less stiff) body was attached. This approach is still used for trucks and some sports utility vehicles (SUVs). But most modern passenger cars use *unibody* construction, where the whole body acts together as a structure, and no separate frame is used. A typical body-in-white is shown in Figure 8-1.

The main criteria by which the body-in-white's structural performance is judged are static stiffness, dynamic stiffness, crashworthiness, and mass. These four criteria are codependent. Static stiffness is a measure of how the body will deform under quasi-static loads. A statically stiff vehicle is desirable as it will be more durable and less prone to squeaks. Dynamic stiffness refers to the vehicle's vibration characteristics. It affects the occupants' perceptions of the vehicle's noise, vibration, and harshness (NVH). Crashworthiness is the term used to measure a vehicle's performance under impact conditions. A good design will be neither too strong (in which case the occupants would undergo very high accelerations) nor too weak (in which case the vehicle would absorb too little energy and not protect the occupant compartment from deformation). Low mass is desirable for performance (acceleration and handling), fuel economy, and material costs.

8.3 History

In the early days of the twentieth century, every automobile was a prototype. There was no "true" design process. Vehicles were generally improved variants of previous vehicles, with modifications designed to improve performance, to reduce cost or to fix problems that had been observed during the previous vehicle's operation. The hand-built methods of early automotive production meant that each vehicle was unique. Testing a vehicle would not necessarily have given feedback that could be used to improve the design of other vehicles of the same nominal design.

But as mass-production techniques were developed during the first quarter of the twentieth century, it became possible to build thousands of nearly identical vehicles. This enabled development testing to be performed to ensure that design faults could be identified and fixed before the vehicles were sold to the consumer. This process led to long vehicle development programs, as there was a significant delay between a change to the design and the feedback from testing of that design.

During the last 30 years of the twentieth century, Computer Aided Engineering (CAE) allowed a vehicle's performance to be predicted before physical prototypes were built and was used to assess each iteration of the design as it was developed. The results from the CAE analyses were then used as a basis for modifications to the design. Physical testing was only performed at the end of the development process and was used to sign off or approve the design. Although CAE can be a faster and more responsive tool than building and testing prototypes, this approach still required long vehicle development programs as the analysis tended to wait for complete design information before it could find faults in the design.

As CAE techniques have become more advanced, their contribution to the vehicle development process has increased. In the current approach, design and analysis operate in parallel. The analysis is continuously updated to include new design information, and feedback from the analysis is continuously used to improve the design. This requires significantly more analysis than has typically been used in the past but leads to more useful design and to a reduction in development times.

Analysis tools are now so refined that the first zero-prototype vehicle development programs are being undertaken, and within the next three years this practice will become commonplace. Computer Aided Engineering is seen as a risk reduction tool, and without the chance of catching faults on a physical prototype, there is greatly increased emphasis on the accuracy and reliability of analysis processes. We predict that there will also be an increasing trend for designing structures that are inherently analyzable, so that the risks of analysis-only development are minimized. This is the reverse of current practices where analysis techniques in the car industry have had to become ever more sophisticated to predict complex behaviors. In this way, the process of designing a car will become more like that of designing a building.

There are many areas of the vehicle design where analysis is replacing test. Three examples are the modeling of exterior aerodynamics, the combustion process, and noise perceived at the driver's ears. But structural analysis (in particular, the prediction of vehicle stiffness and vehicle impact performance) is one area where the ability to replace physical testing with analysis has brought great benefits to the automotive design process.

8.4 The Role of Analysis in the Automotive Design Process

The objective of the engineering design team is to minimize the mass of the body-in-white, to maximize its stiffness, and to optimize its strength.

In Preconcept Design (the earliest stages of vehicle development), the vehicle "package" begins to be defined. The vehicle's "hard points" are set (e.g., length, wheelbase; location and configuration of the major power train components such as engine and transmission; location of occupants). Competitive benchmarking is performed, and targets are set for vehicle structural performance (body static stiffness, full vehicle dynamic stiffness, impact performance). The design team starts to lay out the vehicle "architecture" (e.g., location of main structural members). Engineering analysis has typically been limited to hand calculations, previous experience, and perhaps very simple analytical models. Recent advances in the speed of generation of full shell models of bodies-in-white without full 3D design data means that analysis has become a more important contributor to this stage of design.

During the Concept Design stage, the engineering design team creates a structure to fit inside the exterior and interior surface shape and appearance. Structural design is typically the slave of the exterior and interior design. The traditional approach here was "design (draw) it, then analyze it." Analysis now tends to lead the design. Analysis cannot be used to solve problems automatically, but it can be used to inspire and guide.

Once there is confidence that the targets (for many attributes, not just structural) can be met, then the Production Design is finalized. The automotive structural engineer has to compromise to accept the requirements of manufacturing, production, cost, and so on.

The two main areas of automotive structural analysis are stiffness and crashworthiness.

8.5 Stiffness Analysis

The objective of stiffness analysis is to achieve specific targets for structural performance while minimizing mass. A stiff vehicle will give better ride and handling, will have better NVH characteristics, and will be more durable.

Stiffness models generally increase in complexity during the vehicle development process. Initial models typically analyze the body-in-white only. As the design develops, the models become more detailed and include a representation of all components in the vehicle that affect its dynamic response.

8.5.1 Software for Stiffness Analysis

Stiffness analysis is performed using an implicit linear elastic finite element program such as MSC.Nastran (*MSC.Nastran 2005 Release Guide* 2005). The theory behind the software is not presented here, but it is important to understand a few of the assumptions made by the software.

Linear elasticity means that the response (displacement, velocity, or acceleration) is proportional to the applied load (force, moment, etc.). If a force of 1 N produces a displacement of 1 mm, then a force of 10 N will produces a displacement of 10 mm. No yielding or hysteresis can be accounted for. (NASTRAN can be used for nonlinear analysis, but automotive stiffness analysis assumes linear elasticity.)

Small displacements are assumed. One effect of this of relevance to automotive analysis is that flat panels are understiff when loaded normal to their surface. In a physical panel with a load perpendicular to its surface, membrane action stiffens the panel after a small amount of displacement. But in an analysis where small displacements are assumed, no membrane action occurs.

8.5.2 Creation of a Stiffness Model

8.5.2.1 Geometry and mesh creation. A finite element mesh is created for all components in the body-in-white using a mesh preprocessor such as HyperMesh (owned by Altair Engineering) or ANSA (owned by BETA CAE Systems). The geometry is created from computer-aided design (CAD) surface data from a CAD design software program such as CATIA (owned by Dassault Systèmes), or Unigraphics (owned by UGS). If the mesh preprocessor is unable to import surfaces directly from the design software, then translation is performed via the initial graphics exchange specification (IGES) format. Fillets, holes, and other features of radius less than 5 mm are removed from the geometry file. Fillets and features are replaced with extended faces that meet with a discontinuity in the surface.

Body-in-white panels are modeled as shell (2D) elements. The mesh density and quality affect the accuracy of the model. The mesh should have as many quadrilateral elements (as opposed to triangles) as possible. A nominal element edge length of between 5 mm and 40 mm (depending on the level of accuracy required) is used. Element density increases close to joints, flanges, and other stress-raising features. There must be at least two elements across a flange. There must be at least four elements across the side of a box section, and at least twelve elements around an irregular section (e.g., the A-pillar). A coarser mesh will be suitable for prediction of static body stiffness and global vehicle modes. A finer mesh is used if stresses need to be predicted accurately (for durability analysis) or to predict higher frequencies.

The quality criteria in Table 8-1 can be violated, but this will have an effect on the accuracy of the model, especially when close to joints, flanges, or other stress-raising features.

The mesh is usually created on the midsurface of each component, although this is not strictly necessary when building a mesh for stiffness analysis. If the same mesh will also be used for impact analysis, it is important that there is a small gap between neighboring panels to ensure that there are no initial penetrations in contact surfaces. This will be discussed in greater detail later.

Table 8-1. Typical Element Quality Criteria

Warpage	$<10°$
Skew	$<50°$
Aspect ratio	<5
Minimum edge length	>5 mm
Jacobian	>0.70
Quad min. angle	$>45°$
Quad max. angle	$<135°$
Triangle min. angle	$>30°$
Triangle max. angle	$<120°$

Until a few years ago, it was necessary for neighboring panels to have similar meshes along weld lines to allow them to be welded together using node-to-node rigid body element (RBE2) weld elements (to be discussed later). This required neighboring panels to be meshed simultaneously or meshes to be subsequently adjusted to suit the welds. New techniques use mesh-independent welding, which means that panels do not have to have matching meshes, leading to large savings of time during the mesh development. The representation of welds will be discussed later.

8.5.2.2 Material properties. For a linear elastic stiffness analysis, the required material properties are Young's Modulus and Poisson's ratio and, for modal analysis, mass density. Nominal panel thicknesses are used (the thickness of the steel sheet before the panels are stamped).

8.5.2.3 Joining. Typical joining techniques for steel panels are spot welds and, less frequently, metal inert gas (MIG) welds or laser welds. Aluminum body panels may be joined by adhesive bonding or self-piercing rivets.

A NASTRAN RBE2 element was the traditional feature used to model a spot weld. An RBE2 element connects a single node on each of two or three panels. It acts as a rigid link, transferring forces and moments between the panels. This is also known as point contact welding.

Current analysis techniques use area contact welding in which the spot weld nugget is modeled explicitly as a solid element. The "corners" of the spot welds are connected to their parent panels by RBE3 linear constraint equations. Area contact welding is generally favored, as it tends to be more accurate than point contact welding. The area contact method allows independent meshes to be used on adjacent panels.

8.5.2.4 Miscellaneous components. The windshield and any other fixed glass (e.g., rear quarter windows) are usually included in a body-in-white analysis. The adhesive bonding between glass and body-in-white panels is modeled using solid elements. Other adhesives (e.g., between roof bows and roof panels) are included in the model in a similar way.

Instrument panel (IP) beams and any bolted-on subframes and struts are also included in a body-in-white analysis if they are designed to contribute to the body's stiffness.

8.5.3 Body-in-White Static Analysis

The two static load cases most frequently applied to a body-in-white stiffness model are bending and torsion.

For analysis of the body's bending stiffness, vertical forces are applied to the midpoint of the vehicle rockers or sills. The forces are applied to the rocker webs (vertical surfaces) using RBE3 elements to avoid local panel deformation. A NASTRAN RBE3 element spreads load without providing local stiffening. Single point constraints (SPC) restraints are used to prevent global motion of the body and are applied to the suspension attachment points. The analysis uses the minimum set of restraints required to prevent rigid body motion of the body. This provides a conservative prediction of stiffness.

The bending stiffness, k_b, is calculated from the applied force (F) and the average vertical displacement (dz) of the two load points:

$$k_b = \frac{F}{dz}. \qquad (8\text{-}1)$$

For analysis of the body's torsional stiffness, a couple is applied to the front shock-towers. SPC restraints are applied to the suspension attachment points to prevent rigid body motion of the body.

The torsional stiffness, k_t, is the ratio of the applied Moment (M) and the rotation of the vehicle at the load points (θ):

$$k_t = \frac{M}{\theta}. \qquad (8\text{-}2)$$

The bending and torsional stiffnesses are an objective measure of the stiffness of the structure. These are the primary outputs from a stiffness analysis, but the engineer uses analysis to develop an understanding of the structural behavior of the body. This is achieved through inspection of the mode of deformation, displacement analysis, and strain energy density plots. The engineer looks to see how the body's behavior differs from the ideal. For bending, the ideal structure is a beam. When simply supported at the suspension mounting points and loaded by point loads at the centers of the rockers, a simple beam will deform in the shape of a cubic. For torsion, the ideal structure is a tube. When held at the rear suspension mounting points and loaded by a couple at the front suspension, a tube will twist uniformly along its length.

By animating the deformed shape, it is possible to identify areas of the structure that are undergoing excessive local deformation. Figure 8-2 shows the mode of deformation of a vehicle body under bending. The magnitude of the deformation is unimportant, but the shape allows the structural performance to be assessed.

Displacement analysis is used to determine the location of weak areas. For a bending analysis, the vertical displacements of various points on the rockers and rails are plotted against their longitudinal position. The optimum shape is a

Figure 8-2. A stiffness analysis of a body-in-white undergoing vertical bending.

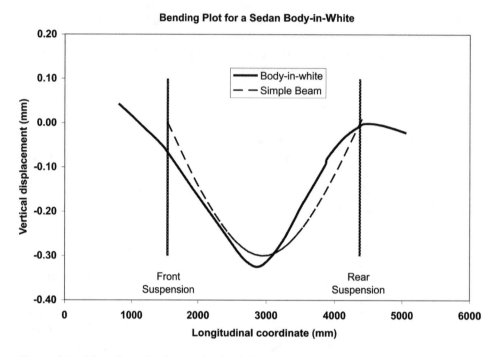

Figure 8-3. A bending plot for a sedan body-in-white.

cubic between the suspension mounting points (see Fig. 8-3). The displacement plot from a bending stiffness simulation is shown on the same figure. Areas of excessive displacement or curvature show weak areas of structure that need to be stiffened.

For a torsional stiffness analysis, the "twists" of the same points are plotted against their longitudinal position. The optimum shape is linear between the suspension mounting points (see Fig. 8-4). The twist plot from a torsional stiffness simulation is shown on the same figure. Areas of excessive slope or curvature show weak areas of structure that need to be stiffened.

Strain energy density is a measure of how hard an element is "working." Strain energy density is the internal energy of an element divided by its volume and is also the integral of stress with respect to strain. In an ideal structure, strain energy density will be evenly distributed around the structure. A contour plot of strain energy density is used to highlight those areas of the structure that are doing most work. These areas should be the focus of attempts to stiffen the structure.

Optimization of the body-in-white with respect to static stiffness is often performed. This will be discussed in a later section of this chapter.

8.5.4 Body-in-White Dynamic Analysis

A body-in-white dynamic stiffness analysis (also known as a modal analysis or a normal modes analysis) is used to determine the mode shapes and natural frequencies

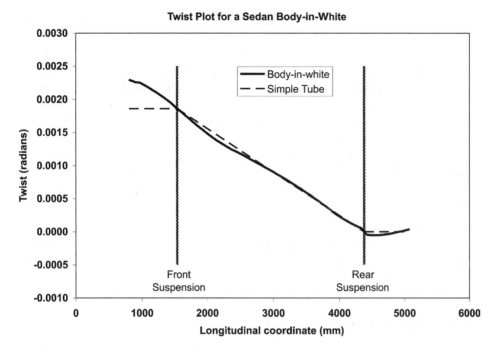

Figure 8-4. A twist plot for a sedan body-in-white.

of vibration of the body-in-white. Although the natural frequencies of the full vehicle are of more interest to the engineer than the frequencies of the body, targets are often also set for the body-in-white alone. This reflects the fact that the body design is always available for analysis or testing earlier than all the additional information (seats, trim, power train, etc.) required for a full vehicle analysis or test.

The dynamic stiffness analysis uses the same finite element model as the static stiffness analysis, but a different solution sequence is used. No single point constraints are applied, and so the body is free to vibrate unrestrained.

The output from the modal analysis is a list of frequencies and a full set of nodal displacements that define the mode shape. The modes are viewed in a graphical postprocessor that animates the mode shape on the screen.

An unrestrained body-in-white should have six rigid body modes at a very low frequency (typically less than 0.1 Hz). These *zero-energy* modes correspond to modes of deformation that have no associated stiffness; zero stiffness implies zero frequency. The six modes represent the three translational degrees of freedom (fore/aft, left/right, up/down) and the three rotational degrees of freedom (roll, pitch, yaw). In practice, the six modes observed in a modal analysis tend to be combinations of these six pure modes. If an analysis gives more than six rigid body modes, then it is likely that a part of the body has been disconnected or the body contains a mechanism. If there are fewer than six modes or if the frequencies of any of the first six modes are above 0.1 Hz, then it is likely that one or more grid points have been restrained.

For a sedan body-in-white, typical targets for the first mode (other than rigid body modes) are 45 to 60 Hz. For a convertible body-in-white, targets would be in the 25 to 30 Hz range. A well-designed body-in-white will have no local modes before the first global modes. A local mode is typically a panel mode (e.g., vertical motion of the floor pan or lateral motion of the quarter panel). The first two modes of the body-in-white are typically global torsion and global bending, where the mode shapes are similar to the modes of deformation under static torsion and static bending respectively. Other common modes are front end lateral (typically caused by poor moment connections at the front of the vehicle), front end nod (caused by a reduction in stiffness of the main vehicle rails as their height reduces as they sweep down to pass under the floor), and rear-end matchboxing, a lozenging mode typically found in minivans and SUVs.

The evaluation techniques used for static stiffness analysis, discussed previously, are also used for dynamic stiffness. Optimization of the body-in-white's dynamic stiffness is common, often in parallel with optimization of the static stiffness. This will be discussed in a later section.

8.5.5 Trimmed Body Dynamic Analysis

As mentioned previously, the natural frequency of vibration of the vehicle has a major effect on NVH, ride quality, and handling. Although a modal analysis of the body-in-white is a useful tool, the full vehicle must be modeled if the modal performance of the vehicle "as driven" is to be predicted.

An intermediate step is to perform a *trimmed body* analysis. A trimmed body is the body-in-white plus closures (doors, hood, deck lid) and all components that are fixed to the body (e.g., steering column and steering wheel, instrument panel, seats, seat belt hardware, spare wheel, fuel tank, radiator, and battery). Seats and closures are generally modeled explicitly using shell elements, but the additional components are generally represented as lumped masses attached rigidly to the body at the appropriate locations. The trimmed body model does not include the engine, transmission, exhaust, suspension, or wheels, which are "soft-mounted" to the body. Their response and the response of the body are decoupled.

The static stiffness of the trimmed body is generally not analyzed, as it will be similar to the static stiffness of the body-in-white. The natural frequencies of the trimmed body will be lower than the frequencies of the body-in-white, as the mass is greater.

A trimmed body analysis may often have a seventh rigid-body mode—rotation of the steering wheel about the axis of the steering column. Any local modes (e.g., battery bounce, bumper pitch) indicate that attachments need to be stiffened. The first global bending and torsion modes are of primary interest. As the only masses missing from a trimmed body analysis are masses that are not coupled to the body, the global bending and torsion modes will be unchanged when the full vehicle is analyzed.

8.5.6 Full Vehicle Dynamic Analysis

A full vehicle model includes representation of the whole vehicle. Exhaust, suspension, wheels, and power train are all added to the trimmed body model. Dynamic analysis only is performed.

This model is used to assess the modal alignment of the vehicle. It is important that the various modes of vibration of a vehicle do not overlap. The lowest frequency modes are suspension modes such as wheel hop or wheel tramp. Engine rigid body modes (the engine vibrating on its mounts) have the next highest frequencies. The full vehicle modal model can predict all these modes. The modes with the next highest frequencies are the modes of the engine at idle (typically 600 to 750 rpm, corresponding to 10 to 25 Hz depending on the engine cylinder configuration). The global torsion and bending modes should therefore exceed 25 Hz. (An additional consideration is the first order wheel vibration. A wheel with tire diameter of 600 mm rotates at 24 Hz at 100 mph and at proportionally lower frequencies at lower speeds.) Additional modes (e.g., steering column vertical motion) are also predicted.

The output from the full vehicle dynamic analysis allows all the modal frequencies to be plotted on a modal alignment chart. An attempt should be made to separate any modes that are within 2 Hz of each other.

8.5.7 Point Mobility or Forced Response Analysis

While dynamic stiffness analysis is a useful tool for predicting the natural modes of vibration of the body or vehicle, it cannot predict the amplitude of the vibration. In order to predict this, a forced response analysis is required. A forced response analysis predicts the response (displacement, velocity or acceleration) of the structure over a range of frequencies when it is excited at a particular location by a sinusoidal input.

First, a modal analysis is performed to determine the mode shapes and frequencies. Next, the mode contributions for each input point at each frequency are calculated, and finally the modal displacements are superposed to calculate the response over the frequency range required.

Damping is a critical component of this calculation method. Without damping, the response would grow to infinity for every natural frequency of the structure. Damping may be applied in one of three ways. Firstly, modal damping may be used. Modal damping is the rate of decay of a particular mode. It is not a physical parameter, and therefore needs to be measured in test. The second type of damping is structural damping, where a damping factor is specified for each material in the vehicle model. The third approach to adding damping to the model (that may be used in conjunction with the second) is to model discrete damper elements. These apply forces between grid points in proportion to their relative velocity. A discrete damper would be used to model a component such as an engine mount.

This solution technique (SOL111 in NASTRAN) is computationally expensive. If the response needs to be predicted up to a certain frequency, then all modes up to a frequency two or three times higher need to be calculated. This requires that the finite element model have a sufficiently fine mesh to predict those modes accurately. Typical calculations today cover the range 0 to 300 or 400 Hz and involve several thousand modes.

8.5.8 Optimization

As discussed previously, the vehicle design process has minimum mass as a strong objective. All the types of stiffness models described previously may be used with an optimizer to minimize mass, maximize performance, or meet a combination of mass and performance.

There are two common forms of optimization—shape optimization and gauge optimization. Gauge optimization (choosing the particular set of panel gauges that provide maximum performance and minimum total mass) is the better developed and more widely used technique. This is a very powerful tool for the analysis engineer.

An optimization analysis requires design variables, design constraints, and a design objective.

Typical design variables are panel gauges. For each panel in the body-in-white, the engineer considers the minimum and maximum allowable gauges. Manufacturing (e.g., the ability to stamp the panel), production (e.g., the ability to weld the panel), and crashworthiness all need to be considered. For a front longitudinal rail, the gauge might already have been optimized for impact performance and so the engineer might not allow the gauge to vary. For other panels, typical limits might be a minimum panel gauge of 0.7 mm and a maximum panel gauge of 2.5 mm.

A typical design objective would be:

- Minimize the body-in-white mass

  A typical set of design constraints would be:

- The first body-in-white frequency must be above 50 Hz.
- The second body-in-white frequency must be above 55Hz.
- The two frequencies must be separated by at least 4 Hz.
- The body-in-white torsional stiffness must be above 15,000 Nm/deg.
- The sum of the thicknesses of parts A, B, and C must not exceed 6.0 mm (to permit spot welding).

The user specifies a series of design parameters to control the optimization (e.g., the number of optimization loops, the amount that a variable may vary between loops, convergence criteria). The optimizer then performs a series of linear static and dynamic analyses and modifies the design variables in order to satisfy the global objective (e.g., minimum mass) while meeting all the design constraints.

If a solution can be achieved without violating any design constraints, output from the analysis is the optimum value for each design variable. A confirmation analysis is then performed, with the panel gauges set to common values. (Sheet steel is generally available in thicknesses of 0.7 mm, 0.8 mm. 0.9 mm, etc. It is not feasible to specify that an automotive panel should be stamped from 0.923 mm steel.)

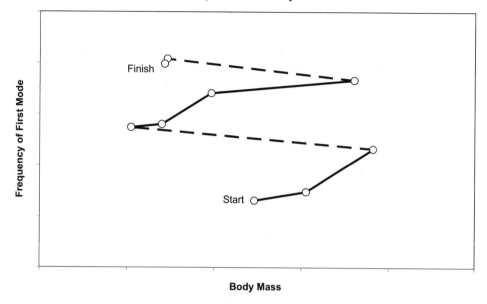

Figure 8-5. A typical design optimization history.

Optimization is a powerful tool, but it should only be applied when engineering judgment has already been used to produce a sound structure with no obvious weak areas. A useful measure during the development of static stiffness performance is *efficiency*, defined as the ratio of static stiffness to mass. The torsional efficiency is therefore the ratio of body-in-white torsional stiffness and body-in-white mass. If the efficiency of the body can be elevated above the target efficiency (even if the mass exceeds the target), it is very likely that optimization can be used to meet the targets for both stiffness and mass.

Figure 8-5 shows how a vehicle's lowest natural frequency might vary with mass during a typical design program. Design changes based on engineering judgment tend to give improvements in performance, but at the expense of an increase in mass (solid lines on the chart). Gauge optimization (dashed lines on the same chart) can then be used to reduce the mass.

8.6 Impact Analysis

During the vehicle development process, a major objective is to ensure that the vehicle will protect its occupants during impact. Analysis is used to support the achievement of this objective.

8.6.1 Impact Scenarios

A variety of impact conditions is considered. These include:

- legislative impacts, mandated by the U.S. and other governments,

- standard, nonlegislative impact tests, often performed by consumer groups, and
- "due care" conditions, performed to give broad confidence in safety levels.

The major legislative tests in the United States are Federal Motor Vehicle Safety Standards (FMVSS). These include FMVSS208 (front impact—30 mph impact into a rigid barrier) and FMVSS214 (side impact—impact of a moving deformable barrier into the side of a vehicle). Performance targets for the tests are defined in terms of the injury criteria of the crash test dummies seated in the vehicle.

Nonlegislative load cases include the New Car Assessment Program (NCAP) 35 mph impact into a rigid barrier and the Insurance Institute for Highway Safety (IIHS) 40 mph offset impact into a deformable barrier.

In due care testing, the manufacturer repeats some or all of the standard tests with different occupants and with seats and steering wheel in a variety of positions.

8.6.2 Software for Impact Analysis

Vehicle impact analysis is performed using a nonlinear transient finite element program such as LS-DYNA (*LS-DYNA Keyword User's Manual Version 970* 2003). Nonlinearity is required to model plasticity and large displacements. The explicit time-integration technique is used.

In the explicit method, each node's acceleration is calculated from the total force applied to the node (internal forces from element stresses; external forces from applied forces and interaction with other bodies) divided by the nodal mass. Accelerations are then integrated to give velocities, and velocities are integrated to give displacements. Element strain rates are calculated from velocities, and element stresses are calculated from element strains. Stresses are then used to recalculate forces and the solution is advanced through time.

This explicit time-integration technique is only stable if the time increment (time step) is less than the Courant time step for each element. (The Courant time step of an element is, very crudely, its shortest length divided by the sonic speed. For most automotive metals, the sonic speed is approximately 5,000 ms[-1], and so the Courant time step for a 5 mm square element is approximately 1 µs.) The explicit method does not require matrix inversion, and the computationally expensive part of the solution algorithm is the calculation of element stresses and strains. Impact analysis software therefore tends to use simple element formulations and a relatively large number of inexpensive time steps.

Contact surfaces are one of LS-DYNA's most important features. They are used to prevent components from passing through each other. A contact surface is comprised of a series of segments. A segment is a shell element or the face of a solid element. At each time step, the contact algorithm looks to see if nodes at the corners of the segments have penetrated other segments. If penetration has occurred, then a force is applied to the penetrating node to push it back out of the surface. An equal and opposite reaction force is spread over the nodes of the penetrated segment. As very small time steps are used, penetrations tend to be small, and nodes on contacting segments tend to find an equilibrium position with a small penetration. Penetration between each node on the surface and each segment on the surface is checked at every time step. While the algorithms are essentially simple, implementing them in a way that allows rapid solution times while obtaining reliable results is very challenging.

Another important attribute of the explicit nonlinear finite element program is its ability to predict the buckling or other unstable behavior of structures. Just as small imperfections in a straight column will cause it to buckle one way or another under compression, so rounding errors in the storage of data will cause the analytical representation of the same column to buckle in different ways. This is important for impact analysis where the compressive buckling of components (e.g., the longitudinal rails in a front impact) is common and frequently leads the analyst to consider the spread of results from a family of similar models rather than a single result. Where computer resources allow, a family of 50 or so models is generated with small random variations in key input parameters. This method is known as stochastic simulation.

8.6.3 Creation of an Impact Model

There are two levels of analysis—"vehicle only" and "vehicle with occupants." Use of a model with the vehicle only is a more mature technique and is easier to perform. But, as mentioned previously, the performance targets for the FMVSS front and side impact tests are based on occupant injury criteria. If a true prediction of the performance of a vehicle in the test is to be made by analysis, then the analysis should include a representation of the occupants and (in order to ensure that loads are transferred to the occupants correctly) the seats, airbags, and seat belts. The use of fully integrated models including vehicle structure, occupants, and restraint systems is becoming more common.

8.6.3.1 Geometry and mesh creation. As was the case for the stiffness analysis described previously, the accuracy of an impact model is a function of the size (number of elements) of the model. In the discussion of stiffness models, we mentioned that a more refined mesh (i.e., more elements) allows the geometry of the vehicle's panels to be modeled more accurately. This will lead to increased accuracy. In an impact analysis, the mesh density must also be sufficiently fine to allow the mode of deformation to be represented by the elements available. Mesh refinement is, however, limited by the computer resources available to run the models. At present, smallest element sizes are commonly limited to 5 mm in critical areas, but the trend is for ever-decreasing element sizes. In future, elements as small as 1 mm may be used.

Because of the requirements for fine meshes in areas of the body that are expected to undergo gross plastic deformation, it was a common occurrence ten years ago to use different meshes for modeling different impact load cases. A model for front impact would have a fine mesh at the front of the vehicle and a coarse mesh at the rear (or even a single rigid element representing the rear half of the vehicle). A model for side impact of the same vehicle would have a fine mesh on the impact side of the vehicle, and a coarse mesh on the nonstruck side. The savings associated with the reduced analysis times of the separate models outweighed the additional costs associated with building these models. But today it is generally more efficient to use a single model for all the different impact load cases.

8.6.3.2 Material properties. In a linear elastic model, a linear relationship between stress and strain is assumed. But in a vehicle impact, yielding occurs and the relationship between stress and strain is more complex. The following data is generally required for each different material grade used in the vehicle design:

- Young's Modulus and Poisson's ratio for preyielding and unloading behavior,
- the yield stress,
- a stress-strain curve showing the postyield relationship between stress and strain, and
- the effect of strain rates on yield stress and postyield behavior.

As for stiffness analysis, nominal panel gauges are generally used (e.g., the gauges of the blank panels before they are stamped). But after stamping, a panel has nonuniform thickness, exhibits nonuniform work-hardening and has nonuniform residual stresses. Recent work has investigated the effect of these on the impact performance of a vehicle (Keer et al. 2001). Initial conclusions tend to be that the effect of residual stamping parameters is generally small (the strengthening effect of work hardening tends to balance the weakening effect of thinning) but sufficiently important to merit further investigation. For hydroformed components, the effect of residual forming properties can be more substantial as, in some areas of a hydroformed section, the section can exhibit work-hardening *and* thickening.

8.6.3.3 Joining. Panels are joined using spot weld elements, which may include one of two failure modes—brittle failure and ductile failure. In the brittle failure model, failure of the spot weld occurs when:

$$\left(\frac{\max(f_n, 0)}{S_n}\right)^n + \left(\frac{|f_s|}{S_s}\right)^m \geq 1 \tag{8-3}$$

where f_n and f_s are the normal and shear forces in the spot weld element and S_n, S_s, n and m are input parameters. Care must be taken to ensure that a spike in the transient oscillation of a spot weld force does not cause the spot weld to fail prematurely. The ductile failure model allows the spot weld element to deform plastically; the failure criterion may still be based on forces, but the ductility of the element ensures robustness against transient oscillations.

8.6.3.4 Modeling of additional components and subsystems. Although the vehicle body-in-white is the most important contributor to vehicle crashworthiness, additional components need to be modeled if the performance of the body-in-white is to be accurately predicted. These include power train (engine and transmission), suspension, wheels and tires, occupants (crash test dummies), seats, interior trim, and restraint systems (airbags and seat belts). The modeling of these components

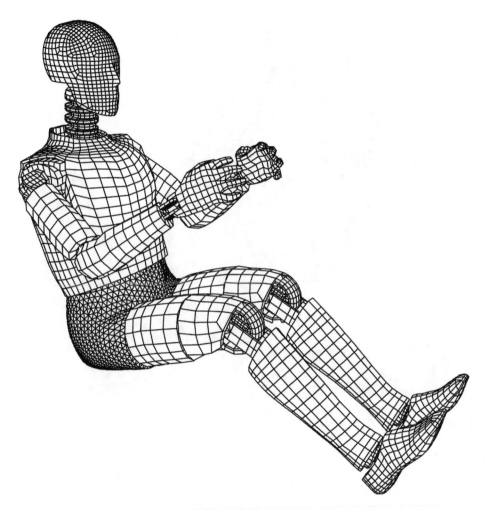

Figure 8-6. The FT-Arup Hybrid III 50th percentile dummy model.

requires as much attention as the modeling of the vehicle body-in-white, but a detailed description is outside the scope of this chapter.

An advanced LS-DYNA finite element model of the Hybrid III 50th percentile crash test dummy is shown in Figure 8-6. This model is part of the family of dummy models developed by First Technology Safety Systems (FTSS), the manufacturer of crash test dummies, and Arup (Moss et al. 1997).

Figure 8-7 shows an image from the deployment of a driver's side airbag, also modeled in LS-DYNA. The initial condition is a finite element model of the airbag in its folded position. The airbag inflator is modeled analytically. As it introduces gas into the airbag, so pressure is developed and the bag inflates. In this instance, the Hybrid III 5th percentile "small female" dummy is located close to the airbag.

8.6.3.5 Component interaction. Contact between components is a major way that load is transferred through the vehicle, especially in the engine compartment.

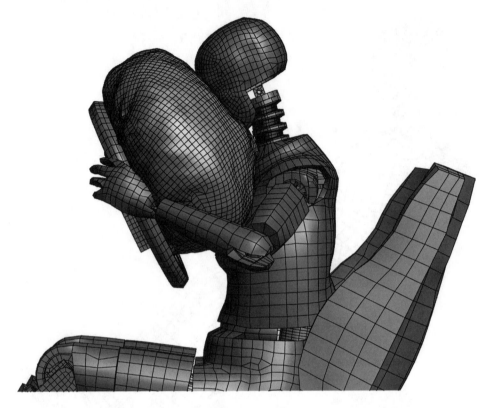

Figure 8-7. Airbag deployment with "Out of Position" 5th percentile dummy.

Loads are applied to occupants by airbags, seats, seat belts, and knee bolsters. All these interactions can be allowed for in an LS-DYNA analysis by the use of contact surfaces. Whereas a few years ago it was necessary to define the individual contact sets, now it is customary to define a single contact surface for the whole model. No node may penetrate any segment. While this is computationally expensive, it is easy to define and does not require all possible contact interactions to be determined in advance.

It is important for the successful use of contact surfaces that the model contains no initial penetrations. Initial penetrations (where a node on one panel is located inside a neighboring panel) can occur when poor geometry is used or when the mesh is created on the design surface instead of the midsurface of the component.

8.6.3.6 Assessing the accuracy of an LS-DYNA analysis. Analyses performed in LS-DYNA or other similar programs need to be assessed for accuracy before their results can be postprocessed. The most important consideration is conservation of energy. The total energy at any time should equal the total energy at the start of the analysis plus the external work performed up to that time.

$$ total\ energy\big|_t \ = \ total\ energy\big|_{t=0} \ + \ external\ work\big|_t. \tag{8-4} $$

Total energy is the sum of kinetic energy, the internal energy of deformable parts and discrete elements such as springs and dampers, contact surface energy (as a node penetrates a segment, it does work against the contact force pushing against it), and other miscellaneous energy terms. External work is the energy added to the system by external forces and pressures. Energy imbalances can be caused by misbehaving contact surfaces or other nonphysical numerical errors.

If the energy balance of the system appears correct, then the user should review the mode of deformation of the vehicle. An experienced user will be able to spot any unusual modes of deformation that might have been caused by modeling mistakes or incorrect input data.

The transmission of force through the structure should also be monitored. During the early stages of a front impact analysis, the barrier force should be approximately equal to the sum of the forces in the main longitudinal rails. As other load paths are developed (e.g., barrier to upper rails or barrier to radiator to engine to firewall), the load distribution will change but should still be checked to ensure that the distribution is realistic and in line with expectations.

One aspect of impact analysis not always considered sufficiently deeply is the lack of repeatability of the behavior of unstable structures. Two physical tests of nominally identical buckling structures would not be expected to give identical results. If an analysis model is run twice on the same computer, it *will* give identical results. (LS-DYNA can be made to give different results if the model is run on a parallel machine with a different number of processors and if the nodal forces are not always summed in the same order.) While it would be frustrating to an analyst to receive two sets of answers to the same problem, it is important for the user to understand the natural variation of results to be expected if small changes are made to the input. Consider, for example, the situation where the gauge of a B-pillar reinforcement is increased from 1.0 mm to 1.2 mm. If the Thoracic Trauma Index (TTI) of the dummy in a side impact analysis decreases by 2 g, is the reduction caused by the strengthening of the B-pillar or is it just a natural variation caused by rounding errors and instabilities? Without this knowledge, the user cannot use analysis to lead the design.

8.6.4 Typical Impact Models

The objective of this chapter is to describe typical modeling techniques rather than to discuss their application as a design tool. The following discussion of typical models and their results is therefore brief.

8.6.4.1 **Front impact.** A typical front impact model is shown in Figure 8-8. This is a simulation of the US NCAP 35 mph impact into a rigid barrier.

For front impact, the success of the crash test is defined in terms of occupant injury criteria and vehicle deformation. The objective is to minimize occupant accelerations and to minimize the amount of structural deformation in the occupant compartment. The first objective (minimizing occupant accelerations) is achieved by minimizing the vehicle's acceleration during the time that the occupant is fully

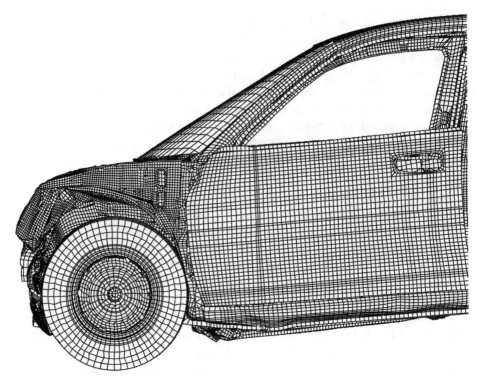

Figure 8-8. Simulation of the 35 mph US NCAP impact test.

engaged in the restraint systems (e.g., airbags, seat belts, and knee bolsters). For a rigid barrier test, the average acceleration from 40 ms to 70 ms is a typical measure. The average vehicle deceleration will be minimized if the vehicle decelerates from its initial velocity over as long a time as possible. This suggests that the vehicle crush should be as great as possible, which is in conflict with the second requirement that there should be minimal structural deformation in the occupant compartment. The optimal performance is achieved when the vehicle crushes as much as possible forward of the occupant compartment, with as constant an acceleration as possible.

8.6.4.2 Side Impact. A typical side impact model is shown in Figure 8-9. This is an analysis of the FMVSS214 side impact test. A barrier with an aluminum honeycomb face, representing the front end of another vehicle, impacts the side of the test vehicle.

For side impact, the objective is to minimize occupant accelerations and (for the European side impact test) occupant deformations. This is achieved by minimizing the intrusion velocity of the side of the vehicle and by making the side structure retain a vertical profile as it intrudes into the passenger space.

8.7 Conclusion

Computational analysis will continue to grow in importance as a design tool. The drive for this will be the desire to avoid the dependence on costly and time-consuming

Figure 8-9. Simulation of the FMVSS214 side impact test.

prototypes. The enabler for this will be the availability of more powerful computers and ever-improving analysis tools and techniques. Some manufacturers are beginning to achieve zero-prototype vehicle development, typically for niche vehicles based on an existing platform, but within a few years it will be standard practice for all vehicles. Analysis will truly lead the vehicle design process in the future.

This will require an increase in the size of impact and stiffness models. The size of impact models has continued to increase. In 1990, a frontal impact model might have included 50,000 elements. In 2005, 1,000,000 elements is more typical. As computers become more powerful, more elements will be used. This will give greater accuracy. More parametric studies will be analyzed to understand the sensitivity of impact results to small changes in initial conditions. The definition of "the performance of a vehicle structure" will no longer mean single values of stiffness, crash deformation, and so on. It will be extended to include the statistical range of such performance indicators. Stochastic analysis (the use of a family of 50–100 models generated with small random variations in important input parameters) will become standard practice for investigating the statistical spread of performance. The robustness of a design (minimizing the variations in performance and the risk of unacceptable performance) will become a key objective in optimization of vehicle structures.

As physical prototyping is eliminated, the risks associated with errors or inaccuracies in the analysis will become much greater. To counter this, as well as continuously improving and validating the analysis methods, designs that are inherently predictable by analysis will be preferred. In the future, the design-analysis relationship that underlies the development of a new vehicle will become similar in principle to that underlying the development of a building today: calculation is the only way to demonstrate the performance of the design, and the design and development process must recognize that fact.

References

Keer, T., Dutton, T., Sturt, R., Richardson, P., and Knight, A. (2001). "The effect of forming on automotive crash results." *Paper 2001–01–3050*, Int. Body Engrg. Conf. and Expo., Detroit, Mich.

LS-DYNA keyword user's manual version 970. (2003). Livermore Software Technology Corp., Livermore, Calif.

Moss, S., Huang, Y., Keer, T., and Shah, B. (1997). "Development of an advanced finite element model database of the Hybrid III crash test dummy family." *Paper 971042*, SAE World Congress and Exhibition, Detroit, Mich.

MSC.Nastran 2005 release guide. (2005). MSC Software Corp., Santa Ana, Calif.

9

Composite Aircraft Structures

D. W. Kelly, M. L. Scott, and R. S. Thomson

9.1 Introduction

Aircraft structures are typically of stressed skin configuration consisting of a thin skin supported on a subframe of transverse and longitudinal members as shown in Figure 9-1. A monocoque structure with only an outer shell is efficient in carrying tensile and shear loads but inefficient for compression loads for many of the shapes found in aircraft. The subframe in Figure 9-1a is therefore added to support the skin. In the fuselage it consists of circular frames and longitudinal stiffeners connected to the skin by closely spaced fasteners. This subframe defines the panel size and hence controls buckling of the skin. Similar stressed skin configurations are found in lightly loaded components of the wing and tail structures. The subframe has the primary goal of supporting the skin against buckling but, in an efficient design, contributes to all loads carried by the structure.

The fuselage cross section is typically circular for a pressurized aircraft to minimize bending as a noncircular shape deforms under pressure loads. For nonpressurized aircraft a rectangular cross section is often used to achieve a more useful cargo space. The internal space is optimized by allowing the longitudinal stiffeners or longerons to pass through cutouts in the frames as shown in Figure 9-1a. Shear clips between the frames and stiffeners stabilize the stiffeners against buckling. This configuration can lead to fatigue in the frames particularly in pressurized aircraft that are subject to one cycle of tensile load for each flight. The frames can be mounted above the longitudinal members to eliminate the cutouts, but this reduces the internal usable volume.

The main elements of the wing are the wing spars. The location of resultant lift on the wing changes with angle of attack but is typically near the quarter chord measured from the leading edge. The front spar is therefore the main structural member

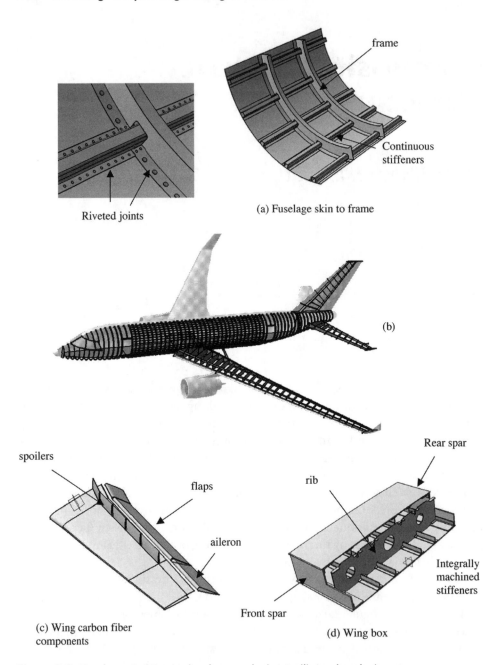

(a) Fuselage skin to frame

(b)

(c) Wing carbon fiber components

(d) Wing box

Figure 9-1. Fuselage structure, wing box, and wing trailing edge devices.

to resist bending and is located near or just in front of the quarter chord. The rear spar creates a box to provide torsional stiffness and provides mounting points for the flaps and control surfaces that are located across the span of the trailing edge. The control surfaces comprise approximately 25% of the chord. High lift devices at the leading edge allow the wing to achieve high angles of attack on landing without flow separation. Typically the spars are *I* sections assembled from separate shear webs and flanges to increase damage tolerance. Such assemblies can increase damage tolerance by not providing a continuous path for crack growth.

The wing shape is optimized to generate lift and the leading and trailing edge devices require hinges that limit their effectiveness as structural members. Therefore the design engineer is forced to work within a defined outer profile and the useful structure is limited to the region between the front and rear spars. For highly loaded sections the optimal structure is a closed box following the outer profile and limited forward and aft by the spars. The closed box provides the required torsional stiffness necessary to limit changes in the angle of attack of the aerofoil. If you watch a wing flex during takeoff you will notice significant bending made evident by movement of the wing tip. The tip on a large commercial jet moves several feet as the lift load is transferred from the wheels to the wing. However there is no noticeable twist in the wing. This reflects the relative significance of torsional stiffness to bending stiffness. The lifting performance of the wing is not significantly affected by bending. Military and aerobatic aircraft, on the other hand, require sharp response to control input and will be designed for greater bending as well as torsional rigidity.

Ribs are normally oriented normal to the front spar to aid construction. They maintain the shape of the wing, transfer loads between the spars, support load introduction, such as from the engine mounts and control surface hinges, and resist compression that arises from the geometry change as the wing bends.

The wing carry-through structure and the main frames of the fuselage at the wing fuselage intersection form a stiff center section for the aircraft. The main undercarriage is located here to take advantage of this extra stiffness and strength. This region also carries the highest bending loads in both the fuselage and the wing. Locating the engines and much of the fuel in the wings relieves the forces and moments that are transferred to the fuselage. Because of this inertia relief it is often not obvious whether the wing root shear and bending moment are higher when the aircraft is fully laden with fuel or when in a low fuel configuration.

Design criteria therefore include torsional and bending stiffness as well as strength. Structural vibrations can be particularly critical if energy can feed into the structure from the airflow. Design requirements usually demand that torsional vibration frequencies are well above bending frequencies and control surfaces are mass balanced to remove coupling between rotation of the control surface and bending of the wing. Mass balancing to locate the center of gravity of control surfaces near the rotational axis of the hinges addresses part of the problem but leaves the designer to eliminate aerodynamic coupling. Stiffness must then be maintained in the control linkages and useful levels of damping help to eliminate unwanted vibrations.

The wing and fuselage are therefore closed cell structures to maintain torsional stiffness and stiffeners are located remote from the bending axis to give bending stiffness. To reduce weight, local buckling of the skin is usually allowed before design ultimate load if the subframe can maintain the integrity of the structure. The aircraft is not expected to encounter these loads in normal flight and the disruption to the airflow is therefore not significant.

The alloys of aluminum used in aircraft construction get their properties from heat treatment and so riveting is used for joining rather than welding. Aluminum, known as the "light alloy," has been the material of choice during the first century of flight but is now being challenged by composite materials. Resistance to fatigue and corrosion, high ratios of strength and stiffness to weight, and high formability during manufacture are properties of carbon fiber composites that are leading to this change. The problems associated with fatigue in aluminum structures, often originating at fastener holes, has driven the aerospace industry towards a damage tolerance philosophy for design and operation. Structures must have sufficient residual strength to tolerate the presence of cracks or corrosion until they are detected by inspection during regular maintenance and are repaired. This has required an integrated approach to design that balances the requirements for low operating costs (reduced maintenance), good fatigue performance (requiring low design stress levels in tension components), and weight (that affects fuel efficiency).

Airframe analysis has traditionally been based on resolving the structure into components that carry different loads (Bruhn 1965). Classical analysis considers the structure to consist of shear-carrying-only panels framed by booms that carry the direct stresses. The effective area of the booms includes all the area of the skin in tension, but only the fraction of the area supported by the subframe in compression. These classical methods of airframe analysis supported the design of aircraft up to and including the Boeing 747 designed in the 1960s before large-scale finite element modeling became possible. Even today, large modern finite element models of wing and fuselage structures can follow this historical approach. Models consisting of shear panels surrounded by beam elements, with lumped areas to represent the direct stress carrying capability of the structure, permit a simple interpretation of the load flow from the forces in the beam elements. Stiffeners can still be lumped into the plate elements when the analysis relies on the tension and compression stiffness of the structure such as in bending of the main wing box. When the analysis involves local bending of panels, or buckling and postbuckling analysis, full plate shell models of panels and stiffeners are required.

This chapter deals with the analysis of composite aircraft structures. Composite panels often have significantly increased bending stiffness compared to aluminum panels and the drive toward efficient manufacturing processes has reduced the number of components in the supporting subframe (Nui 1992). These changes have driven the design toward lean semimonocoque configurations. The parallel growth in computing resources has allowed composite structures to be modeled with shell elements with plates representing stiffeners. Attention now has to be placed on effective modeling of assemblies and joints and ensuring the insight available in the approximate models is retained in the new modeling strategies.

The continuing development of computer systems is also promoting the integration of electronic resources in the design office. Computer Aided Design (CAD) and Finite Element Analysis (FEA) are being married by integrating FEA codes into CAD systems. In the past the industry has coped with difficulties associated with transfer of geometry data between the designers, the stress analysts, and those focused on manufacturing. Good housekeeping was required to ensure CAD, FEA, and manufacturing models were updated in parallel as the design evolved. In the modern environment CAD systems such as CATIA V5 (owned by Dassault Systèmes) and UGS NX (owned by UGS Corporation) are providing an integrated platform for design, analysis, and manufacturing, working from a single database. In addition rule-based tools are being developed to monitor the design process. Knowledge Based Engineering (KBE) is a technology that enables the knowledge and experience of engineers to be captured and deployed. KBE and automation are releasing engineers from many of the time-consuming and repetitive tasks, allowing them more time to concentrate on refining the product. Rules and checks embedded in the design tools ensure the design conforms to best practice. For example, the edge distance for a rivet installation in a panel is a function of the rivet diameter. However best practice might require the edge distance is defined for the next larger fastener above that specified by panel loads. This permits rivets to be drilled out and replaced if repairs are required. Such rules and checks improve productivity and limit the costs involved in downstream correction of mistakes.

While algorithms that define the topology of structures remain an active research area (Suzuki and Kikuchi 1991), algorithms that vary the thickness of structural components to minimize weight and satisfy strength, stiffness, and frequency constraints are now robust enough to be a standard tool for the structural engineer. In this respect, design and analysis in aircraft structures have strong similarities with the modeling of other thin shell structures such as automobiles and maritime structures, and the detail will not be repeated here. Instead this chapter will follow the modern development in aircraft structures toward use of carbon fiber composite materials.

9.2 The Design Environment

The U.S. Federal Aviation Regulations (FAR) and the European Joint Aviation Requirements (JAR) are the regulations applied to aircraft structures. These regulations define limit loads as the highest loads expected in normal operation of the aircraft resulting from both maneuvers and gusts. Ultimate loads are equal to 1.5 times the limit loads. The structure must show no detrimental effect for limit loads, and must briefly sustain ultimate loads. A structure loaded above limit load will need to be inspected and possibly replaced after the load event. Typically an aircraft experiences a load factor of 1 in steady level flight and a load factor of 2 in a turn banked at 60°. Limit load factors of the order of 3 cover maneuvers and turbulence for commercial transport aircraft, with higher values applying to fighter and sports aircraft. Buckling in the skin that may impair the airflow may be forbidden at normal operating load factors, but permitted at the load factors between limit and ultimate. Ultimate loads are normally only encountered in extreme events such as a crash.

The design process for a composite component is depicted in the flow chart in Figure 9-2. The developers of CAD systems such as CATIA V5 and UGS NX are working toward integrated environments for design, analysis, and manufacture, and these are being extended to include features specific to composite structures. CATIA V5, for example, is configured as a number of "work benches" that include tools for surface creation and conceptual design and surface and solid modeling.

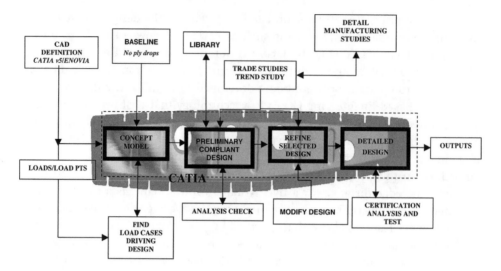

Figure 9-2. Design process flow.

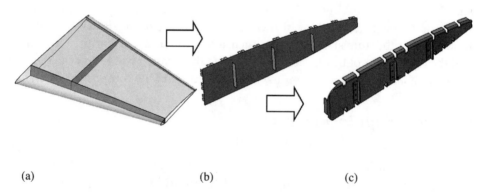

(a)   (b)   (c)

Figure 9-3. Rib design from concept to preliminary design to detailed design.
(a) Conceptual design. (b) Preliminary design. (c) Detail design.

For composite structures a composite workbench is included where each surface has attributes that define the ply stack in the laminate and zones over which the same laminate is used. Interfaces include direct links to manufacturing tools and finite element solvers including both proprietary products from the system developer but also embedded systems such as MSC.NASTRAN (owned by MSC Software Corporation).

Increasing sophistication in the structural model is depicted in Figure 9-3. The process progresses from conceptual design, to preliminary design, and finally to detailed design. The model for conceptual design of composite components usually consists of simple surfaces. The model does not include the detail of stiffeners, ply drops, and assembly. At this stage of the design process the analysis is focused on defining the configuration of the structure. As the design progresses, detail is added including the design of the laminate, stiffening against buckling and the method of joining. Joining can require flanges on the stiffener and bonded or riveted connections. All of these contribute to the weight and cost and make more difficult the estimation of weight and cost early in the design process. In addition there are details of assembly, such as those already discussed, that demand that stiffeners on the wing skin are continuous, requiring cutouts in the rib.

9.3 Finite Element Analysis

The design team involved in a development of an aircraft structure will develop finite element models that reflect the different stages of the design process. The modern trend is to develop full interactive computational fluid dynamics (CFD) and structural models to supplement wind tunnel testing, but the computing load associated with the CFD analysis limits the structural model to a coarse simulation. Lifting line and lifting panel approaches to the aerodynamic analysis remove the need for a full 3D discretization of the air flow and permit sufficiently refined structural models to capture the aeroelastic response of the structure. This enables engineers working on aeroelastic problems sufficient accuracy to model static divergence of the lifting surfaces and control reversal and the dynamic interaction at the critical flutter speeds. These analyses require the full structure of the aircraft as flutter modes can involve the interaction of the wings, fuselage, and tail structures and can be driven by rotation of the control surfaces about their hinges. Mass distribution and damping can be as significant to flutter as stiffness, and mass balancing needs to be included in the models to identify the correct coupling of the bending and rotational vibrational modes in the structure. Aeroelastic analysis is best left to the expert as small changes in the phase relationships can dominate the energy transfer.

Static structural modeling of the entire airframe in sufficient detail to capture the stress distribution in structural components is now possible, but the time taken for the analysis imposes significant delays on the design process. Design teams focused on the conceptual design of the aircraft configuration, such as the arrangement of spars and ribs in the wing, will therefore develop models where the stiffened plates are represented by plate elements and beams represent the stiffeners. Ribs will be simple surfaces with no detailed modeling of joints. If required, plate elements can be replaced by shear panels and direct stress carrying areas included in beams. This aids the interpretation of the load flow and load paths in the structure. Recently the algorithms developed to define topology have become a new tool for determining the optimal configuration of internal members. An overpopulated model of the structure with additional ribs and spars can be subjected to a design analysis that will select the main load bearing members and provide fresh insight for the designer.

Stiffness is more difficult to capture than stress distribution. Lightweight design means many aircraft structures have dominant primary load paths and classical analysis based on statics will return useful results. For stiffness, however, the stiffness of the joints is very important and models are tuned from previous test results.

Accurate modeling of fastener behavior is also important in the modeling of crash events and bird strike. These simulations require full transient dynamic modeling using explicit FEA. Aluminum exhibits significant plastic deformation, and crash modeling of aluminum structures is very advanced. However, composites fail in a brittle fashion for some load cases and absorb significant energy in others. Fiber failure is brittle, but in failure modes that involve significant matrix cracking the energy absorbed is high. Accurate modeling of collapse of composite structures and response to impact events can require large 3D models and is the subject of current research.

Cocured structures have joints that are bonded or a continuous configuration, and failure can be initiated by delamination of the layered material. Modeling of these events typically requires use of fracture mechanics approaches. The microscope reveals a complex multiphase domain of fibers and resin populated by defects with failures initiating in the resin, in the fibers, and at the fiber resin interface.

Full plate and shell models and geometrically nonlinear analysis are required to accurately predict the buckling and postbuckling behavior of the structure. Buckling of the skin affects the airflow but is permitted for load factors outside the normal operating environment of the aircraft. The accurate prediction of postbuckling stiffness and collapse of the structure is one of the modern challenges for researchers in structural analysis. The curved panel structures are also imperfection-sensitive, and dynamic buckling can precede the buckling predicted by static algorithms.

Efficiency in turnaround of analyses in the design process requires local as well as global analysis. For example, secondary structures such as trailing edge flaps must conform to the bending profile of the wing. Boundary conditions to force this compatibility can be applied at the hinges when modeling the secondary structures.

In addition, the matrix of load cases for aircraft structures can grow to thousands and so insight is required to extract the dominant load cases for a particular component. The loads on the airframe depend on the position of the center of gravity (cg), which depends on the loading of the aircraft, but also the fuel load, and on the phase of the flight. Control surface movement and maneuvering imposes bending and torsional loads on the structure that also depend on the position of the cg. So the load case matrix builds rapidly as different control movements of the aileron, rudder, and elevators can be applied with different cg positions and during different maneuvers. Superimposing a range of gust and landing loads causes the matrix to grow. Design teams learn the load cases that dominate the design of their components and whether their structures are strength or stiffness driven. The full load case set is left to the global model. This is an obvious application for the development of future KBE tools and automation. Programs need to be written to sort through multiple load cases in the result files to identify which load cases are driving the design.

Postbuckling stiffness was an important consideration in the case study described in Section 9.7, and so the panels and stiffeners were modeled with plate elements. A significant requirement was the prediction of the buckling load and it was checked by a full-scale test. For composite structures, typical finite element models consist of plate elements that have anisotropic material properties defined from the individual plies that make up the laminate. Laminate strains can be recovered on a ply-by-ply basis to predict ply failure. However, resin-based failures and failures of the interface between the fiber and resin, including delamination and growth of disbonds, require knowledge of strains in the resin. Full 3D analysis is then required and may include the residual stresses that result from the curing process. The resin hardens at elevated temperature, and the cooling process to room temperature creates residual stresses due to the difference in thermal expansion coefficients for the fiber and resin. The resin strains are further exacerbated by the strain concentration that occurs around the fibers under transverse loading. Failure prediction can therefore require micromechanical modeling (Gosse and Christensen 2001).

9.4 Design and Analysis of Composite Structures

Traditionally the use of composites in commercial aircraft has been restricted to wing and tail trailing edge structures such as the flaps, ailerons, and rudders identified in Figure 9-1c (Hart-Smith 1986, 1993). However, as confidence is growing in the ability of industry to design, manufacture, and maintain high quality components, composite materials are finding increasing application. For example, the Airbus A380 has approximately 16% of its structural weight composite. In the Boeing 787 this will increase to 50%. In this chapter a case study will consider the aileron of a medium-sized jet. The discussion will focus on design and analysis and will include comparisons with full-scale test results.

The introduction of fiber reinforced materials has been absorbed directly into FEA with classical laminate theory (Daniel and Ishai 1994) providing in-plane and bending stiffness for plate and shell elements. The laminate is defined in the finite element pre-processor and strains required for failure indices can be determined on a ply-by-ply basis. Three-dimensional representation may be required to give accurate through-thickness stresses. Brick elements are then implemented with, typically, a layer of bricks representing a group of plies with the same orientation.

The introduction of fiber reinforced materials has, however, added the design of the material at a sublaminate level as a new dimension in the design process. The number of plies in a panel defines the thickness, but the orientation of each ply in Figure 9-4a, the use of unidirectional tape or cross-ply fabric, and the specification of the fiber and resin materials add additional variables in the design process. The designer must also comply with rules for best practice that are often linked to manufacturing processes. The list of rules includes the orientation of surface plies for damage tolerance, the number of consecutive plies with the same orientation, and limits on the radii of curves. A concurrent engineering approach is therefore required so that the manufacturing process is defined early in the design process. A commonly quoted figure is that the initial 5% of the design process fixes 70% of the project cost (Salamone 1995). It becomes increasingly expensive to rectify mistakes as the design process progresses.

The reward for use of carbon fiber composites in appropriate applications is a reduction in weight, sometimes by as much as 25% over aluminum structures. This improvement usually comes with a penalty in cost and the designer must assign a value to weight reduction for a composite structure to be cost competitive. Costs are reducing as new manufacturing processes are developed, but the current figure is of the order of $500 per kilogram for commercial aircraft structures (Baker et al. 2004).

An advanced application of fiber-reinforced materials uses groups of fibers, called tows, steered along trajectories in the structure. This approach is very effective when load cases are clearly defined such as for filament winding of pressure vessels and fuselage structures. The tows follow optimal paths to carry load or provide stiffness. The number of fibers in a tow is typically from 1k (1,000) to 12k (12,000). Fibers are not continuous but average lengths, typically of the order of 5 to 10 cm, are long enough for efficient load transfer. An application considered later in this chapter redirects load around a cutout in a panel. In the future, the introduction of carbon nanotubes—using carbon atoms arranged in an optimal cylindrical configuration—promises significant further weight reduction. How to use these superfibers is currently challenging designers and structural engineers.

In this modern arena it has become apparent that automated procedures that attempt a push-button solution to the overall design problem are restricted by the common occurrence of local minima in the design space. These local optima trap the gradient-based search methods. This problem is hugely compounded by the design of the laminate where there are often multiple ply configurations that satisfy the constraints at the same weight (Gürdal et al. 1999). Optimization algorithms based on static and dynamic optimality criteria that define fully stressed or maximum stiffness configurations partly overcome these problems. Genetic algorithms (GAs) that offer some hope of identifying global solutions are replacing traditional gradient-based search methods particularly for laminate design. For the global structure, Design of Experiments approaches (Montgomery 1991) sample the design space using KBE tools. Presented as trade studies based on automated redesign and analysis, they allow the design team to explore the design space and answer "what-if" questions while still meeting project deadlines.

In parallel with these developments is an increasing sophistication in failure prediction. Compatibility is assumed to exist between the plies so that maximum strain criteria on a ply-by-ply basis are easier to implement than stress criteria. However, composites comprise different components, and the way failure can manifest itself includes failure of the fibers as shown in Figure 9-4b, cracking in the resin, and failure at the interface between fiber and resin. Interlaminar failure drives the analysis toward 3D models that can predict through-thickness stresses. Micromechanical failure models require modeling of the stresses in the resin between the fibers and include residual stresses from the curing process. The resulting global/local modeling spans the global dimensions of the structure from the order of meters in the panels to nanometer dimensions at the fiber resin interface.

Classical tension cycling fatigue is not a significant problem in fiber-reinforced structures as the fibers tend to bridge the crack and prevent its growth. However this form of cyclic degradation is replaced by damage growth in compression as delaminations grow due to "brooming" of the laminate. These delaminations are difficult to detect when the laminate is not loaded. Damage tolerant design has therefore become the basis of modern aerospace design using composites. Barely Visible Impact Damage, BVID, is defined as the smallest level of surface damage that can be identified by visual inspection. The impact that causes this level of damage can cause subsurface delaminations. The design philosophy is that this damage will not affect the compression strength and will not grow under normal service loads. At the

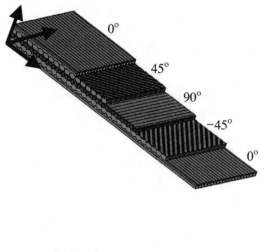

(a) Laminate structure

(b) Micromechanical failure in tension and compression

Figure 9-4. Structure of the material in carbon fiber reinforced composites. (a) Laminate structure. (b) Micromechanical failure in tension and compression.

macrolevel the philosophy of damage tolerant design for damage that exceeds BVID is to ensure damaged structures continue to perform their function until the damage has grown sufficiently to be found and repaired during regular maintenance. If new developments in condition monitoring achieve their goals, the regular maintenance of the airframe could be replaced by continuous monitoring using an array of built-in sensors.

In addition composites are sensitive to environmental effects and properties are defined for both a hot-wet condition and a cold-dry condition. Typically the hot-wet test coupons that provide the material properties are saturated with moisture and heated to 70°C. The resin dominated properties can be significantly reduced in these conditions. At the other extreme is a cold condition as low is −55°C for which thermal cracking can occur due to the different thermal expansion coefficients for the components in the composite. While not immediately critical, these microcracks permit moisture ingress and can significantly reduce the useful life of the structure.

9.5 Laminate Design

The design of the laminate is usually treated as a subproblem once the results of initial finite element models become available. These initial models use typical (average) properties and thickness-based optimization can be implemented to identify efficient designs. Panel loads are then recovered from the finite element model. Algorithms using classical laminate theory can then be used to apply these loads to representative elements to determine the in-plane strains and curvatures and define the orientation and number of plies required in the laminate. The algorithms ensure compliance with limiting strains defined from coupon tests.

Optimization of the laminate is difficult because several laminates can provide the required stiffness with a similar number of plies. Genetic Algorithms are particularly suited to this problem (Nagendra et al. 1996). The Genetic Algorithm is an optimization technique based on a simulation of Darwin's theory of evolution of natural species. From a randomized population of stack sequence, the GA evolves (selection, cross over, mutation) based on several fit goals in order to get a population of acceptable stack sequences. The best is then chosen as the sequence that achieves compliance with goals, while simultaneously achieving the minimum number of plies and the maximum margins of safety.

The laminate design rules defined earlier can be accommodated in the measure of fitness that drives the GA. These rules include permissible orientations of the plies (for example, 0°, 90°, and ±45°), fixed directions for the surface plies, and the maximum number of plies in a group with the same orientation. The objective of minimum weight is reflected if the number of plies is minimized. Laminates can be designed for in-plane strength and for panel buckling. Bending (and hence buckling) stiffness is achieved by aligning plies near the surface with the direction of the bending load and by including core materials to define a sandwich structure.

The design of the ply configuration for a complex panel includes multiple zones, picture frame stiffening near the joints between different components, and ply drops in more lightly loaded regions (Adams et al. 2004).

9.6 Load Paths and Load Flow Trajectories for Fiber Placement

The concept of load paths and load flow is widely used by design engineers to describe the way a structure carries applied loads from the point of application to the point of reaction in the structure. Designers may identify a path formed by statically determinate sets of members in a structure that can carry the load. In this way they ensure that the structure can support the applied load. A structure is redundant if there is more than one of these paths.

Load paths can be mapped using results from an FEA (Kelly et al. 2001). Contours are plotted aligned with total stress vectors defined by:

$$V_x = \sigma_{xx}i + \tau_{yx}j + \tau_{zx}k$$
$$V_y = \tau_{xy}i + \sigma_{yy}j + \tau_{zy}k \; . \tag{9-1}$$
$$V_z = \tau_{xz}i + \tau_{yz}j + \sigma_{zz}k$$

Here τ_{xy} is the shear acting on the surface whose normal is in the x-direction, directed in the positive y direction. The forces acting on a plane with normal given by $\vec{n} = n_x i + n_y j + n_z k$ are obtained by integrating the total stress vectors

$$F_x = \int V_x \cdot \vec{n} \, dA$$
$$F_y = \int V_y \cdot \vec{n} \, dA \tag{9-2}$$
$$F_z = \int V_z \cdot \vec{n} \, dA$$

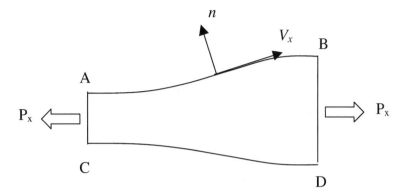

Figure 9-5. Contours for a path along which P_x is constant.

where the dot indicates the vector dot product.

The load path for a force in a given direction is a region in which the force in that direction remains constant. For example, if the path in Figure 9-5 is to define a region in which the force P_x remains constant, the requirement is to determine the curved contour forming an edge along which the normal and tangential edge loads make no contribution to force in the x-direction. This requires that there is no contribution to the x-force on sides AB and CD. On AB this requires

$$F_x\big|_{AB} = 0$$

$$or \quad \int_A^B V_x \cdot \vec{n} \, dA = 0.$$

This is achieved if the normal to the surface is perpendicular to V_x, as the dot product is then zero. Alternatively, this is achieved if the surface tangents are parallel to the vector V_x as indicated in Figure 9-5.

Separate load path contours are defined for the X-force, the Y-force, and the Z-force. In general any coordinate axis can be used so a path can be defined for a force in any arbitrary direction. However, because the theory relies on applying equilibrium to a free body diagram of a section of the structure, the path for each of the orthogonal components is distinct.

Figure 9-6a gives an example that is easily interpreted. The pin-jointed structure is subject to a vertical load. Because of the pin joints, the vertical load is transmitted to the supports as defined in Figure 9-6b. In addition to these Y-direction forces, a secondary set of loads is created at the supports to carry the bending moment. These loads are transferred through the structure as shown in Figure 9-6c. The theory defined above can be applied to this case but is usually reserved for two- and 3D domains. For the truss the flow of the load can be tracked by taking sections and applying equilibrium.

Applications of load flow analysis in composite structures include the identification of efficient subframe configurations that reflect the increased stiffness of the laminate skins. Fiber placement allows the paths of fiber bundles to be controlled in the manufacturing process and load paths rival principal stress trajectories as

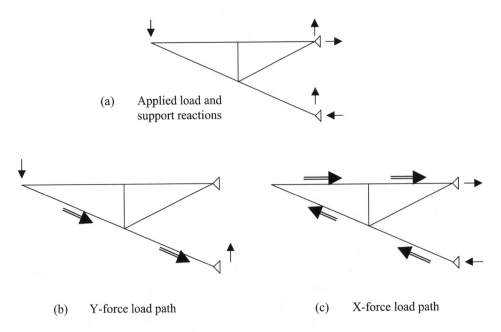

(a)   Applied load and
      support reactions

(b)   Y-force load path   (c)   X-force load path

Figure 9-6. Load paths for a simple truss structure.

a method for design of these paths. Both load paths and principal stress trajectories limit the shear stress in the weaker matrix in the highly loaded regions of the structure. The fibers also act to transmit shock loads across laminates increasing the importance of identifying hard points in the structure particularly for shock loading.

9.7 Aileron Analysis and Design: Case Study 1

The following case study considers the design of a composite aileron for a medium-sized jet aircraft in which buckling is allowed in parts of the structure. Stiffness and damage tolerance are the main design drivers.

9.7.1 Rationale

Carbon fiber reinforced composites in honeycomb sandwich panels have been widely used in control surfaces for commercial transport aircraft for well over two decades. Typical design configurations for components, such as ailerons, flaps, and spoilers, incorporate carbon fiber/epoxy ribs and spars that are mechanically fastened to honeycomb sandwich skins. Poor impact resistance, water ingress, skin/core delamination, and costly repair of damaged skins are the most common causes of problems raised by operators of aircraft with this type of construction. The most promising concept to replace sandwich structures for relatively lightly loaded components, such as control surfaces, is thin, stiffened skins that are designed to buckle below limit load levels (Schneider and Gibson 1985; Hart-Smith 1986, 1993; Hawley 1986; Ashizawa and Toi 1990).

The design and manufacturing technology associated with composite/honeycomb sandwich panels was considered the baseline for the new composite control surface design presented here. The structure studied represents the aileron of a midsized, commercial, jet transport aircraft. In order to demonstrate the weight and cost saving potential of thin buckling skins with integral blade stiffening, a complete aileron has been designed, incorporating many innovative features. Central to this task was the selection of a one-piece composite manufacturing process that enabled the blade stiffened skins, rear spar, and ribs to be cocured.

The primary objectives for the new design were as follows: to reduce the mass of the control surface box structure by at least 30%; to reduce manufacturing costs; to achieve or exceed existing requirements for quality, durability, and damage tolerance; and to achieve existing requirements for stiffness and environmental degradation. Full-scale tests were undertaken to validate the design.

9.7.2 Aileron Design

9.7.2.1 Aileron configuration. The overall dimensions of the aileron are 2.8 m × 0.4 m, with a local sweep angle of 16 degrees, conforming with the sweep of the wing as shown in Figure 9-7. There are four hinges supporting the aileron through its front spar. A 1.7 m long balance tab is located at the trailing edge extending from the inboard end of the structure. The aileron is actuated by a single control rod next to the inboard hinge.

9.7.2.2 Load cases. The loads acting on the aileron were divided into three groups: distributed surface pressures resulting from the air loads, secondary pressure loads resulting from the tab and leading edge, pressure loads from the structure that supports balance weights ahead of the hinge line, and induced bending loads from bending of the wing. The design ultimate pressure profiles resulting from air loads applied to the top and bottom skins of the aileron are shown in Figure 9-8. They were assumed to be distributed evenly, with 50% on the upper and lower surfaces.

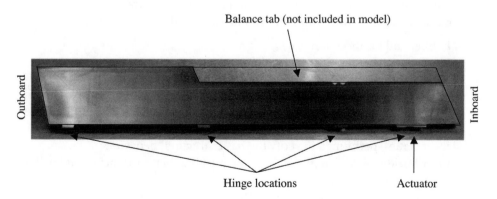

Figure 9-7. Aileron configuration.

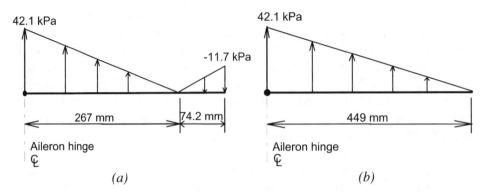

Figure 9-8. Chordwise pressure distribution (ultimate air load) in (a) the tab region and (b) the no-tab region.

9.7.2.3 Design environment. CATIA was the CAD and manufacturing environment chosen for the project and is used extensively in the aerospace industry. The CATIA design database was used as the sole design authority and was transferred digitally between the design center and the manufacturing center. The aileron was modeled as a 3D solid in CATIA. Midsurface data from the CATIA solids was then translated into the MSC.PATRAN and MSC.NASTRAN environment for analysis purposes. The 3D model was also used to generate the numerically controlled (NC) machine files for milling all tooling.

9.7.2.4 Material data. In the design of the aileron, two carbon fiber/epoxy material types were used. Unidirectional tape was used for the majority of the structure while plain weave fabric was used to reinforce particular regions. The allowable strains for the tape and fabric materials are presented in Table 9-1. To satisfy damage tolerance requirements, the allowable strains under hot/wet/damaged condition were used for design purposes.

9.7.2.5 FEA. The aim of the design process was to determine the most efficient structure to satisfy the various design criteria. The effect of the design iterations was determined through the use of FEAs through the predicted strains, buckling behavior, and stiffness. The primary design variables included the following:

- Number and location of ribs.
- Number and location of stiffeners.
- Layup and thickness of skins, spars, stiffeners, and ribs.
- Local thickness increases at points of load introduction.

   Using MSC.NASTRAN, the following FEAs were conducted:

- Linear static—performed to verify the basic structural behavior.
- Linear buckling—to determine the basic buckling behavior including buckling load and mode shapes.
- Geometric nonlinear—to predict the full postbuckling behavior of the aileron, including the overall buckling behavior, strain levels, and stiffness.

Table 9-1. Hot/Wet/Damaged Material Properties for Carbon/Epoxy Tape and Fabric

Property	Carbon/Epoxy Tape	Carbon/Epoxy Fabric
Ply Thickness (mm)	0.094	0.21
Longitudinal Modulus (GPa)	125	57.2
Transverse Modulus (GPa)	9.0	57.2
In-Plane Shear Modulus (GPa)	4.6	4.8
Poisson's Ratio	0.34	0.06
Longitudinal Tension Allowable Strain ($\mu\varepsilon$)	5250	5250
Longitudinal Compression Allowable Strain ($\mu\varepsilon$)	4050	4050
Transverse Tension Allowable Strain ($\mu\varepsilon$)	10750	5250
Transverse Compression Allowable Strain ($\mu\varepsilon$)	25810	4050
Shear Allowable Strain ($\mu\varepsilon$)	7950	7950

The number and orientation of plies were adjusted, accounting for manufacturing constraints, until no regions experienced a negative margin of safety at design ultimate load (DUL) in the geometric nonlinear analysis.

9.7.2.6 Finite element model. The FEA model, shown in Figure 9-9, consisted of 13,835 four-node shell elements, 26 three-node shell elements, and 211 two-node beam elements. The composite materials were represented using shell elements with a layered composite property definition. In the skins, a mesh refinement giving seven shell elements between stiffeners was used to accurately represent the buckling behavior. The ribs were represented by shell elements while some stiffeners were modeled using beam elements to improve computational efficiency. Appropriate laminate properties were calculated and applied to the beams. The mesh at one end of the aileron was further refined to give greater detail of the structural response close to the actuator, as shown in Figure 9-10.

The air loads were applied as distributed pressure loads on the top and bottom skins. The wing bending loads were applied via enforced displacements at the hinge points.

9.7.2.7 Design summary. The baseline honeycomb structure was redesigned, while maintaining the stiffness, by using thin buckling skins incorporating multiple, cocured blade stiffeners. Since similar front and rear spars to the baseline control surface were used, bending stiffness was not significantly altered. To match the torsional stiffness of the baseline control surface, a total of 9 ribs and 12 stiffeners

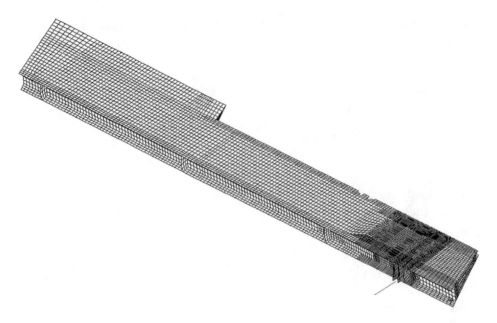

Figure 9-9. The aileron FEA model.

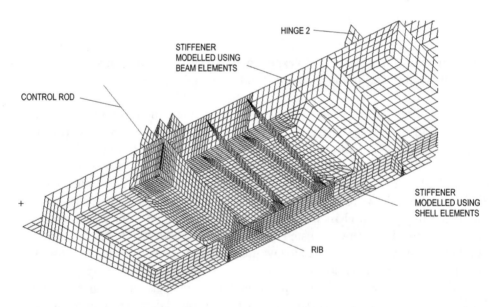

Figure 9-10. Internal structure and refined mesh region near actuator.

were incorporated, which were positioned chordwise along the span as shown in Figure 9-11. The thin skins consisted of two pairs of ±45° plies to resist the torsional loading, two 0° (spanwise plies), and one 90° (chordwise plies) to transfer the air loading to the surrounding structure, as shown in Figure 9-12. Local skin buildups were incorporated at the front and rear spars and around major hinge ribs.

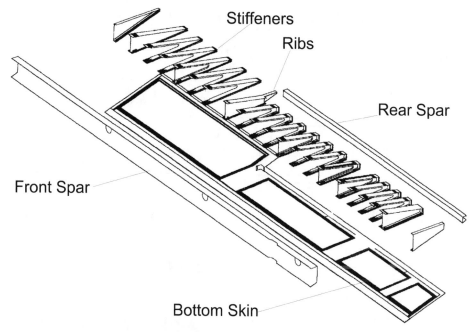

Figure 9-11. Aileron configuration (top skin not shown).

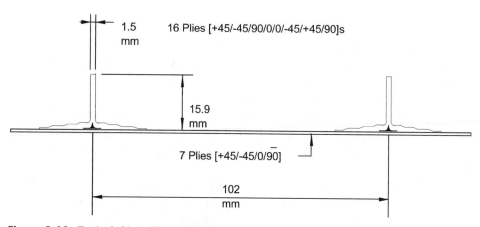

Figure 9-12. Typical skin-stiffener interface.

To achieve significant weight savings over the honeycomb sandwich baseline structure, emphasis was placed on a thin-skinned postbuckling design. Buckling was only permitted in the skins of the box structure. The front and rear spars, which carry the bending loads on the aileron, were not permitted to buckle nor cripple at ultimate load. Likewise, ribs were designed to be nonbuckling. Some sections of the upper and lower skins, however, were reinforced to prevent buckling to allow attachment of a fairing to the lower skin for the tab control rods. Multiple chordwise blade-stiffeners were incorporated to stiffen the thin, buckling skin.

Table 9-2. First Five Predicted Mode Shapes from the Linear Buckling Analysis

Buckling Mode	% DUL	Mode Shape (Top Skin)
1	19.8	
2	20.7	
3	21.6	
4	22.4	
5	24.1	

9.7.2.8 Predicted behavior. The first five eigenvalues from the linear buckling analysis, as a percentage of DUL, and the corresponding mode shapes are shown in Table 9-2. The analysis predicted a lowest eigenvalue of 19.8% of DUL, which reflects a buckling ratio of 5.1 at ultimate load and 3.4 at limit load. The first ten buckling modes occurred in the top skin primarily in the center region of the aileron and were strongly influenced by overall bending of the aileron. No buckling was predicted in the bottom skin due to the tensile component induced from overall bending.

The geometric nonlinear analysis predicted the maximum deflection at DUL of 50.0 mm at the outboard tip, as shown in Figure 9-13. Complex compression/shear buckling of the top skin was experienced. The predicted tip deflections on the control surface for both linear and nonlinear analyses are presented in Figure 9-14. The tip deflection of the aileron is a function of the torsional stiffness of the structure, which is affected by the buckling of the skins, and the bending stiffness of the ribs, stiffeners, and rear spar. Up to the point of buckling, the tip deflection is the same for linear and nonlinear analyses.

9.7.2.9 Mass breakdown. A breakdown of the mass of the aileron is presented in Table 9-3. The total predicted mass of 4.47 kg compares with a mass of 7.06 kg for the baseline honeycomb stiffened aileron, which represents a weight saving of 37%.

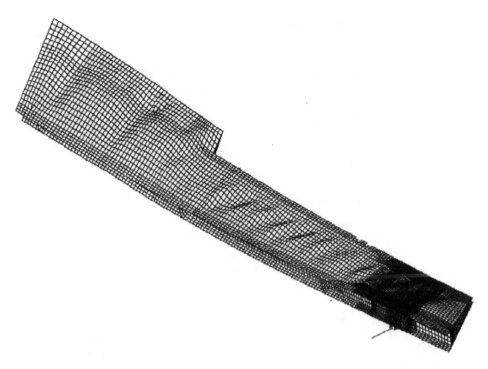

Figure 9-13. Predicted deformed shape at DUL from nonlinear analysis (deformation exaggerated).

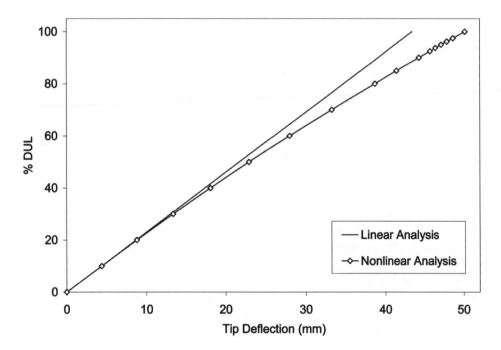

Figure 9-14. Tip deflection variation with load for linear and nonlinear analyses.

Table 9-3. Aileron Mass Breakdown

Item	Mass (g)
Skins (top & bottom)	2,120
Ribs	288
Front Spar	1,095
Rear Spar	205
Rib Post	100
Stiffeners	282
Enclosure Ribs	98
Fasteners	278
Total	4,466

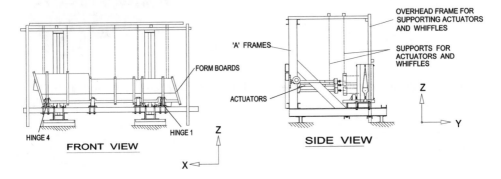

Figure 9-15. Schematic of aileron test arrangement.

9.7.3 Full-Scale Aileron Test

The aileron design was substantiated by means of a full-scale static test, the results of which were used to validate the analysis and design procedure.

9.7.3.1 Structural test arrangement. The test article was mounted in a purpose-built test rig as shown in Figure 9-15. The test article was positioned with its chord in the vertical plane to enable observation of the top and bottom skins under load as extensive buckling was anticipated. The test article was mounted to the test rig at the four hinge points and a link from the control surface to the test rig at the actuator location resisted all hinge moment. The two middle hinge fittings on the test rig were capable of being displaced normal to the chord line to enable wing bending displacement to be applied to the test article. These two hinge fittings also incorporated a load cell each to measure the normal-to-chord hinge reaction load during the tests.

The aileron was extensively instrumented with gauges to measure strain and transducers to measure displacements. Two shadow moiré grid panels were placed over the bottom skin in the test, one at the inboard end beside the actuator rib and one outboard where the tab region ends to view the buckling mode shapes.

9.7.3.2 Load application. The simulated air loads were applied to the test article through nine form boards, one located at each rib. Application of the loads at the ribs permitted the skin to buckle freely. These form boards were loaded using six electric screw jack actuators and a whiffle tree arrangement where necessary.

Displacements for the wing bending case were also based on the same aileron design load case and determined by applying induced bending loads at the middle two hinges in the FEA model and observing the displacement of these hinges. These displacements were then applied during the test, and were 7.5 mm and 9.9 mm at Hinges 2 and 3 respectively.

9.7.3.3 Test procedure. The tests were conducted in the following order to ensure they progressed from least to most severe: limit air loads only, limit air loads and wing bending displacement, ultimate air loads only, ultimate air loads and wing bending displacement. No failure of the test article occurred during the tests and only results from the final load case have been presented.

As the means of load introduction used in the test differed significantly from the design load cases, an analysis was performed with loading that simulated the test situation to enable an accurate comparison between FEA and experiment.

9.7.4 Results and Comparison with Analysis

9.7.4.1 Buckling behavior. Buckling was observed to initiate during testing at approximately 48% DUL, although some bays did not buckle until over 70% DUL. This was followed by a change in buckling mode at 89% DUL. In the FEA, buckling was predicted to initiate at 45% DUL. No significant mode change was predicted to occur.

A comparison of the general buckling pattern experienced during test and that predicted by the FEA at DUL is presented in Figure 9-16. A photograph taken at DUL during testing is shown in Figure 9-16a, while the rendered displacement result from FEA is shown in Figure 9-16b. Buckling patterns are clearly visible over the entire aileron and the comparison between the analytical and experimental buckling modes is generally very good.

9.7.4.2 Torsional behavior. The displacements measured at the trailing edge were compared with the FEA. While the comparison of the raw displacements was reasonable, it was more appropriate to compare the relative displacement or twist that the aileron experienced as this eliminates the influence of test rig deflections. A summary of the values at a DUL are presented in Table 9-4. These results demonstrate the excellent comparison between test and the FEA with only 2% difference at DUL.

9.7.5 Outcome

In this case study, the design and validation of a carbon fiber composite aileron of a midsized, jet transport aircraft has been described. The design features cocured,

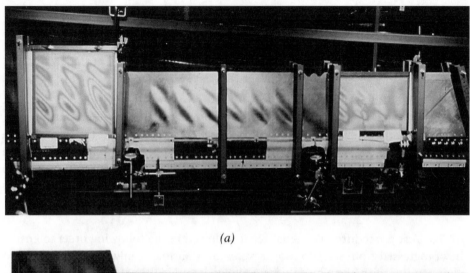

(a)

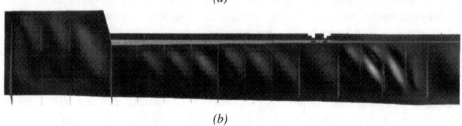

(b)

Figure 9-16. Comparison between the predicted and actual buckling behavior at DUL.

Table 9-4. Comparison between Test and FEA of Trailing Edge Displacements at a DUL

| | Actuator Rib | | Outboard Rib | | Relative | |
	Displacement	Angular Rotation	Displacement	Angular Rotation	Twist	Error
Test	12.4 mm	2.14°	68.1 mm	8.71°	6.57°	—
FEA	4.8 mm	0.84°	60.5 mm	7.74°	6.69°	2%

thin-skinned, postbuckling structure with chordwise ribs and blade-stiffeners. The design was performed in CATIA interfacing with the FEA tools MSC.PATRAN and MSC.NASTRAN in which linear static, linear buckling, and geometric nonlinear analyses were conducted to optimize the design. Weight savings of 37% over the baseline honeycomb stiffened design were realized.

The design process required the balancing of the stiffness of the skin and subframe so that an efficient stressed skin construction was obtained. Figure 9-17a shows

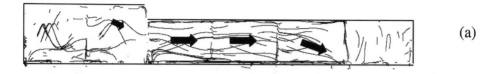

(a)

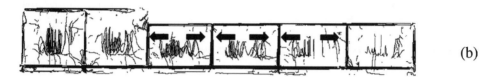

(b)

Figure 9-17. Load flow in the aileron according to relative stiffness of the skin and subframe.

the load flow in a typical structure with a stiff skin. The model indicates load flowing to the hinge comprising the torsional restraint provide by an actuator and returning to a second hinge via the front spar. The design in Figure 9-17b has reduced skin thickness allowing the ribs to become effective in transferring load to the front spar.

The successful validation of the postbuckling composite aileron demonstrates the great potential of this design for the next generation of transport aircraft.

9.8 Panel with Opening: Case Study 2

This second case study is directed toward the detailed design of fiber architecture for a panel with stress concentration. A current development in fiber-reinforced composite structures is the use of fiber placement to utilize the high tensile properties of individual fibers. Fibers in the form of tows are either placed by robots onto tacky surfaces along predefined trajectories or stitched onto dry preforms that are later injected with resin. Applications include hole reinforcement such as around cutouts in the fuselage or in load introduction components such as hinges and fittings. The aim of the analysis in this application is to define the path to be followed by the fiber.

The plot in Figure 9-18 shows trajectories that are tangent to the load paths for horizontal and vertical loads traversing near a cutout in a panel subjected to a shear load. A particle tracing algorithm has been borrowed from fluid mechanics visualization codes such as Techplot (owned by Amtec Engineering) to plot a trajectory parallel to the vectors in Eq. 9-1. In fluid mechanics the particle trace is determined by the velocity vectors and typically a fourth order Runge-Kutta algorithm is used.

An application reported in Li et al. (2002) achieved a 37% improvement in the ultimate shear load with the addition of steered fiber that contributed 3% to the weight of the panel. In this application the tows were continuous. An additional

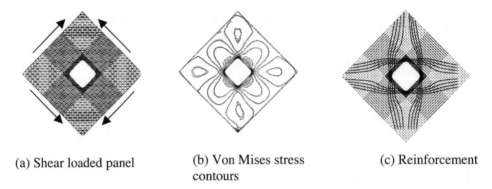

(a) Shear loaded panel

(b) Von Mises stress contours

(c) Reinforcement

Figure 9-18. Fiber steering embedded in a composite component.

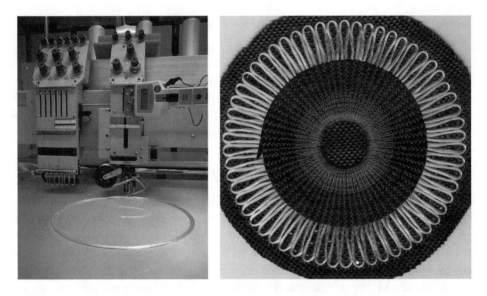

Figure 9-19. Stitching tows onto a preform and typical fiber architecture for a load bearing hole. Reprinted with the permission of Hightex Verstarkungsstrukturen GmbH, Dresden.

level of optimization would have dropped tows remote from the corners but this was beyond the capability of the robotic laying procedure that was available for manufacture of the panels. In this application the load paths for the tensile and compressive loads parallel to the horizontal and vertical axes are indistinguishable from trajectories defined using the principal stress vectors. Figure 9-19 shows a tow placement machine and fiber architecture typical of a load-bearing hole. The tows are stitched onto a dry preform prior to injection of resin to create the component.

9.9 Conclusion

In this chapter, the application of the finite element method to the design and analysis of aerospace structures has been reviewed. The key driver in the design of these structures is minimizing structural weight while meeting the performance and regulatory requirements. The discussion has focused on the analysis of thin-walled structures of stressed skin configuration. The integration of CAD, analysis, and manufacturing and the emergence of KBE and automation as standard tools have been discussed. The loading on the airframe is provided by the fluid flow and while computational fluid dynamics (CFD) is advancing rapidly, integrated CFD/FEA systems are yet to be applied to global analysis of aerospace structures.

The main development in the last decade has been the emergence of carbon fiber composite structures as the preferred material for design. This chapter has therefore focused on the challenges that exist for using computational analysis to support the adoption of this new material. The future development of composite technology will include the exploitation of the fiber architectures of these materials. This is immediately apparent in tensile applications dominated by a single load case. More general applications will depend on the development of new technologies that could, for example, separate the compression and shear carrying functions into different structural components thereby reinventing the stressed skin configuration.

Several topics have not been included in the chapter but this should be seen as a problem of space rather than relevance. For example, the interested reader should investigate the stochastic processes of robust design that are being applied in the industry to avoid imperfection sensitive structures. The fundamental concept is to recognize variability in the design parameters and search for outliers that represent a combination of unlikely events that could lead to failure in a structure. For composite materials an obvious candidate is the final configuration of the plies following the curing process. Curing requires placing the laminate under pressure at elevated temperature to achieve high fractions of fiber relative to resin. In complex geometries the fibers can be displaced and are only located by destructive sectioning and examination. The identification of low probability events will become more significant as we move closer toward zero tolerance to failure.

Acknowledgments

Parts of the postbuckling aileron design case study were originally published in Thomson and Scott (2000), and are reprinted here with permission from Elsevier Publishers, UK.

References

Adams, D. B., Watson, L. T., Gürdal, Z., and Anderson-Cook, C. M. (2004). "Genetic algorithm optimization and blending of composite laminates by locally reducing laminate thickness." *Advances in Engrg. Software*, 35(1), 35–43.

Ashizawa, M., and Toi, Y. (1990). "Unique features and innovative application of advanced composites to the MD-11." *Proc., AIAA/AHS/ASEE Aircraft Design, Systems and Operations Conference, AIAA-90–3217*, American Institute of Aeronautics and Astronautics, Dayton, Ohio.

Baker, A., Dutton, S., and Kelly, D., eds. (2004). *Composite materials for aerospace structures*, AIAA Education Series.

Bruhn, E. F. (1965). *Analysis and design of flight vehicle structures*, Tri-State Offset Co., Ohio.

Daniel, I. M., and Ishai, O. (1994). *Engineering mechanics of composite materials*, Oxford University Press, New York.

Gosse, J. H., and Christensen, S. (2001). "Strain invariant failure criteria for polymers in composite materials." *AIAA J., 2001–1184*, American Institute for Aeronautics and Astronautics, Dayton, Ohio.

Gürdal, Z., Haftka, R. T., and Hajela, P. (1999). *Design and Optimization of Laminated Composite Materials*, Wiley-Interscience, New York.

Hart-Smith, L. J. (1986). "Lessons learned from the DC-10 carbon-epoxy rudder program." *Douglas Aircraft Paper 7734*, Douglas Aircraft Company, Long Beach, Calif.

Hart-Smith, L. J. (1993). "Innovative concepts for the design and manufacture of secondary composite aircraft structures." *Proc. 5th Australian Aeronautical Conf., McDonnell Douglas Paper MDC 93K0081*, The Institution of Engineers, Melbourne, Australia.

Hawley, A. V. (1986) "Ten years of flight service with DC-10 composite rudders—A backward glance." *Proc. SAE Aerotech '86, Douglas Aircraft Paper 7733*, Society of Automative Engineers, Long Beach, Calif.

Kelly, D. W., Hsu, P., and Asadullah, M. (2001). "Load paths and load flow in finite element analysis." *Engrg. Computations*, 18(1,2), 304–313.

Li, R., Kelly, D. W., Arima, S., Willgoss, R. A., and Crosky, A. (2002). "Fiber steering around a cut-out in a shear loaded panel." *Proc., 47th Int. SAMPE Symp.*, B. M. Rasmussen, L. A. Pilato, and H. S. Kliger, eds., Long Beach, Calif., 1853–1861.

Montgomery, D. C. (1991). *Design and analysis of experiments*, Wiley, New York.

Nagendra, S., Jestin, G., Gürdal, Z., Haftka, R. T., and Watson, L. T. (1996). "Improved genetic algorithm for the design of stiffened composite panels." *Computers and Struct.*, 58(3), 543–555.

Nui, M. C. Y. (1992). *Composite airframe structures, practical design information and data*, Conmilit Press Ltd., Hong Kong.

Salamone, T. A. (1995). *What every engineer should know about concurrent engineering*, Marcel Decker Inc., New York.

Schneider, C. W., and Gibson, D. C. (1985). "Design and certification of a composite control surface." *Composites: Design and manufacturing for general aviation aircraft, SAE SP-623*, Society of Automotive Engineers, Wichita, Kan.

Suzuki, K., and Kikuchi, N. (1991). "A homogenization method for shape and topology optimization." *Comp. Meth. in App. Mech. and Eng.*, 93, 291–318.

Thomson, R. S., and Scott, M. L. (2000). "Experience with the finite element modeling of a full-scale test of a composite aircraft control surface." *Composite Struct.*, 50(4), 331–345.

10

Ship Structures

Jeom Kee Paik and Owen F. Hughes

10.1 Analysis of Ship Structures

During the last few decades there have been significant changes in the size and types of ships, but an even bigger change has occurred in the way that the structure of a ship is designed. In the past, the analysis and design of merchant ship structures was done using empirical classification society rules and basic structural theory, the former being based on accumulated experience and the performance records of successful ships. Due to the complexities of modern ship structures and the demand for greater reliability and integrity, the calculation of action effects (e.g., stress, deformation) has been increasingly undertaken by finite element method (FEM), although classical theory of structural mechanics and previous experience still play an important role. The large size of ship structures requires both global and detailed finite element modeling, and interaction between them. Guidelines for analysis and design of merchant ship structures are provided by the classification societies that include American Bureau of Shipping (USA), Bureau Veritas (France), Det Norske Veritas (Norway), Germanischer Lloyd (Germany), Korean Register of Shipping (Korea), Lloyd's Register (UK), and Nippon Kaiji Kyokai (Japan), among others. Recently, Common Structural Rules (CSR) were developed by the International Association of Classification Societies (IACS).

This chapter reviews the characteristics of typical merchant ship structures and the procedures and assumptions involved in their finite element analysis. Current developments in nonlinear finite element analysis of ship structures to determine the load-carrying capacity are also presented.

10.2 Nature and Structural Characteristics of Merchant Ships

The size and principal characteristics of ships are determined by their mission or intended service, and as a result ships are among the largest and most complex mobile structures in existence. The merchant ship of today is designed for a specific type of cargo rather than general cargo. At the overall level a merchant ship structure is similar to a box girder.

In contrast to land-based box girders, however, a ship does not rest on a fixed foundation, but it is supported by buoyant pressures exerted by a dynamic and ever-changing fluid environment. Therefore, ship structures are likely subjected to various types of actions including static, dynamic, and impact actions, all of which vary with operating conditions (e.g., laden or ballast, ship speed, and heading) and sea states, and they must be designed so that all of these actions can be sustained without structural failure. The action variability and the geometric complexity of modern ship structures give rise to difficulties in structural analysis.

Ship structures are composed of a number of structural components that normally serve two or more functions, while their arrangements depend primarily on the type of cargo. For instance, a ship hull consists of deck, side shell, and bottom shell that are stiffened in the lengthwise and girthwise directions by support members at certain spacings. Besides being the principal structural components, they must also be watertight. Likewise, bulkheads are designed not only to provide water (or oil) tight boundaries between the internal compartments, but also to contribute to overall strength.

In the subsequent sections, the nature and structural characteristics of some typical merchant cargo ships are described.

10.2.1 Bulk Carriers

Bulk carriers are designed for carrying dry bulk cargoes such as iron ore, coal, grain, bauxite/alumina or phosphate rock, which are loaded directly into the cargo space without any intermediate form of containment.

The size of bulk carriers is affected by various factors such as storage capacity, water depth at the berth, and regularity of the demand for the cargo. This large variety of influential factors has created various types of bulk carriers, as measured by their displacement (total weight, including cargo). Principal types are handysize bulkers (25,000–50,000 tonnes), handymax bulkers (35,000–50,000 tonnes), Panamax bulkers (50,000–100,000 tonnes, and able to pass through the Panama Canal with principal dimensions limited to a length of 289.5 m, a beam of 32.3 m and a draft of 12.04 m), Capesize bulkers (100,000–180,000 tonnes) and very large bulk carriers (VLBCs) (over 180,000 tonnes). For dense cargoes such as iron ore, the cargo is usually loaded in every alternate hold.

Together with a double bottom, the bulk carrier structures typically have triangular wing tanks at the upper two corners and triangular hopper tanks at the lower two corners, as shown in Figure 10-1. The shape of the wing tank is influenced by the angle of repose of the cargo, while the shape of the hopper tank is to allow the cargo to slide toward the centerline as the cargo hold becomes nearly empty. Side frames are attached at inner side shell between the bottom of the wing tank and the top of the hopper tank. To make loading / unloading of bulk cargoes efficient, transverse bulkheads have corrugated sections in order to have a smooth surface on

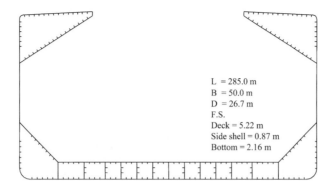

L = 285.0 m
B = 50.0 m
D = 26.7 m
F.S.
Deck = 5.22 m
Side shell = 0.87 m
Bottom = 2.16 m

Figure 10-1. Schematic representation of midsection of a 170,000 DWT single-sided bulk carrier.

both sides. They also have upper and lower transverse box structures with sloping sides, called stools, whose primary purpose is to reduce the span of the bulkhead between deck and inner bottom. Also, bulk carriers have relatively large hatches.

10.2.2 Oil Tankers

Tankers are designed for the transport of crude oil or oil products. Most crude oil carriers are very large, carrying a cargo of over 200,000 tonnes, while oil product carriers are commonly small or medium. For tankers with large breadth, longitudinal vertical bulkheads extend continuously from bow to stern. Depending on the arrangement of these longitudinal bulkheads, the ship classification societies usually define three standard types of tankers:

- Type A are tankers with two longitudinal bulkheads. Typical examples of Type A tankers are very large crude oil carriers (VLCCs) and shuttle tankers with cross ties, and smaller tankers without cross ties.
- Type B are tankers with one (centerline) longitudinal bulkhead. Typical examples of Type B tankers are Suezmax tankers (designed to be able to pass through the Suez Canal), Aframax tankers, and shuttle tankers.
- Type C are tankers without longitudinal bulkheads. These are usually small tankers.

In typical tanker structures, oil-tight transverse bulkheads divide the ship into several cargo holds. Following the Exxon Valdez grounding in 1989, all new oil tankers must have a double bottom and double sides, in order to minimize oil outflow in collision and grounding, as shown in Figure 10-2. Ballast water must be stored in segregated (special purpose) ballast spaces rather than in empty cargo tanks in order to prevent pollution when deballasting. Transverse girders are placed below deck structures extending from inner side shell to an adjacent longitudinal bulkhead, or between two longitudinal bulkheads, or between two inner side shells. For large tankers, horizontal cross ties at the mid-depth of the ship are sometimes placed at the same locations.

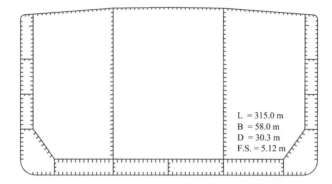

Figure 10-2. Schematic representation of midsection of a 313,000 DWT double hull tanker with two side-longitudinal bulkheads.

10.2.3 Container Ships

Container ships are designed for carrying containers, which are of standard width (8 ft), height (usually 8 ft) and length: 20 ft, 40 ft, 45 ft, and 48 ft, the first two being the most common. Combinations of sizes are usually expressed in terms of Twenty-Foot Equivalent Units (TEU). The size of a container ship is normally determined by the operating route and the trade pattern. There are currently five sizes of container ships: feeder ships (100–2,000 TEU), sub-Panamax ships (2,000–3,000 TEU), Panamax ships (about 4,800 TEU and able to pass through the Panama Canal), post-Panamax ships (larger than Panamax), and very large container ships (over 9,000 TEU).

About 60% of containers are loaded in cargo holds under the deck, with cell guides to hold them in place, and the rest are loaded on the deck, with steel lashing rods to hold them in place.

A special feature of container ship structures is a very wide hatch opening as shown in Figure 10-3, which implies that torsional strength is of a primary concern in their analysis and design. Double bottom and double sides provide closed cells

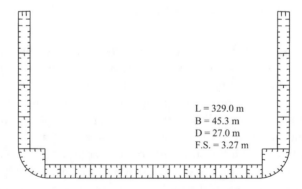

Figure 10-3. Schematic representation of midsection of a 9,000 TEU container.

in the hull cross section to increase the torsional rigidity of the hull, while they also provide space for ballast water. In some container ships, closed box structures are placed in the longitudinal and transverse directions to further increase the torsional rigidity of the hull.

10.3 Ship Structural Actions

Ship structures are subjected to various types of actions, which may be grouped according to their time duration: static actions, low-frequency dynamic actions, high-frequency dynamic actions and impact actions (Paulling 1988).

Static actions are those arising from the weight and buoyancy of the ship. Low-frequency dynamic actions occur at frequencies that are sufficiently low compared to the frequencies of the vibratory response of the ship hull and its parts so that the resulting dynamic effects on the structural response are relatively small. Such actions include hull pressure variations induced by waves or oscillatory ship motions, and inertial reaction forces resulting from the acceleration of the mass of the ship and its cargo or ballast water.

High-frequency dynamic actions have frequencies that approach or exceed the lowest natural frequency of the hull girder. A typical example is wave-induced springing (flexural vibration of the hull girder) that may occur when the natural period of the hull girder overlaps the period of shorter components of the encountered waves. Since springing occurs at a higher frequency than that of ordinary wave-induced bending, it increases the number of stress cycles during the ship's lifetime, thus increasing the hull girder fatigue damage.

Impact actions are those whose duration is even shorter than the period of the high-frequency dynamic loads. Examples of impact loads are slamming and green water impact on deck. Slamming causes a sudden upward acceleration and deflection of the bow and excites hull girder flexural vibration in the first two or three modes, typically with a period in the range of 0.5 to approximately 2 sec. This slam-induced vibration is termed *whipping*.

In ship structural analysis and design, the most common actions are the static and low-frequency dynamic actions, the latter usually being treated as static or quasi-static actions. The high-frequency dynamic actions and impact actions can be important in specific design cases.

Since the characteristics of ship structural actions vary significantly depending on loading/unloading, operating conditions and sea states, all potential conditions during the ship's lifetime must be taken into account in the analysis and design of ship structures. Flooding and damaged conditions should also be considered.

Information about estimating ship structural actions can be found in textbooks such as Paulling (1988) and Hughes (1988). For calculating the design actions of merchant ships the classification societies and IACS provide simplified formulae or guidelines, while a direct calculation of ship structural actions is in principle recommended for various patterns of loading and operational conditions.

As the most primary hull girder action component, for instance, the total vertical bending moment M_t of ship hulls can be defined as the extreme algebraic sum of still-water moment and wave-induced moment, taking account of dynamic action effects, as follows

$$M_t = k_{sw} M_{sw} + k_w (M_w + k_d M_d) \qquad (10\text{-}1)$$

where k_{sw}, k_w and k_d are action combination factors for still-water bending moment M_{sw}, wave-induced bending moment M_w, and dynamic bending moment M_d, respectively, the last arising from either slamming or whipping.

M_{sw} is taken as the maximum value of the still-water bending moment resulting from the worst load condition on the ship in both hogging and sagging. A detailed distribution of the still-water moment along the ship's length can be calculated by a double integration of the difference between the weight force and the buoyancy force, using classical beam theory, that is, regarding a ship hull as a beam.

M_w is taken as the mean value of the extreme wave-induced bending moment that the ship is likely to encounter during its lifetime. For the safety and reliability assessment of damaged ship structures, a short-term analysis using hydrodynamic strip theory (so called because the hull is idealized as a series of short prismatic sections, or girthwise strips) may be used to determine M_w when the ship encounters a storm of specific duration (e.g., 3 hours) and with certain small encounter probability. On the other hand, a long-term analysis will be employed to determine M_w for newly built ships, and the design formula provided by IACS is normally used. In calculating M_w, a second order strip theory may need to be used in order to distinguish between sagging and hogging wave-induced bending moments.

M_d is taken as the extreme dynamic bending moment in the same wave condition (e.g., sea states) as the wave-induced bending moment while the effect of ship hull flexibility is accounted for in the computation of M_d. For whipping in very high sea states, M_d is normally ignored, while for slamming, it is approximately taken as $M_d = 0.15 M_w$ for tankers in sagging, but $M_d = 0$ in hogging (Mansour and Thayamballi 1994).

While external pressure actions imposed on the ship hull in seaways can be calculated in terms of sea water heads, the internal pressure actions must be determined for each fully loaded cargo hold and ballast tank as caused by the dominating ship motions (pitch and roll) and the resulting accelerations.

The longitudinal bending stresses σ_x acting on a typical merchant ship hull under vertical bending moment can then be estimated by classical beam theory, as follows

$$\sigma_x = \frac{M_t}{Z} \qquad (10\text{-}2)$$

where $Z = I/y =$ section modulus, $I =$ moment of inertia of the hull cross-section, $y =$ distance from the neutral axis to the point of stress calculation.

Using classical theory of structural mechanics alone, it is not straightforward to calculate the action effects of a ship hull under combined actions that include local actions, global shear forces and torsional moments as well as vertical bending moment. The use of the FEM is essential in this regard.

10.4 Ship Structural Action Effect Analysis

For convenience of finite element analysis of complex ship structures, the response of ship structures under applied ballast/cargo loading and sea conditions may be classified into the following five levels:

- global structure (or hull girder),
- cargo hold (or hull module),
- grillage,
- frame and girder, and
- local structure.

For each action case the resulting action effects at each level are calculated by the FEM and are then combined appropriately, using correlation factors relevant for that action case. The response at each level provides the boundary conditions for the next lower level analysis. At each level the finite element modeling is slightly different for different ship types.

10.4.1 Types of Analysis

10.4.1.1 Static or dynamic. The analysis at each structure level may need to include a dynamic structural analysis, depending on whether that level of structure is subjected to any significant rapidly varying actions, that is, actions for which the shortest component period is the same order of magnitude or shorter than the longest natural period of that level of structure.

At the hull girder and cargo hold levels, a wave-excited dynamic analysis is usually not required for most merchant ship types, while a calculation of hull girder natural frequency is nearly always performed. For relatively flexible ships including some container vessels or naval ships that are susceptible of springing, however, a dynamic analysis may be required at hull girder and cargo hold levels.

At the principal member and local structure levels, a vibration analysis may be required if there are some significant and unavoidable sources of excitations (e.g., propellers, machineries). In most cases the only requirement is to calculate the natural frequencies and to design the structure so as to avoid resonance.

10.4.1.2 Deterministic or probabilistic. For most types of merchant ships for which the hull girder actions are already well established and their characteristic values are available, only the deterministic analysis is required. However, the probabilistic analysis should be used for ships if these are not well established such that characteristics values of wave-induced hull girder actions do not exist. If a probabilistic analysis is required, it is usually required only at the hull girder level because the most uncertain action is normally the wave action.

At the lower structure levels, the characteristic values of action effects are obtained from the hull girder level analysis and thus the analysis at these lower levels is usually deterministic. On the other hand, for limit state analysis that deals with structural failures, the situation is reversed because the principal source of uncertainties associated with structural failures arises at the principal member level or local structure level. Therefore, the limit state analysis is probabilistic at the lower structure levels, while it is deterministic at the higher structure levels.

10.4.1.3 Linear or nonlinear. Wave response is dynamic, probabilistic, and nonlinear in nature. For simplicity, a linear analysis is often used under several simplifying assumptions: (a) the irregular wave surface of the ocean can be represented as the linear sum of a large number of individual regular waves of different heights and frequencies, (b) the hydrodynamic forces on a ship hull can be obtained by considering each transverse section of the ship separately and combining the results linearly, and (c) the wave force acting on each section is linearly proportional to the difference between the local wave height and the ship's still water plane.

The accuracy of the first two assumptions is usually satisfactory, while the third is valid for ships which are approximately wall-sided in the water plane region. If this is not so or if there is any other source of nonlinearity, a nonlinear method of response analysis should be used.

10.4.2 Extent of Modeling

It is important to realize that whenever parts of ship structures (i.e., not the entire ship structure) are modeled and analyzed, the results of the finite element analyses strongly depend on the boundary conditions of the model. The more local the model, the stronger is this dependency. The extent of analysis for each model must be large enough so that the structural area of interest will be relatively unaffected by approximations in the boundary conditions.

10.4.2.1 Global structure model. The global structure model is used to investigate the response of the overall ship hull and its primary strength members to both still water and wave-induced hull girder actions. At this level the primary concern is the overall stiffness and the global or nominal stresses of primary strength members along the entire ship length rather than local or detailed stresses. The resulting nominal stresses in primary structural members must be smaller than the allowable stresses defined by the classification societies. The nominal stresses are also utilized for checking the ultimate or failure strength of primary support members (e.g., girders, frames, transverse webs) based on yielding and buckling.

Figure 10-4 shows a typical global model of a ship hull. Figure 10-5 represents a zoomed-in picture of the global model for a container ship. For container ships, which have large hatch openings, the torsional response is governed by the structural arrangement and the load distributions over the entire length of the ship. Therefore, the finite element model is usually the full length of the ship. The model can be of half breadth, but only if the particular finite element program being used can accommodate large openings and asymmetric actions. In addition to deck house structures, all longitudinal members and all primary transverse members (e.g., bulkheads, cross decks, transverse webs) are included in the model. Structural members that do not contribute to the global strength of the ship may be omitted, but their masses should be included in the model.

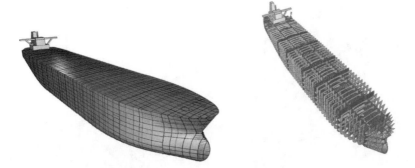

Figure 10-4. Examples of the global hull girder model.

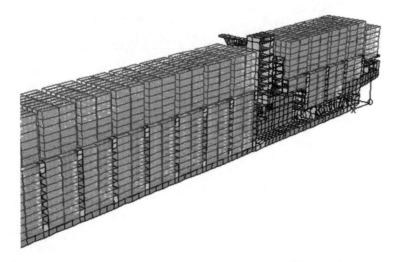

Figure 10-5. A zoomed-in picture of the global finite element analysis model for a container ship structure.

10.4.2.2 Cargo hold model. The cargo hold model is used to examine the response of the primary strength members in the middle portion of the hull girder (i.e., several cargo holds) under the action of internal cargo (or ballast water) and external water pressure. The boundary conditions are obtained from the global structure analysis, or are defined by the ship classification society guidelines.

Figures 10-6 and 10-7 show typical finite element models for the cargo hold level analysis of conventional bulk carrier and tanker structures, respectively. The extent of the cargo hold analysis depends on the ship type, the loading conditions, and the degree of symmetry of the hull structure in the longitudinal and transverse directions.

While some classification societies recommend taking two cargo hold lengths (i.e., $1/2 + 1 + 1/2$) at amidships or in the forward half, others recommend taking three hold lengths (i.e., $1 + 1 + 1$), as those shown in Figures 10-6 and 10-7. A half

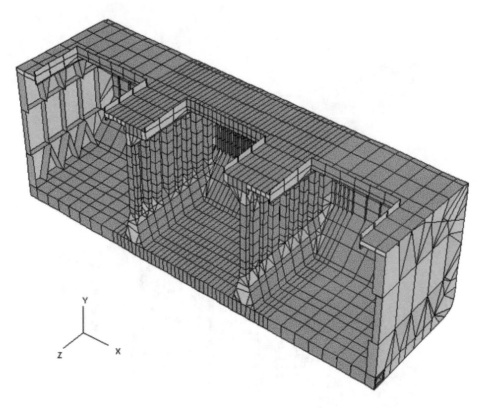

Figure 10-6. Example of the finite element model for the cargo hold analysis of a conventional bulk carrier structure.

breadth model may be used, but again only if the finite element program being used can correctly model asymmetric actions.

10.4.2.3 Grillage model. The grillage model is used to investigate overall or local strength behavior of a continuous plated structure supported by both longitudinal girders (or stiffeners) and transverse frames, under a lateral pressure, or other action that is normal to the plane of the grillage. Examples are double bottoms, bulkheads, or decks. Figure 10-8 shows a typical example of the grillage model. The main purpose of a grillage analysis is to determine how the lateral action is transferred to the boundaries of the grillage and to calculate the action effects.

10.4.2.4 Frame and girder model. The frame and girder level is related to the response of 2D or 3D frame structures such as transverse web frame systems or longitudinal girder systems, including the flanges that are provided by the associated plating. Figures 10-9a and 10-9b show examples of the 2D and 3D frame and girder models, respectively. The purpose of these frame and girder analyses is to examine the bending and shear behavior in the plane of the structure web, and also torsion, for which fine mesh modeling is required. This level of analysis may be included as part of the cargo hold analysis. Alternatively, it may be undertaken separately with the boundary conditions obtained from the cargo hold analysis.

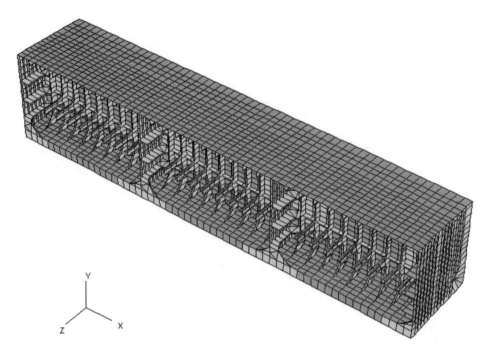

Figure 10-7. Example of the finite element model for the cargo hold analysis of a conventional tanker structure.

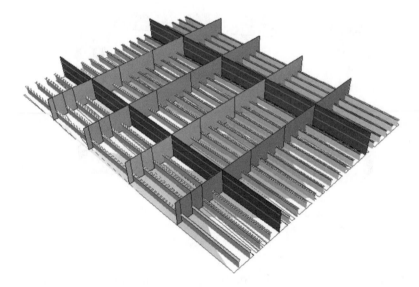

Figure 10-8. Example of the grillage model.

10.4.2.5 Local structure model. The local structure model is used to investigate the response (e.g., stresses) of local or special structural members and of structural details. An example of a local member is a laterally loaded plate stiffener with its connecting brackets, subject to relative deformations between end supports. In ship structures,

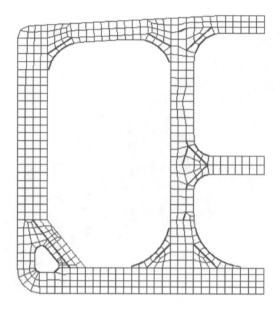

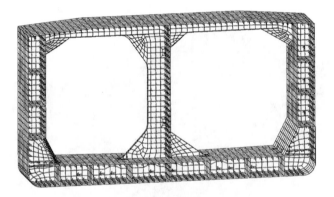

Figure 10-9. (a) Example of the 2D transverse frame model. (b) Example of the 3D frame and girder model.

fatigue analysis and design is usually done using the S-N approach ($S =$ fluctuating stress, $N =$ associated number of cycles). This approach can be used for stresses at three progressively more detailed levels: nominal stress, hotspot stress, and notch stress, and each level requires a different set of S-N curves, while for ship structures the hotspot level is the most common.

Figure 10-10 shows an example of a local finite element model to examine the stress concentration in discontinuities.

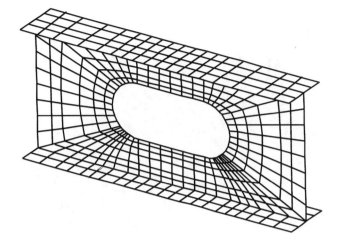

Figure 10-10. Example of the local structure level model for a cutout.

10.4.3 Types of Finite Elements

In structural modeling, the types and numbers of finite elements must be selected so that they will be able to accurately portray the stiffness and stresses of the structure to be analyzed, while they may also depend on the analysis levels noted above. For this purpose it is important to better understand the characteristics of the various element types that are available in current finite element computer programs. For numerical analysis of merchant ship structures the following types of elements are typically employed:

- Truss element, which is a 1D element with axial stiffness alone without bending stiffness
- Beam element, which is a 1D element with axial, shear, bending, and torsional stiffnesses
- Membrane stress element (or called plane stress element), which is a 2D element with membrane stiffness in the plane, but without out-of-plane bending stiffness
- Plate-shell element, which is a 2D element with membrane, out-of-plane bending, and torsional stiffnesses
- Solid element, which is a 3D element
- Boundary and spring element
- Point or mass element

A finite element model usually involves several types of elements. In this case, one should be careful to achieve compatibility in the displacement functions of the elements and smooth transfer of the boundary actions and stresses, particularly when both membrane stress and plate-shell elements are involved, because this combination is error prone.

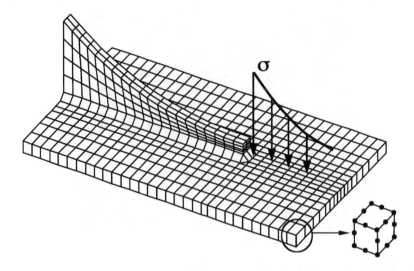

Figure 10-11. Example of higher order solid elements for local structure level analysis.

It is usually very important to understand whether the effect of out-of-plane bending can be ignored or not in the strength analysis of structural members. For the upper four levels of modeling, plate-shell and beam elements are normally recommended. Beam or truss elements are often used to model the face plates of primary members such as vertical webs and horizontal stringers of transverse bulkheads.

The local level usually uses solid elements, and these may be higher order elements with midside nodes as shown in Figure 10-11. Modeling with finer meshes of the plate-shell elements is also considered. If a local structure to be analyzed is subjected to predominantly in-plane actions, it may be idealized by a 2D model using membrane stress elements.

Point or mass elements may be used to represent the corresponding nodal forces due to ballast water or cargoes. For analysis of container vessel structures, containers are sometimes modeled by mass elements. The equivalent concentrated forces in terms of container size, mass, and acceleration are applied at contact nodes to the ship hull.

10.4.4 Geometric Idealization and Meshing

Idealization of structural geometry must be undertaken to relevantly represent the global and local stiffness of the structure. The mesh fineness is in principle determined depending on the types and characteristics of elements to be used so that it should represent the actual stiffness and distribution of lateral pressure actions, resulting in action effects with sufficient accuracy.

10.4.4.1 Global structure model. For finite element analysis at the global structure level, a coarse mesh modeling extending over the entire hull length may be adopted as long as it relevantly reflects the overall stiffness and global stress distribution along the ship length as shown in Figures 10-4 and 10-5.

All primary longitudinal and transverse members (e.g., bottom shell, side shell, deck plating, bulkhead plating, stringers, girders, and transverse webs) are best modeled by quadrilateral plate-shell elements. Triangular plate-shell elements are used where necessary, but their stresses must be converted from element to member coordinates. Longitudinal members may be given their total or as-built scantlings, while transverse members are given their net (minimum rule) scantlings, which excludes a corrosion margin. Support members that do not involve a deep web may be modeled by beam or truss elements.

Stiffened panels and grillages may be modeled as an assembly of plate-shell elements and beam elements. Alternatively, they may be modeled by anisotropic plate-shell elements. For deck house structures, only primary members are included in the model while local stiffeners may be omitted if desired. Small secondary members and structural details that do not affect the overall stiffness and strength of the ship hull may also be omitted; for example, brackets at frames, and sniped short stiffeners that prevent local buckling and small cutouts. However, large cutouts should either be directly modeled or approximated by using an equivalent reduced plate thickness.

Items that do not contribute to structural stiffness such as hatch covers and main/auxiliary engines are modeled by concentrated or distributed mass elements in the relevant locations.

10.4.4.2 Cargo hold model. At the cargo hold level, the finite element model is usually coarse mesh, with several fine mesh submodels as shown in Figures 10-6 and 10-7. The model must relevantly reflect the stiffness and stresses of principal strength members such as plating, support members (e.g., stiffeners, girders, frames) and brackets. Girder webs with cutouts may be modeled by noncutout webs with reduced equivalent thickness. The net scantlings (i.e., without corrosion allowance) are used.

Four-noded plate-shell or membrane elements together with two-noded beam or truss elements are typically employed for finite element modeling of the cargo holds. Higher order elements (e.g., six- or eight-noded elements) may be employed when a coarser mesh model is used. Decks, shell, inner bottom, and longitudinal bulkhead plates are modeled by plate-shell elements so that lateral pressure actions can be applied. As previously noted, mixing membrane and plate-shell elements in a 3D model is not recommended. Face plates of primary members such as vertical webs and horizontal stringers of transverse bulkheads can be modeled by beam or truss elements.

10.4.4.3 Grillage model. The grillage structure is typically modeled by beam elements with attached effective plating as shown in Figure 10-8. A fine mesh of the beam elements is normally needed to accurately reflect the deflection behavior of the plate-beam combinations under lateral action.

10.4.4.4 Frame and girder model. As shown in Figure 10-9, one plate-shell element is typically used to model plating between stiffeners. Three or more elements are required over the height of web frames or girders so that the stresses of plate webs should be readable without interpolation or extrapolation.

The mesh should be fine enough to represent the large openings in the web frame. One or more elements are needed for webs and flanges of corrugated sections. The mesh size of large brackets may be equal to the stiffener spacing or smaller. At the bracket toes, the rectangular mesh may be terminated at the nearest node of the primary structure so that the last triangular shape element may be omitted.

10.4.4.5 Local structure model. As shown in Figure 10-10, a fine meshing which provides a "good" aspect ratio (i.e., length/breadth ≈ 1.0) of the plate-shell elements is generally required to reflect the behavior of the local structure under large deformations. Three four-noded plate-shell elements are typically used for the web height of the stiffeners and for plate flanges. Beam elements may be used to model the face plates, while plate-shell elements may be employed for effective flanges of curved areas.

At least three plate-shell elements are typically used to model plating between stiffeners, but much more than three plate-shell elements are normally required over the height of web frames or girders. When the structural details are modeled by plate-shell elements, the appropriate mesh size is typically of the order of plate thickness, t (i.e., a "$t \times t$ mesh"), to examine hotspot stresses.

10.4.5 Conditions at Boundaries and Supports

Relevant conditions at the boundaries and supports must be provided by suppressing or prescribing translational displacements and rotations to prevent the rigid body motion of the model or to model the interaction between neighboring and considered structural areas along the model boundaries or to represent the constraints of existing supports. In principle, the conditions at boundaries or supports should not cause unrealistic constraints in terms of displacements or rotations.

For the global level analysis, sufficient degrees of freedom (normally six) are constrained to prevent the rigid body motion of the model. The translational supports should be located away from the areas where the stresses are of interest. Forces in the constrained nodes (i.e., translational supports) may be eliminated by generating balanced forces. The global structure model should typically be balanced to within 1% of the vessel's displacement. The remaining unbalanced forces at the constrained nodes should be eliminated by distributing equivalent inertia forces throughout the global analysis model. Alternatively, rigid body motion of the global structure model can be handled by using distributed spring elements over the hull surface which represent the reaction forces to submersion of the ship hull.

For lower level analyses, symmetric boundary conditions can often be applied considering the symmetry related to structural arrangements and load application. Also, the boundary conditions may be prescribed based on the load effects obtained from a higher level. The (translational or rotational) displacements or forces may be applied at the boundaries of the model.

To explain the boundary conditions of the finite element model, a more detailed example is herein considered for the cargo hold analysis of a bulk carrier structure as shown in Figure 10-12.

As shown in Figures 10-12 and 10-13, the model extends to the midlengths of holds 3 and 5. Cargo hold 4 is empty (as are holds 2 and 6), and holds 3 and 5 are full. All primary strength members including frames and stiffeners are included in the modeling. While one plate-shell element is used between longitudinal stiffeners in the transverse and vertical directions, three plate-shell elements are used between bottom floors. For beam elements, the attached plating has a reduced or effective breadth to allow for shear lag. Elements in the center plane of the ship have a half thickness.

Because of the half breadth model, symmetric conditions are generally applied with regard to the center line. However, the boundary conditions at the ends of the model are not very easy to come by. If the model is subjected to uniform lateral actions alone, as shown at the top of Figure 10-14, symmetric conditions can also

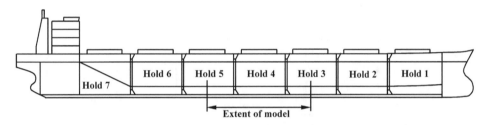

Figure 10-12. General arrangement of a typical bulk carrier.

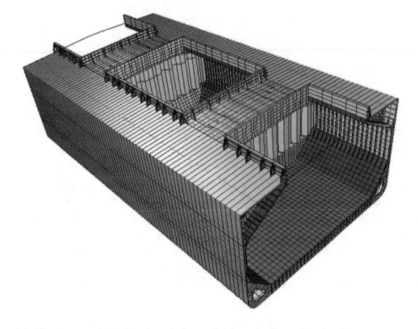

Figure 10-13. A cargo hold level analysis model for a bulk carrier structure.

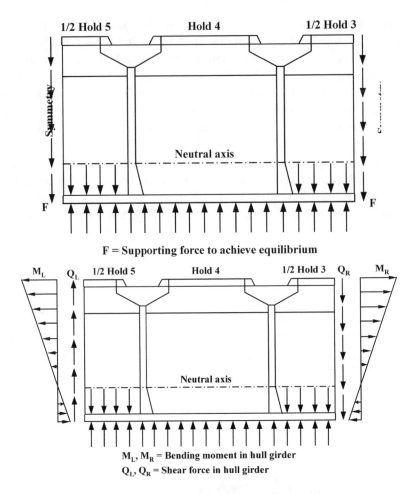

Figure 10-14. Boundary conditions of a cargo hold analysis model.

be applied with reasonable certainty even at the ends, and the additional stresses due to global hull girder bending may be superimposed on the results by the former model analysis.

On the other hand, the ends of the model can shift and rotate if hull girder bending and shearing forces are applied as shown at the bottom of Figure 10-14. In this case, distributed displacements or forces which can be obtained from the results of the global structure analysis may be prescribed over the cross section at the ends, assuming that a plane cross section of the hull remains plane.

Alternatively, it may be considered that the cargo hold model is supported in the vertical direction by vertical springs along the intersections of the side and the transverse bulkhead, between the inner side and the transverse bulkhead, and between the longitudinal bulkhead and the transverse bulkhead. The spring constants are uniformly distributed along the corresponding intersections. Instead of application of the vertical springs, vertical forces may be applied along the intersections mentioned above, but the displacement of one nodal point at each intersection is additionally fixed in the vertical direction to remove the rigid body motion.

When the mid cargo hold is empty and the fore and aft cargo holds are full, as in Figure 10-13, longitudinal compressive forces may be generated in the cargo hold model because of the boundary conditions at both ends. To remove these fictitious compressive forces, therefore, counteracting horizontal forces are applied at one point around the bottom part of one end of the model. The plane where the concentrated horizontal forces are applied at the point must remain in plane so that the nodes of that plane must move linearly in the longitudinal direction.

The boundary conditions for the cargo hold model under hull girder loads are of course different from those under local loads mentioned above. For hull girder vertical bending, some classification societies suggest that one end of the cargo hold model is completely fixed, and the other end deflects as a rigid plane. Other classification societies use a simple support at each end of the cargo hold model.

For vertical shearing forces, the symmetric boundary conditions are applied at both ends. For the cargo hold model covering a half breadth of the ship under vertical shearing forces, symmetric boundary conditions are applied along the centerline. For the cargo hold model covering a full breadth of the ship under vertical shearing forces, the intersections between the transverse bulkhead and the longitudinal center girder at inner bottom are fixed in the transverse direction to prevent the rigid body motion.

10.4.6 Load Application

For the global structure level analysis, various standard hull girder action conditions (e.g., vertical still water bending, vertical wave-induced bending, horizontal bending, wave-induced torsion) and their combinations are considered. Ballast water or cargoes are distributed into the corresponding nodal points using mass elements. Local pressure distribution of the tanks need not be considered for the global structure level analysis. Static and hydrodynamic external water pressure actions are applied to the external plate-shell elements which form the envelope of the ship hull.

As shown in Figure 10-15, the magnitude of pressure actions acting on the transverse bulkheads are calculated for the worst cases, when the ship is at a wave crest and a wave trough, either from the classification society rules or directly. These pressure actions are applied as equivalent nodal forces at the related nodes.

The sectional forces and moments will also be applied at both ends of the cargo hold model, as shown in Figure 10-16. Once the sectional forces, Q_L, and moments, M_L, at the left end of the model are specified, the corresponding sectional forces, Q_R, and moments, M_R, at the right end of the model may be determined to satisfy the equilibrium, as follows

$$Q_R = Q_L + \sum_i P_i, \quad M_R = M_L - Q_L \ell - \sum_i P_i(\ell - x_i) \qquad (10\text{-}3)$$

where P_i = vertical forces acting at the distance, x_i, from the left end of the model.

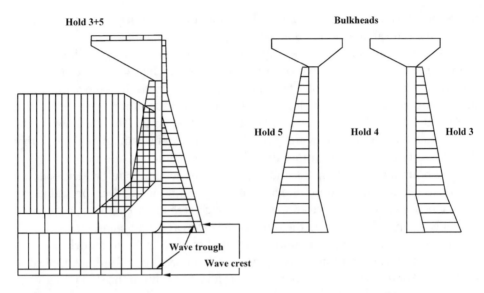

Figure 10-15. Example of pressure actions acting on the cargo hold model.

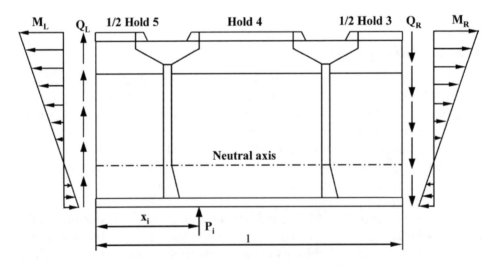

Figure 10-16. Sectional forces and moments at the ends of the cargo hold analysis model.

A set of the sectional forces and moments are selected from those obtained for various load application, operating conditions and sea states, which give maximum hogging and sagging moments in the cargo hold area and maximum shearing forces at the bulkhead locations.

For local load application, all cargo/ballast loading conditions that can potentially cause extreme stress states should be considered. Bulk carriers often have combinations of full and empty cargo holds, and flooding conditions are also investigated as part of the cargo hold analysis.

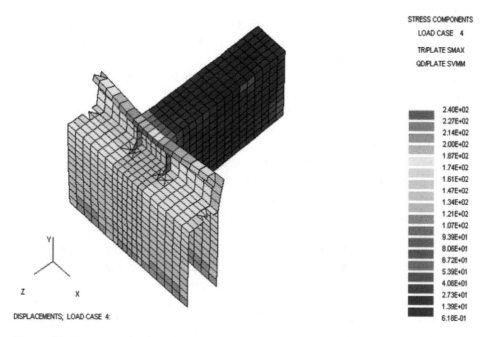

STRESS COMPONENTS

LOAD CASE   4

TRI/PLATE SMAX

QD/PLATE SVMM

	2.40E+02
	2.27E+02
	2.14E+02
	2.00E+02
	1.87E+02
	1.74E+02
	1.61E+02
	1.47E+02
	1.34E+02
	1.21E+02
	1.07E+02
	9.39E+01
	8.06E+01
	6.72E+01
	5.39E+01
	4.06E+01
	2.73E+01
	1.39E+01
	6.18E-01

DISPLACEMENTS; LOAD CASE 4:

Figure 10-17. A sample plot of deformations and stresses obtained by a finite element analysis.

10.4.7 Assessment of the Computed Results

Various color graphics techniques are used to display the resulting deformations and stresses. Figure 10-17 shows a typical example of such displays. In this way, areas where the deformations or stresses exceed permissible limits are readily identified. The areas of stress concentration observed by a higher level analysis will be further analyzed by the finite element model at the local structure or structural details level.

Related to the element stresses, some extrapolation or interpolation may be necessary to examine the stress distribution around the areas of interest which do not have nodes or elements. Using the calculated stresses, buckling and yielding checks can be made in accordance with the relevant strength criteria. Based on the computed results, the stress ranges can also be determined for analysis of fatigue strength.

10.5 Simplified Nonlinear Finite Element Analysis of Ship Structures

Traditionally, procedures and criteria for structural design of merchant ships are based on allowable stresses and buckling strength adjusted by a simple plasticity correction. During the last two decades, however, the emphasis in design of land-based structures, offshore platforms, and naval ships has been moving from the allowable stress design to the ultimate limit state design. This is because the ultimate limit state is a much better basis in determining the real safety margin of any economically designed structure (Paik and Thayamballi 2003, 2006). Therefore, the development of procedures and criteria for ultimate limit state design of merchant ship structures is one important area of future challenge.

There are two different prescriptive methods for the safety assessment of ship structures against a survival condition. One method is based on the load-carrying capacity (ultimate limit state) and the other is based on the energy absorption capacity. Ultimately, however, application of risk-based methods should be a way to go (Paik and Thayamballi 2006).

The first method is used in normal loading situations where the magnitude of deformation is small (of the order of plate thickness). In this case, the structural safety is related to the margin between the ultimate load-carrying capacity and the extreme applied load during the structure's lifetime.

The second method is used in accidental loading conditions such as ship collisions or grounding, where the deformations are much larger than the thickness of structural members. Here the structural safety is related to the margin between the energy absorption capacity and the initial kinetic energy loss during the accident. For both methods of safety evaluation it is essential to precisely analyze the nonlinear behavior of the structures either until the ultimate limit state is reached or until the initial kinetic energy is entirely absorbed.

Under extreme or accidental actions, the structures can be involved in highly nonlinear response associated with yielding, buckling, crushing, and fracture of individual structural components. In finite element analysis of a time-independent nonlinear problem, the stiffness equation is expressed by $\{R\} = [K]\{U\}$ where $\{R\}$ = the load vector, $\{U\}$ = the displacement vector, and $[K]$ = the stiffness matrix which is a nonlinear function of $\{U\}$. For a given load $\{R\}$, one must determine nodal displacements $\{U\}$ by solving the stiffness equation.

Quite accurate solutions of the nonlinear structural response can be obtained by application of the conventional FEM. However, a disadvantage of the conventional FEM is that it requires enormous modeling effort and computing time for nonlinear analysis of large structures. Therefore, most efforts in the development of new nonlinear FEMs have focused on reducing modeling and computing times.

The most obvious way to reduce modeling effort and computing time is to reduce the number of degrees of freedom so that the number of unknowns in the finite element stiffness equation decreases. Modeling the object structure with very large-sized structural units is perhaps the best way to do that. Properly formulated structural units or super elements in such an approach can then be used to efficiently model the actual nonlinear behavior of large structural units.

Ueda and Rashed (1974, 1984), who suggested this idea, called it the idealized structural unit method (ISUM). Unlike the conventional nonlinear FEM, however, ISUM models a basic structural member making up the structure as one ISUM unit with a few nodal points. In conventional finite element modeling, a huge number of fine meshes are normally required to obtain precise nonlinear behavior, while for ISUM analysis the structure is modeled by only a few superelements. Thus with

ISUM the size of the numerical problem is greatly reduced, giving large savings in modeling and computing times.

Since ship structures are composed of several different types of structural members such as beam-columns (support members), rectangular plates, and stiffened panels, it is necessary to develop various ISUM units for each type of structural members in advance. The nonlinear behavior of each type of structural members is idealized and expressed in the form of a set of failure functions defining the necessary conditions for different failures which may take place in the corresponding ISUM unit and a set of stiffness matrices representing the nonlinear relationship between the nodal force vector and the nodal displacement vector until the limit state is reached. The ISUM units (elements) so developed are used in the framework of the nonlinear matrix displacement procedure applying the incremental method. In terms of simplicity of element geometry, the method can be termed FEM, but the method can be termed complex FEM in terms of sophistication of element characteristics.

Figure 10-18 shows the procedure for developing an ISUM unit. Thus far the following ISUM units have been developed for the purpose of ultimate strength analysis of ship structures: a deep girder element (Ueda and Rashed 1974, 1984), a plate-beam combination element (Paik and Thayamballi 2003), a rectangular plate element (Ueda et al. 1984, 1986a, 1986b; Bai et al. 1993; Masaoka et al. 1998; Fujikubo et al. 2000; Paik and Thayamballi 2003, 2006), and a stiffened panel element (Ueda et al. 1984, 1986a, 1986b; Bai et al. 1993; Paik and Thayamballi 2003). For analysis of the internal mechanics in ship collisions and

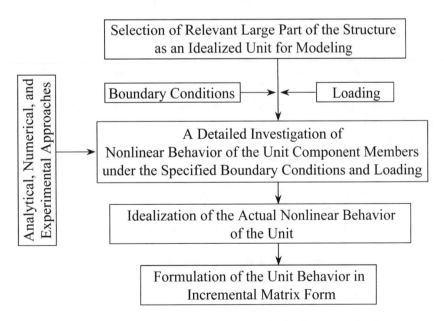

Figure 10-18. Procedure for the development of an idealized structural unit.

grounding, the rectangular plate element, the stiffened panel element, and the gap/contact element are available (Paik and Pedersen 1996; Paik and Thayamballi 2003). The ISUM has been successfully applied to nonlinear analysis of ship structures in overloading (Paik et al. 1996, 2001, 2002b; Wang et al. 2000; among others) and in collision and grounding (Paik et al. 1999, 2002a; Brown et al. 2000; among others).

It should be noted that individual developers of the ISUM units may of course employ somewhat different approaches from each other to idealize and formulate the actual nonlinear behavior of the structural members. In the following, ISUM modeling technique for analysis of the ultimate strength and the internal collision/grounding mechanics is outlined.

10.5.1 Structural Modeling

Steel-plated structures are typically composed of several different types of structural members such as support members (or beam-columns), rectangular plates, and stiffened panels. In ISUM modeling, such members are regarded as the ISUM units, as shown in Figure 10-19. It is important to realize that an identical structure may be modeled in somewhat different ways by different analysts, but it is of course always the aim to model so that the idealized structure behaves in the (nearly) same way as the actual structure.

The beam-column unit has two nodal points, as shown in Figure 10-20, one at the left end and the other at the right end. Each node is located where the beam is connected to another member. The nonlinear behavior of the beam-column unit is expressed by three translational degrees of freedom at each nodal point. For 3D analysis, however, six degrees of freedom must be considered (Paik and Thayamball 2006).

A rectangular (unstiffened) plate can be modeled as one rectangular plate unit as shown in Figure 10-21, while a stiffened panel can be modeled as one stiffened panel unit as shown in Figure 10-22. The nonlinear behavior of the rectangular plate or stiffened panel unit is formulated by three degrees of freedom at each of the corner nodal points. Again, six degrees of freedom must be considered for 3D analysis (Paik and Thayamball 2006).

A larger support member such as a deep girder, in which local web buckling can occur, may be modeled as an assembly of the panel unit and the beam-column unit, where the web is modeled as one plate (or panel) unit and the flange is modeled as one beam-column unit. Even though a plate is not exactly rectangular, it may be modeled as an equivalent ISUM rectangular plate unit of average length and breadth.

In a ship-ship collision accident, the shape of the striking ship bow is normally not uniform. The gap and contact conditions between the striking and the struck ships are thus modeled by means of the one or two dimensional gap/contact units.

Depending on the purpose of analysis, one may use different types of the ISUM units with different nonlinear characteristics as shown in Figure 10-23. For an ultimate strength analysis, the ISUM beam-column units need to take into account flexural or lateral torsional buckling and yielding; the ISUM plate units need to consider buckling and yielding; the ISUM stiffened panel units need to consider plate buckling, lateral torsional buckling of stiffeners, buckling of stiffener web, and gross

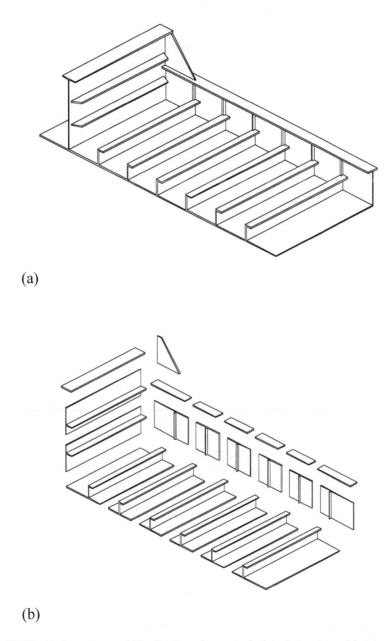

(a)

(b)

Figure 10-19. Various types of idealizations for a steel plated structure. (a) A typical steel plated structure. (b) Structural idealization as an assembly of plate stiffener combination units. (c) Structural idealization as an assembly of plate stiffener separation units. (d) Structural idealization as an assembly of stiffened panels.

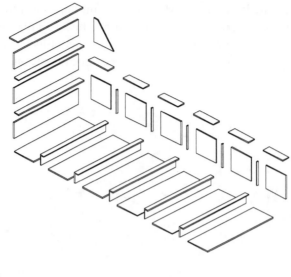

(c)

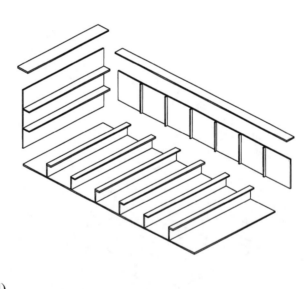

(d)

Figure 10-19. *continued.*

yielding. The effects of combined actions and initial imperfections should also be accounted for.

For a collision or grounding mechanics analysis, the units should accommodate yielding, crushing (folding), and rupture behavior as shown in Figure 10-24 and also the dynamic action effects.

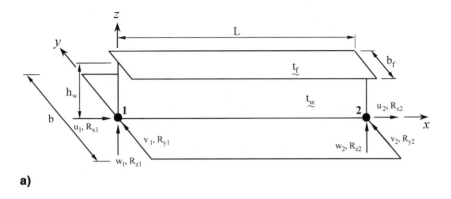

a)

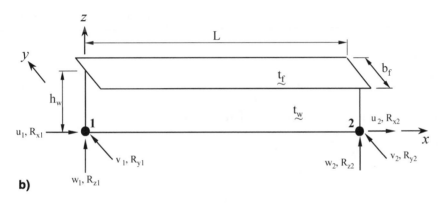

b)

Figure 10-20. (a) The ISUM beam-column unit with attached plating (•, nodal points). (b) The ISUM beam-column unit without attached plating (•, nodal points).

For an elaborate description of the idealized structural unit method in terms of formulating the nonlinear behavior, Paik and Thayamballi (2003, 2006) may be referred to.

For the ultimate strength analysis of ship structures under extreme hull girder actions, the same extent of the cargo hold level analysis previously noted in Section 10.4.2 or more approximately a hull section between two adjacent transverse frames can be taken. For the collision analysis of ships where the bow of the striking ship collides with the side structure of the struck ship, the model would approximately consist of a cargo hold between two transverse bulkheads where the collision occurs. The rectangular plate or stiffened panel units are employed for the purpose of modeling side girders, transverse webs, and inner/outer shell plating. The 1D or 2D gap/contact units can be used for the purpose of modeling the contact conditions between the side shell of the struck ship and the bow of the striking ship.

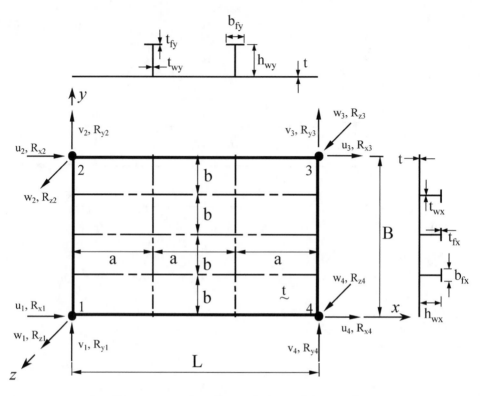

Figure 10-21. The ISUM rectangular plate unit (•, nodal points).

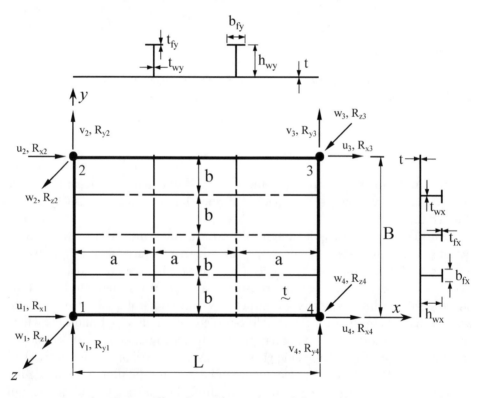

Figure 10-22. The ISUM stiffened panel unit (•, nodal points).

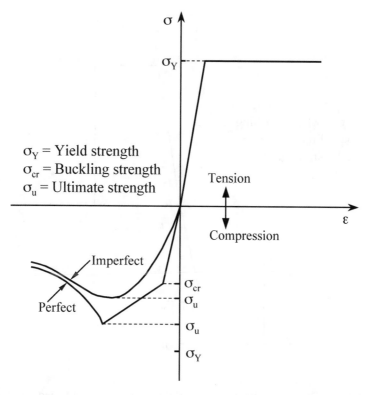

Figure 10-23. Idealized stress-strain behavior of the ISUM plate or stiffened panel unit for the ultimate strength analysis.

10.5.2 Modeling of Postweld Initial Imperfections

Postweld initial imperfections in the form of initial deflection and residual stress may affect stiffness and strength of structures, and so they need to be included in a nonlinear analysis, particularly for the ultimate strength. The postweld initial imperfections are included in the formulation of the ISUM units for the ultimate strength analysis as parameters of influence.

10.5.3 Boundary Conditions

Figure 10-25 shows an example of the boundary conditions applied to a cargo hold of a bulk carrier under vertical bending and sectional shear. It may be noted that the degrees of freedom at the nodal points at one end of the hull module are restrained in both ship length and depth (vertical) directions. At the centerline a symmetry condition is applied such that the degrees of freedom of those nodal points are restrained in the transverse direction. When the hatch end beams of the ships are restrained (fixed) by stiff hatch coamings or supporting members, the degrees of freedom at the nodal points at hatch ends may also be restrained

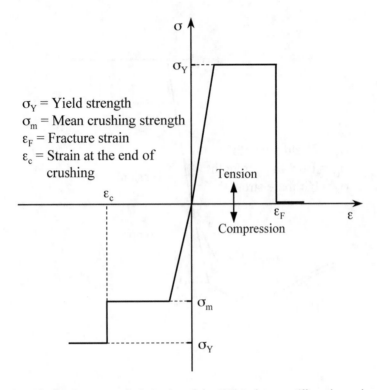

Figure 10-24. Idealized stress-strain behavior of the ISUM plate or stiffened panel unit for structural crashworthiness analysis.

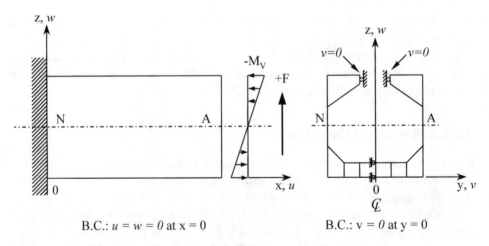

Figure 10-25. Example of the boundary and loading conditions for the ultimate strength analysis of a bulk carrier under vertical bending and shearing forces.

in the beam direction. Otherwise they will be left free. For analysis of the internal mechanics in ship collisions or grounding, the relevant boundary conditions can also be applied.

10.5.4 Load Application

To solve the nonlinear stiffness equation, the incremental method is normally employed in ISUM so that the loads are applied incrementally through the nodal points. For the ultimate strength analysis of ship hulls, nodal displacements or forces are increased incrementally at the unrestrained end of the hull module. The bending moment can be generated by applying linear axial displacements with respect to the neutral axis of the hull section, which itself can vary at each incremental loading step due to any structural member failure.

To produce a hull girder shearing force, a uniform shear displacement can be applied at the nodal points in the corresponding direction. The ultimate strength of a ship hull under a combination of various hull girder load components may be calculated numerically either by applying one load component incrementally until the ultimate limit state is reached with the other load components at a certain constant desired magnitude or by applying all load components simultaneously until one of the load components reaches its ultimate load capacity.

10.5.5 Application Examples

The ISUM theory has been automated within nonlinear Analysis of Large Plated Structures using the Idealized Structural Unit Method (ALPS/ISUM) program (Paik and Thayamballi 2003, 2006). The ALPS/ISUM program has three main modules for different purposes, namely, ALPS/GENERAL (2006) for simulating the progressive collapse behavior of general types of steel plated structures such as box girder bridges and offshore platform decks under extreme actions, ALPS/HULL (2006) for the progressive collapse analysis of ship hulls under extreme hull girder actions, and ALPS/SCOL (2006) for simulating the structural crashworthiness of steel plated structures under crushing actions arising from accidents such as collisions or stranding.

Two examples of the ISUM application to the analyses of ultimate strength or internal collision mechanics of ship structures are now shown. A physical test for investigating the progressive collapse characteristics under vertical sagging moment was undertaken on a welded steel frigate ship structural model with 1/3 scale to the original ship dimensions (Dow 1991). ALPS/HULL is now used to analyze the progressive collapse behavior of the Dow test model and the results are then compared with the experimental results. In the ALPS/HULL computations, both sagging and hogging cases are considered.

Figure 10-26 shows the ALPS/HULL model for the test structure. For simplicity, the hull module between two transverse frames is taken as the extent of the present analysis, though it is not difficult to take the entire structure as the extent of the ISUM analysis if needed. The test structure is idealized as an assembly of plate-stiffener separation units;

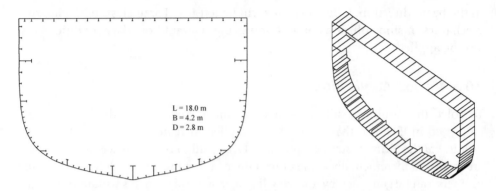

Figure 10-26. Midship section and ALPS/HULL model for the progressive collapse analysis of the Dow frigate test structure (L = length, B = breadth, D = depth).

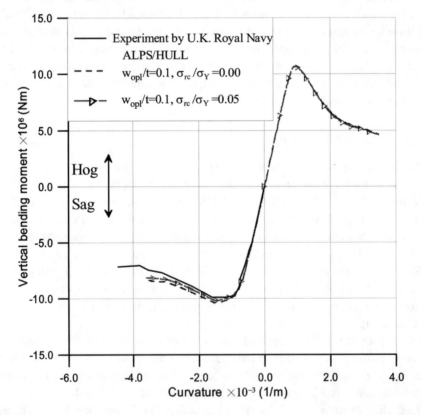

Figure 10-27. Progressive collapse behavior of the Dow test structure under vertical moment.

plating between stiffeners is modeled using the ISUM rectangular plate unit and stiffeners without attached plating are modeled using the ISUM beam-column unit. The webs of deep girders in bottom structures are also modeled using the ISUM rectangular plate units, while their flanges are modeled using the ISUM beam-column units.

Figure 10-27 shows the progressive collapse behavior of the Dow test structure under sagging or hogging moment, as obtained by ALPS/HULL. The Dow test result

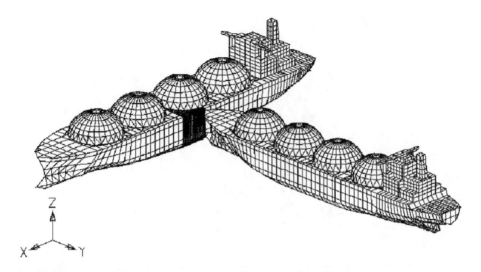

Figure 10-28. An example of the MSC/DYTRAN model for the LNG carrier to LNG carrier collision accident.

for sagging is also plotted. It is seen from Figure 10-27 that ALPS/ISUM provides quite accurate results when compared with the experiment.

For the second example, a ship to ship collision accident is now analyzed by ALPS/SCOL, where the midside structure of a spherical type Liquid Natural Gas (LNG) carrier carrying 135,000 m³ of liquefied natural gas is struck by the bow of another LNG carrier of the same size. Both the striking and the struck ships are fully loaded. This collision case is denoted by "SS-F." Other scenarios, with different loading conditions and types of striking ship, are given in Paik et al. (2002a). In the current example, the striking ship's speed is at 50% of full speed (i.e., 11.25 knots or 5.77 m/sec) at the beginning of the accident, while the struck ship is stationary, alongside a pier, and thus cannot move sideways.

Two computer programs are used to obtain the collision force penetration curves of the struck ship, namely, MSC/DYTRAN (1998) and ALPS/SCOL. The former program is based on the conventional nonlinear FEM and the latter is based on the ISUM theory for analysis of structural crashworthiness due to ship collisions. Figures 10-28 and 10-29 represent examples of the structural modeling for MDC/DYTRAN and ALPS/SCOL, respectively.

In the MSC/DYTRAN model, the entire ship structures are included using fine meshes around the impact location while coarse meshes are used at the other parts away from the collision location. In the ALPS/SCOL model, a quarter of the mid cargo hold of the struck ship, including the LNG tank together with the striking ship bow, is taken as the extent of the analysis; the ISUM plate and stiffened panel units are employed to model the struck structures, and the 2D gap/contact elements are used to represent the gap and contact conditions between the struck ship side and the striking ship bow. In both models, it is assumed that the striking ship bow is a rigid body which does not dissipate the strain energy.

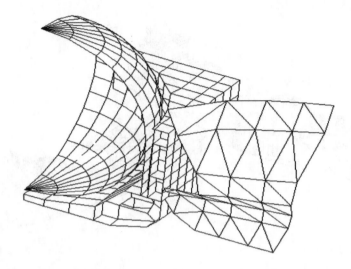

Figure 10-29. An example of the ALPS/SCOL model for the LNG carrier to LNG carrier collision accident.

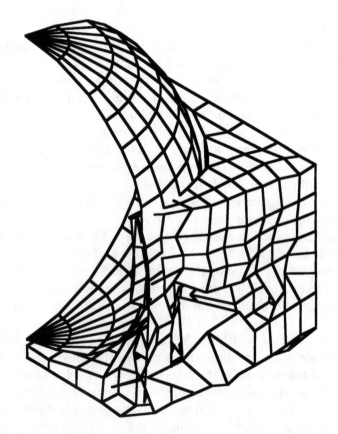

Figure 10-30. Deformed shape of the ALPS/SCOL model immediately after the struck LNG cargo tank starts to fracture.

The fracture strain of the structural members in both ship hull and cargo tank structures is supposed to be 10% in the ALPS/SCOL model, which uses ISUM units, while it is taken to be 20% in the MSC/DYTRAN model, which uses fine meshes, the former being more related to the characteristics of the weld metal fracture and the latter being related to those of the base metal fracture.

The computations were continued until the struck LNG carrier reaches the accidental limit states of the following two conditions, namely, (1) the striking ship bow penetrates the boundary of the struck ship cargo tank, and (2) the LNG cargo tanks start to rupture. In reality, the energy dissipation capability of the struck LNG carrier structure may be evaluated at the earlier state of the two limiting conditions. Figure 10-30 shows the deformed shape of the ALPS/SCOL model. Figure 10-31 compares the collision force penetration curves of the two different methods. Figure 10-32 shows the absorbed energy penetration curves for the same collision, as obtained by integrating the areas below the collision force penetration curves. As apparent, the ALPS/SCOL predictions correlate fairly well with the more refined MSC/DYTRAN analyses.

Of interest, the computing time required for the ALPS/SCOL analysis was about 4 hours using a Pentium III personal computer, while it was more than 14.4 hours

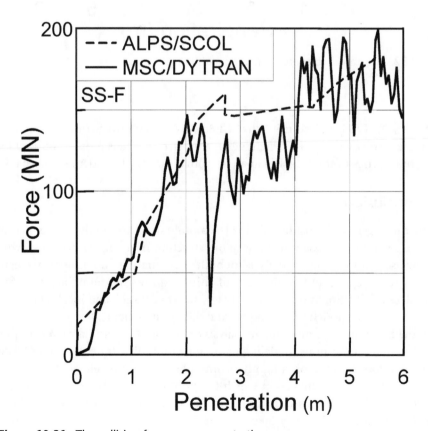

Figure 10-31. The collision force versus penetration curves.

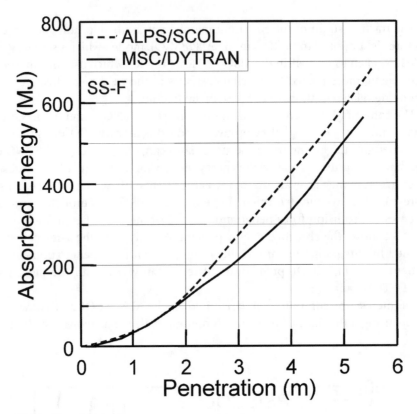

Figure 10-32. The absorbed energy versus penetration curves.

for the MSC/DYTRAN analysis using the SGI Power Challenge XL Series computer. As evident, the former will be of a benefit when a quick estimate is more important for carrying out a series of the collision analyses with a variety of accident scenarios.

10.6 Conclusion

Ships are the largest and most heavily loaded mobile structures in the world. Recent years have seen a big change in the way the structure of a ship is designed, with increasing use of finite element analysis for both stress analysis and ultimate strength analysis. The large size and complexity of ships requires modeling at up to five levels of detail: full ship, hull module, grillage, stiffened panel, and local. These levels involve different modeling techniques and different elements.

Although nonlinear finite element analysis can be used for ultimate strength prediction, it requires enormous modeling effort and computing time. Therefore ISUM has been developed, based on a family of simplified large-scale elements—one for each type of structural member—in which the nonlinear behavior is idealized and expressed in the form of a set of failure functions defining the necessary conditions for different failures which may take place in that member, and a set of stiffness matrices that can be used incrementally to represent the nonlinear relationship between the nodal force vector and the nodal displacement vector. Besides the prediction of ultimate strength under operational loads, ISUM can also be used for collision analysis. An example is given of a VLCC tanker bow colliding with the side of a liquefied natural gas carrier.

References

ALPS/GENERAL. (2006). *A computer program for the progressive collapse analysis of general types of plated structures*, Proteus Engineering, Stevensville, Md.

ALPS/HULL. (2006). *A computer program for the progressive hull collapse analysis of ships and ship-shaped offshore structures*, Proteus Engineering, Stevensville, Md.

ALPS/SCOL. (2006). *A computer program for structural crashworthiness simulation of plated structures*, Proteus Engineering, Stevensville, Md.

Bai, Y., Bendiksen, E., and Pedersen, P. T. (1993). "Collapse analysis of ship hulls." *Marine Struct.*, 6, 485–507.

Brown, A., Tikka, K., Daidola, J. C., Lutzen, M., and Choe, I. H. (2000). "Structural design and response in collision and grounding." *SNAME Transactions*, 108, 447–473.

Dow, R. S. (1991). "Testing and analysis of 1/3-scale welded steel frigate model." *Proc., of the Int. Conf. on Adv. in Marine Struct.*, Dunfermline, Scotland, 749–773.

Fujikubo, M., Kaeding, P., and Yao, T. (2000). "ISUM rectangular plate element with new lateral shape function–longitudinal and transverse thrust." *J. of the Soc. Naval Architects of Japan*, 187, 209–219.

Hughes, O. F. (1988). *Ship structural design, a rationally-based, computer-aided optimization approach*, Society of Naval Architects and Marine Engrs., Jersey City, N.J.

Mansour, A. E., and Thayamballi, A. K. (1994). "Probability based ship design; loads and load combination." *SSC-373*, Ship Struct. Committee, U.S. Coast Guard, Washington, D.C.

Masaoka, K., Okada, H., and Ueda, Y. (1998). "A rectangular plate element for ultimate strength analysis." *Proc., of the 2nd Int. Conf. on Thin-Walled Struct.*, Singapore, 1–8.

MSC/DYTRAN. (1998). *User's manual, version 4.5*, MSc Corporate, Santa Ana, Calif.

Paik, J. K., Choe, I. H., and Thayamballi, A. K. (2002a). "Predicting resistance of spherical-type LNG carrier structures to ship collisions." *Marine Tech. and SNAME News*, 39(2), 86–94.

Paik, J. K., Chung, J. Y., Choe, I. H., Thayamballi, A. K., Pedersen, P. T., and Wang, G. (1999). "On rational design of double hull tanker structures against collision." *SNAME Transactions*, 107, 323–363.

Paik, J. K., and Pedersen, P. T. (1996). "Modeling of the internal mechanics in ship collisions." *Ocean Engrg.*, 23(2), 107–142.

Paik, J. K., and Thayamballi, A. K. (2003). *Ultimate limit state design of steel-plated structures*, Wiley, Chichester, UK.

Paik, J. K., and Thayamballi, A. K. (2006). *Ship-shapped offshore installations: Design, building, and operation*, Cambridge University Press, Cambridge, UK.

Paik, J. K., Thayamballi, A. K., and Che, J. S. (1996). "Ultimate strength of ship hulls under combined vertical bending, horizontal bending, and shearing forces." *SNAME Transactions*, 104, 31–59.

Paik, J. K., Thayamballi, A. K., Pedersen, P. T., and Park, Y.I. (2001). "Ultimate strength of ship hulls under torsion." *Ocean Engrg.*, 28, 1097–1133.

Paik, J. K., Wang, G., Kim, B. J., and Thayamballi, A. K. (2002b). "Ultimate limit state design of ship hulls." *SNAME Transactions*, 110, 285–308.

Paulling, J. R. (1988). *Strength of ships, Chapter IV, Principles of Naval Architecture,* The Society of Naval Architects and Marine Engineers, Jersey City, N.J., 205–299.

Ueda, Y., and Rashed, S. M. H. (1974). "An ultimate transverse strength analysis of ship structures." *J. of the Soc. Naval Architects of Japan,* 136, 309–324 (in Japanese).

Ueda, Y., and Rashed, S. M. H. (1984). "The idealized structural unit method and its application to deep girder structures." *Computers & Struct.,* 18(2), 277–293.

Ueda, Y., Rashed, S. M. H., and Paik, J. K. (1984). "Plate and stiffened plate units of the idealized structural unit method (1st report)—under in-plane loading." *J. of the Soc. Naval Architects of Japan,* 156, 389–400 (in Japanese).

Ueda, Y., Rashed, S. M. H., and Paik, J. K. (1986a). "Plate and stiffened plate units of the idealized structural unit method (2nd report)—under in-plane and lateral loading considering initial deflection and residual stress." *J. of the Soc. Naval Architects of Japan,* 160, 321–339 (in Japanese).

Ueda, Y., Rashed, S. M. H., Paik, J. K., and Masaoka, K. (1986b). "The idealized structural unit method including global nonlinearities—Idealized rectangular plate and stiffened plate elements." *J. of the Soc. Naval Architects of Japan,* 159, 283–293 (in Japanese).

Wang, G., Chen, Y., Zhang, H., and Shin, Y. (2000). "Residual strength of damaged ship hulls." *Proc., Ship Structure Symp. on Ship Struct. for the New Millennium: Supporting Quality in Shipbuilding,* Arlington, Va.

Complex Structural Analysis and Structural Reliability

Robert E. Melchers and Xiu-Li Guan

11.1 Safety of Structures

Perhaps somewhat provocatively, Freudenthal argued some 40 years ago that when viewed as a whole, the theories underpinning structural engineering have developed in a somewhat lopsided manner; ". . . it seems absurd to strive for more and more refinement of methods of stress-analysis, if, in order to determine the dimensions of the structural elements, its results are subsequently compared with so-called working stress, derived in a rather crude manner by dividing the values of somewhat dubious material parameters obtained in conventional material tests by still more dubious empirical numbers called 'safety factors'" (Freudenthal 1961). Fortunately, time has somewhat caught up with this view. However, serious issues remain about estimating structural safety, particularly for complex structures.

Safety factors have given way to the more rational partial factor safety formats and these have been underpinned, increasingly, by structural reliability theory. But this is not the whole story. There is increasing demand for safety assessment outside the conventional code formats. Such assessment may be necessary for design purposes for critical structures or for the assessment of existing structures for remaining life, particularly where the safety or economic considerations are significant. For these purposes good quality tools for structural analysis are required as well as appropriately detailed and philosophically sound tools for making decisions about structural safety and structural performance. In addition, such decision tools demand realistic estimation of reliability and safety, not merely nominal estimates as used in underpinning code development. For realistic and hence typically complex structural, geotechnical, and infrastructure systems, in general this is a major computation task. This follows easily from considering the coupling of reliability analysis tools and complex structural analysis techniques such as finite elements.

This chapter deals with the combination of structural reliability theory and advanced structural analyses tools. The result is, in principle, a very complex and highly demanding computational problem. Several approaches can be taken to render the problem more amenable to solution. All involve some degree of problem simplification and some degree of compromise in the estimated failure probability. These approaches are briefly reviewed and related to the increasing demand for realistic estimates of structural safety.

11.2 Safety Rules for Structural Design

In structural engineering it is customary to think of structural safety being assured through appropriate design rules and through quality assurance procedures. History shows that this approach works well, with relatively few cases of structural failure. When a failure does occur it is typical for detailed investigations to be performed. The result may be modification of design rules or tightening of quality assurance procedures. In principle this is nothing less than a heuristic or "learning from experience" approach. It has always been part of engineering. However it can be expensive, it can cause unnecessary death and injury, and it may have severe consequences for the engineers and builders involved. Presumably there may also be an element of luck involved since not all structures are exposed during their lifetime to the loads that potentially might be applied to them and for which they should have been designed.

Although early engineers were skeptical of structural theory (Heyman 1988) and demanded tests to justify their designs, history shows that theoretical predictions increasingly have become the norm. Modern structural engineering relies heavily on analytical tools to predict likely structural behavior under defined loading conditions. Computer-based analytical tools can deal with very complex structural configurations and increasingly with nonlinear material behaviors. They were backed by much detailed testing of structures, adding confidence that the analytical tools can replicate expected structural behavior.

However, analytical prediction of structural behavior also depends on proper modeling of material properties and of the loads. It depends on the way structural analysis is translated into design and eventually into the finished structure through the documentation, construction, and commissioning processes. These latter aspects are not normally considered in structural analysis. They fall in the domain of structural safety.

The setting of appropriate factors of safety or partial factors has much vexed the profession. This stems from the gradual realization that the traditional approach, which tended to consist of expert committees making decisions about appropriate factors of safety, had no moral or theoretical justification (Pugsley 1955; Julian 1957). As understanding of structural behavior improved, design rules became increasing complex. As a result, there was continual pressure to reduce factors of safety in the face of the generally satisfactory behavior of previous structures. This was fuelled also by economic competition between material providers and between national design codes.

The greater complexity occasioned by the technical improvements to design codes typically had been sweetened by small economic gains for owners (more economic designs, more slender structures, etc.). However, it could be argued that this has been at the cost of a steady erosion of the actual margin of safety or even of the robustness (reserve of strength) of modern structures. There are obvious limits to the continued use of this approach.

The gradual development in parallel of the modern theory of structural reliability offered a way forward, most clearly demonstrated in the U.S. study for a probability-based limit state design code (Ellingwood et al. 1980). This landmark work forms the touchstone of much subsequent code-based development work for safety rules.

Underpinning all modern limit state design rules is the theory of structural reliability (e.g., Madsen et al. 1986; Melchers 1999). Rightly, it has been described as a means "to remove the concept of structural safety from the realm of metaphysics" (Grandiori 1991). It has permitted establishment of a more rational decision framework about structural safety. In turn this means that decisions about safety or risk can be made at a higher plane of abstraction and with a level of detail that was not possible previously. At times it forces analysts to face issues and data requirements that might otherwise have been overlooked. It has also been criticized as demanding information that is not obtainable (e.g., Elms and Turkstra 1992). However, others argue it is better to understand deficiencies in knowledge than to resort to simpler methods that simply hide lack of knowledge (Ellingwood 1999; Melchers 2000). It follows that despite the greater rationality of probabilistic methods, in actual applications a certain level of subjectiveness remains and is probably unavoidable.

High-quality analysis tools and a sound approach to ensuring technical safety at the design stage are not sufficient. As noted, quality assurance in design, documentation, construction, commissioning, and use also are important. This is illustrated by examination of cases of structural failure. Only seldom are design issues involved. The majority of structural failures are associated with:

1. one or more human errors during design, documentation, or construction,
2. organizational error or deficiency during design or construction or use of the structure, and
3. abuse of the structure such as through illegal deliberate loading or overloading.

Extensive and detailed investigations show that the factors involved are largely of a nontechnical nature (Matousek and Schneider 1976; Sibley and Walker 1977; Turner 1978; Nowak 1986). They include: ignorance, carelessness, negligence, mistakes, underestimation, inappropriate delegation, unusual materials, unusual construction, inappropriate organization, financial pressures, industrial relations pressures, political pressures, and sabotage. It has been argued that the greater the technical complexity of a structure, the more likely it is that these factors will be involved in failure through organizational problems (Turner 1978).

It is important to recognize, however, that these factors are present also in structural projects that do not fail. There are other factors that tend to ameliorate them. These include: level of education and training, work environment, complexity reduction, personnel selection, quality-assurance systems, self-checking, independent checking, inspection, monitoring in use, maintenance, and legal sanctions.

It must be concluded, therefore, that the actual failure rates of real structures are governed by a much more complex set of factors than is considered in a structural design and in the design codes that apply. It follows that the nominal probability of failure underpinning modern structural engineering design codes has only a tenuous relationship with the likelihood of failure of a structure as built (Melchers 2002).

Code theory and code calibration is a self-consistent approach, useful for helping to establish rationality in formal safety rules for design (Ditlevsen 1983). It is an essential and logical requirement prior to any attempts to exert greater control over human and organizational influences. But being based on nominal loads and generic structures and on safety rules that have changed every few years (Rusch and Rackwitz 1972), code-based structural reliability theory cannot predict actual failure rates. Something more is needed.

11.3 Structural Safety Assessment and Decision

Reasonably realistic estimates of structural safety are increasingly of interest. One application area is the assessment of existing structures for remaining safe life under deteriorated conditions. Existing road bridges also must meet higher legal load limits for modern transport vehicles. One approach is to consider the existing structure in terms of its ability to comply with current design codes, perhaps with some allowances. Unfortunately, many structures do not comply, despite their apparent intuitive adequacy under current and future expected operational conditions. Scrapping or rehabilitating a structure simply because it does not comply with current design requirements may not be an appropriate decision. The reason for this has to do with the fundamental assumptions made in developing design codes.

Structural design codes apply to a broad class of structures. Code writers have to consider the code possibly being applied to a variety of possible structural types. Thus, code development (calibration) uses generic rather than structure-specific material and structural properties. For this reason alone the failure probability implicit in the code rules does not necessarily give a good estimate of the failure probability of any particular structure designed to the rules.

In addition, the structural reliability theory employed in code calibration usually is simplified owing to technical complications. When coupled with generic rather than structure-specific material and structural properties, the result is that the failure probability estimate becomes a nominal measure, not necessarily giving an accurate indication of actual failure probability for any particular structure. This is an important matter. Increasingly it is recognized that structure-specific safety verification needs to be done with actual material properties, actual loads, and with a realistic rather than a nominal safety format (Melchers 1999; Faber 2000).

The second application area recognizes that most engineered structures are part of larger infrastructure systems. These include transportation facilities, chemical, petrochemical and nuclear power plants, pipelines, and so forth. Increasingly it is the case that regulatory authorities require (and owners often desire) detailed probabilistic risk assessments of such facilities. The criteria for assessment are mainly economic or in terms of the risk to human life.

Apart from certain private sector projects, many major infrastructure projects are assessed on economic criteria, usually based on decision-theoretic tools and criteria such as the net present value of expected lifetime costs for a project, for example (Raiffa and Schlaifer 1961),

$$N(t) = \sum_k B_k - \sum_j C_j - \sum_i p_i C_i \qquad (11\text{-}1)$$

where $N(t)$ is the net present value as a function of the time horizon t selected for the analysis, B_k are the expected benefits, C_i are the expected direct costs, and p_i is the probability of occurrence of the expected cost C_i of the ith consequence, all costs being expressed in consistent (e.g., monetary) terms and discounted over time as necessary. In Eq. 11-1, the probabilities p_i are assumed to have realistic, not notional, interpretation. This is important since even small changes in p_i may lead to considerable changes in $N(t)$. And this may have a very significant effect on the assessment outcome.

Regulatory requirements typically limit the maximum acceptable probability of occurrence of some prescribed adverse outcome. Often this is the risk to human life, usually fatality. The setting of criteria and the philosophy involved has a long and rich history (e.g., Conrad 1980). Table 11-1 shows a typical set of criteria.

An elaboration of the maximum acceptable probability concept is the "As Low As Reasonably Practical" (ALARP) approach increasingly being adopted by regulatory authorities (HSE 1999). In this approach, the regulatory probability limits to be applied may depend on practical and economic considerations. Typically the State,

Table 11-1. Quantitative Maximum Acceptable Individual Fatality Rates (Typical)

Application	Fatality Rate ($\times 10^{-6}$ per year)
Individual "death rate" (background rate as averaged from disease and accidents)	100
Individual accident rate (children only)	0.5
Individual accident rate (children and adults)	1
with about 10 deaths per accident	5–10
with about 100 deaths per accident	10–50
for industrial accidents	50

Note: Based on various data in Stewart and Melchers (1997).

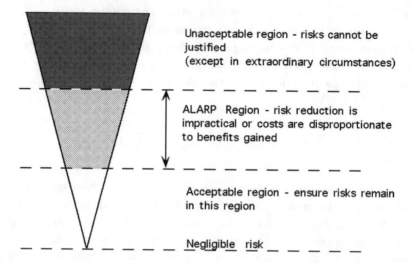

Unacceptable region - risks cannot be justified
(except in extraordinary circumstances)

ALARP Region - risk reduction is impractical or costs are disproportionate to benefits gained

Acceptable region - ensure risks remain in this region

Negligible risk

Figure 11-1. ALARP concept.

through one of its agencies, prescribes these levels (Fig. 11-1). Again, the risk levels are expected to have a realistic meaning. This requires a similarly realistic estimate of the probabilities of failure associated with the project.

11.4 Principles of Structural Reliability Analysis

11.4.1 Introduction

The principal tool for the establishment of the partial factors in modern structural design codes is the theory of structural reliability. It can provide nominal probabilities as used in code calibration work. It can also produce estimates of the more realistic failure probabilities required for assessment purposes. The difference between them lies in two areas. One is the level of refinement of structural reliability theory used in the analysis. The other concerns the type of data and models used for the variables involved. This section describes the fundamental structural reliability problem. It is the key to realistic probability estimates. Section 11.5 describes various simplifications to the theory and indicates how this leads to nominal probability measures. The following section deals with the integration of complex structural analysis and structural reliability theory.

11.4.2 Load Representation

Most structures are required to resist various types of loading, including live loads such as from people and vehicles, as well as earthquake, wind, snow, temperature, and wave loads. Most of these loads can be represented as a stochastic process. Figure 11-2 shows a realization (trace) of such a process. It fluctuates about a mean value. The process is described by the instantaneous probability distribution, shown on the left of Figure 11-2.

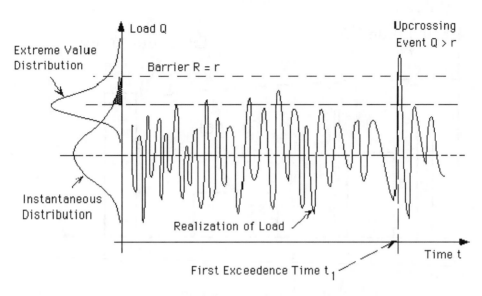

Figure 11-2. Realization of wind-loading modeled as a continuous random process showing an exceedence of structural resistance and time to first exceedence.

Figure 11-2 shows a barrier $R = r$. It describes the resistance of the structure against the load. An upcrossing event $Q > r$ occurs whenever the resistance is exceeded. Of particular interest for the safe life of a structure is the time t_1 to the first occurrence of such an event, shown as the "first exceedence event." For acceptable structural safety, t_1 should be sufficiently long.

Live loads usually are modeled better as a series of pulses. The pulses may represent events such as an office party or a meeting. Each cause short-term, high-intensity loading. Any one pulse may cause the strength of the structure to be exceeded (Fig. 11-3).

Because loads usually are random processes in time, the first exceedence event will be a random variable. The probability that the process will upcross the barrier $R = r$ (i.e., the structural strength) in a given time period is the central question for the estimation of the probability of structural failure. As will be seen below, in its present form it is not an easy problem to solve, and usually it is considerably simplified (with associated loss of accuracy).

Most structures are subject to several loads and the combination of some or all loads acting on the structure may lead to an exceedence event. The structural strength (resistance) in Figure 11-4 is shown as an envelope. The region under the envelope is a "safe" domain and that outside is the failure domain. The first exceedence time is now the time when one of the load processes or the combined action of the two processes out-crosses the envelope of structural capacity.

Design codes use load combination rules. These were originally specified intuitively, but modern limit state design codes have rules that derive from probability theory, albeit very much simplified. They may be used in a limited way for structural reliability

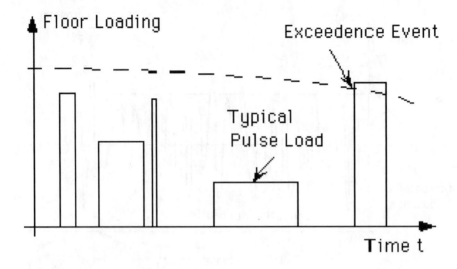

Figure 11-3. Load modeled as a pulse process showing exceedence of structural resistance.

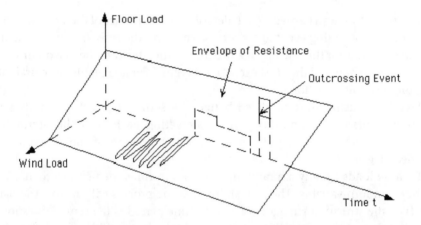

Figure 11-4. Envelope of structural strength (resistance) showing a realization of the vector load process and an outcrossing by the floor loading component.

analyses (Ditlevsen and Bjerager 1984; Ditlevsen and Madsen 1996) but for a proper structural reliability analysis, however, such simplified rules strictly are not correct.

11.4.3 Probabilistic Design Load

Before proceeding, a comment about descriptions such as "the 1,000-year" load is appropriate. Figure 11-2 shows an "instantaneous" probability distribution on the left. This describes the probabilities of occurrence of all possible values of the loading process. A small (shaded) part of the probability density function lies above the level $R = r$. This is the probability that the load w will be greater than this value. If this probability is such that the load will only be greater than this level, on average, once every 1,000 years, the load level, say q_d, is referred to as the 1,000-year load. The 1,000 years is also known as the return period for this load. Equivalently, a load greater than q_d has a probability of occurrence per year of about 0.001 or less. In Figure 11-2, the probability of occurrence of loading greater than q_d is represented by the shaded area (not to scale). Obviously, the shaded zone will be smaller for higher values of q_d, that is, the probability that the load will exceed the strength will be smaller and the return period will be longer.

Sometimes only the maximum value of the load in a given time period (e.g., each year) is recorded. The corresponding probability distribution is then known as the extreme value distribution (Fig. 11-2). A parallel argument applies for the load description in this case. These concepts are discussed in more detail in the literature.

Statements such as "q_d is the 1,000-year load" are useful for description of nominal loads selected for design codes. However, such descriptions provide no information about the likelihood of greater load levels or about load levels that are less likely to occur (and therefore of greater magnitude). Evidently, high load levels are particularly significant for structural safety estimation. It follows that knowledge of the complete upper tail of the load distribution is required.

11.4.4 Strength Uncertainty

When the structure is finally constructed, its actual strength R usually is not known precisely. This is because the designer's intent has to be translated to the finished structure. This is not an exact process. Variations occur in workmanship at all stages and materials have variable properties. As a result, the location of the line $R = r$ in Figure 11-2 should be represented as a probabilistic estimate such as shown in Figure 11-5. The level $R = r$ shown in Figure 11-2 is just one realization of many possible outcomes.

Since the actual strength outcome is uncertain, all reasonable possibilities must be considered in a reliability analysis. More generally, structural and cross-sectional dimensions, workmanship, boundary conditions, and so forth must all be considered. For example, if the strength of a cross section consists of an uncertain material strength M and an uncertain cross-sectional area A, the strength S of the member (in tension, say) is given by

$$S = M \cdot A. \tag{11-2}$$

The probability density function $f_s(s)$ of S can be determined from the corresponding probability density functions for M and A through Eq. 11-2. Unless these density functions are of simple form, this will require numerical integration. A particularly simple approach is to calculate just the first two moments of S.

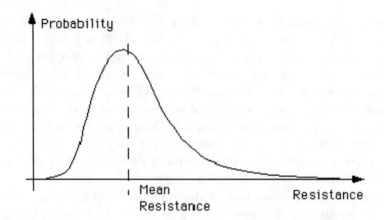

Figure 11-5. Schematic probability density function of material strength.

These are the mean μ_s and the variance, σ_s^2 respectively. Standard expressions for this are as follows:

$$\mu_S = \mu_M + \mu_A \tag{11-3}$$

$$V_S^2 \approx V_M^2 + V_A^2 \tag{11-4}$$

where $V = \sigma / \mu$ is the coefficient of variation.

11.4.5 Probability of Failure—Formulation

The above concepts can be formulated relatively simply. Let $[0,t_L]$ be denoted the design life for the structure. Then the probability that the structure will fail during $[0,t_L]$ is the sum of probability that the structure will fail when it is first loaded and the probability that it will fail subsequently, given that it has not failed earlier, or:

$$p_f(t) \approx p_f(0,t_L) + [1 - p_f(0,t_L)] \cdot [1 - e^{-vt}] \tag{11-5}$$

where v is the outcrossing rate. Already in writing the expression $[1-e^{-vt}]$ in the second term it has been assumed (not unreasonably) that structural failure events are rare and that such events therefore can be represented by a Poisson distribution. This makes expression Eq. 11-5 approximate but not unreasonably so for structural failure estimation. It would not be appropriate for, say, serviceability failure as these would not, normally, be rare events.

The outcrossing rate v can be estimated by assuming that the random load processes continue indefinitely and have a stationary statistical nature (e.g., in the

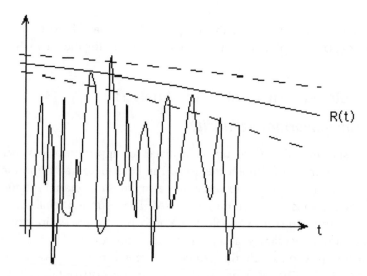

Figure 11-6. Outcrossing when there is *ẋn* structural deterioration with time, i.e., when *R=R(t)*.

simplest case, their means and variances do not change with time). Then the out-crossing rate is estimated from:

$$v = \int_{\text{safe domain}} E\left(X_n^{\dot{Y}} | X = x\right)^+ f_X(x) dx. \tag{11-6}$$

Here $X = X(t)$ represents the vector of random processes and ( )$^+$ denotes the (positive) component that crosses out of the safe domain (the other components cross back in and are of no interest). The term $E(\dot{X}_n | X = x) = \dot{x}_n = n(t).\dot{x}(t) > 0$ represents the outward normal component of the vector process at the domain boundary and is there for mathematical completeness. Finally, the term $f_x(\mathbf{x})$ represents the probability that the process is actually at the boundary between the safe and unsafe domain (i.e., at the limit state). Evidently, if the process is not at this boundary it cannot cross out of the safe domain.

The result Eq. 11-6 can be extended to allow for gradual deterioration or enhancement of the structural strength with time with the result that $p_f(0,t_L)$ and v become time dependent $v = v(t)$ (Fig. 11-6).

11.4.6 Solution Methods

The central problem in structural reliability theory is how to evaluate expressions $p_f(0,t_L)$ and $v = v(t)$ to estimate the expected life and hence the reliability of the structure. Only a few analytical solutions exist for estimating the outcrossing rate (Eq. 11-6). The usual approach is through Monte Carlo simulation, including employment of methods of variance reduction (e.g., Ditlevsen et al. 1987; Mori and Ellingwood 1993; Melchers 1995). These follow the same general principles as outlined in Section 11.5.3

below. However, it is fair to say that the computational demand is extremely high. Partly for this reason, considerable effort has been made to simplify the theory. This will now be discussed.

11.5 Simplified Structural Safety Analysis

11.5.1 Time-Invariant Reliability

The theory sketched above was not recognized for application to structures until after the development of simpler methods. The latter have had important application in code calibration work. However, they also have important limitations arising from the assumptions on which they are based. The result is a reliability estimation approach that hides the time variable in the probabilistic descriptions of loads, now modeled as random variables rather than random processes.

The first step is to assume that structural strength remains essentially constant with time, that is, the line $R = r$ in Figure 11-6 remains horizontal, as in Figure 11-2. This is reasonable provided the structural strength has not been seriously degraded by processes such as corrosion.

The second step is to assume that each load process is stationary, that is, its statistical properties do not change with time. This means that typically the mean of the load process is always the same, as is the variance (and higher moments, in general). Again, this is not an unreasonable assumption for many load processes when they are active. A load can now be considered as a random variable since one way of viewing a random process is as the limit in time of a sequence of random variables. Thus only the probability density function of a load need be considered. However, some care is required in how the load is defined and in treating the combination of several loads.

To see how a load must be defined as a random variable rather than as a stochastic process, consider the simple case shown in Figure 11-7a. For the bar to fail the matter of interest is the maximum value of the load likely to act during the life of the structure and the probability with which this load value will occur.

In principle, this information can be obtained experimentally. The maximum load on the structure would be recorded each year and plotting in a histogram. This

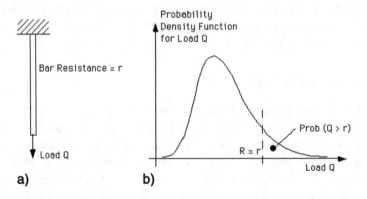

Figure 11-7. (a) Bar in tension under applied load. (b) Probability density function of $f_Q(\ )$ the applied load.

would allow a probability density function $f_Q()$ to be inferred. Since it deals with the maximum values of the loads only, it would be an extreme value distribution (in this case for the maxima). In Figure 11-2 it is shown on the vertical axis. A typical extreme value distribution is also shown in Figure 11-7b.

11.5.2 Time-Invariant Probability of Failure

The probability of failure now becomes the probability that the maximum annual load applied to the structure will be greater than its capacity at any time during the life of the structure. Evidently, this failure event can occur only once (unless the structure is repaired to exactly its former strength). The probability will depend on: (1) the value of the maximum load (a random variable), (2) the probability of occurrence of different possible values of candidate maximum loads, and (3) the actual strength of the structure as expressed by the probability density function for strength.

For a reasonably well-designed structure, failure would not occur under low values of the applied load, but failure is increasingly more likely with greater magnitudes. If the strength of the bar is known to be exactly $R = r$ then the failure event can be expressed as

$$Z = r - Q < 0. \tag{11-7}$$

The probability that the bar will fail is then:

$$p_f = \text{Prob}(Z < 0) = \int_r^\infty f_Q(x)dx \tag{11-8}$$

where $f_Q()$ is the probability density function of the load (Fig. 11-7b). In practice, the actual strength of the bar cannot be known precisely. However, it can be expressed as a random variable R with parameter values such as the mean μ_R and variance σ_R^2 and the probability density function $f_R()$. Previous experience with similar bars and with similar design, documentation, and construction techniques can give estimates for values of these parameters. Eq. 11-8 then becomes:

$$p_f = \text{Prob}(R < Q) = \text{Prob}(Z < 0) = \text{Prob}[G(X) < 0] \tag{11-9}$$

where $G()$ is known as the limit state (or performance) function and $X = (R,Q)$ denotes the (random) vector of loads, resistances, and other factors that may influence the failure event. The expression $G(X) < 0$ represents the condition that the bar fails.

In the simplest case, represented in Fig. 11-7, it follows that Eq. 11-9 becomes:

$$p_f = \text{Prob}(R < Q) = \int_{-\infty}^\infty F_R(x)f_Q(x)dx \tag{11-10}$$

where $F_R()$ is the cumulative distribution function for R, given by

$$F_R(r) = \text{Prob}(R < r) = \int_{-\infty}^r f_R(x)dx. \tag{11-11}$$

Eq. 11-10 may be interpreted (loosely) as follows. Under the integral, the first term, given by (Eq. 11-11), denotes the probability of failure given that the actual load has the value $Q = x$. The second term is the probability that the load takes the value $Q = x$. This is then integrated over all possible values of x.

Interestingly, Eq. 11-10 and 11-11 do not have the initial failure probability $p_f(0, t_L)$ of Eq. 11-5. It has been subsumed into the random variable representation for the extreme load applied at any time during the life of the structure and the assumption that the maximum load is applied *only once* during the lifetime of the structure. The assumption is that the probability of failure is not affected by precisely when in the lifetime this event occurs. This is the time-invariant structural reliability problem.

Since most structures consist of many members, typically there will be a number of random variables describing structural strength. There also may be multiple loads, each describable validly as a random variable. Eq. 11-10 and 11-11 may then be combined to become:

$$p_f = \int \ldots \int_{\Delta_f} f_X(x)dx \tag{11-12}$$

where $f_X(\ )$ is the joint density function of the vector of random variables X. It collects all the random variables in the reliability problem. The region of integration in Eq. 11-12 is $D_f[D_f : G_i(X) < 0, i = 1, \ldots, m]$, known as the *failure domain*. It is specified by failure criteria (such as a maximum allowable stress, or a yield condition) and through a structural analysis process.

Eq. 11-12 is easier to solve that Eq. 11-5 and 11-6. However, because it may be of very high dimensionality, its solution still may not be a simple matter in general. This is particularly the case if the limit state functions $G_i(X) = 0, i = 1, \ldots, m$ describe a complex failure domain, for instance, a failure mode of a complex system.

11.5.3 Numerical Solution Methods—Monte Carlo

The most obvious approach to solving expression Eq. 11-12 is to attempt to use direct numerical integration. For dimensions of X is greater than about 5, this is not normally considered feasible. The next best option is to use Monte Carlo simulation to perform the integration. It works on the basis of random selection of actual values of each of the variables, running the structural analysis for these values and deciding if the structure fails. This is repeated N times. The estimate for the failure probability is then given by the number of failures divided by N.

More precisely, the process can be stated as follows. Eq. 11-12 can be written as

$$p_f = J = \int \ldots \int I[G(x) < 0] f_X(x) dx \tag{11-13}$$

or, discrete form

$$p_f \approx J_1 = \frac{1}{N} \sum_{j=1}^{N} I[G(\hat{x}_j) \leq 0], \tag{11-14}$$

where $I[\ ]$ is an indicator function defined such that $I[\theta] = 1$ if the expression represented by θ is "true" and zero otherwise. $\hat{x}_i$ represents a (the jth) vector of random samples selected from $f_x(\ )$.

To apply Eq. 11-14, a discrete set of values is chosen for each of the random variables in the problem. This set is termed $\hat{x}_j$. For each such set, the limit state function (i.e., $G(\hat{x}_i)$) is evaluated. If it is violated (i.e., if the structure fails) the sum in Eq. 11-14 is incremented by one. Otherwise no action is taken. The process is repeated N times and the sum in Eq. 11-14 counts the number of failures n. The ratio n/N is an estimate of the probability of failure. In theory, as $N \rightarrow \infty$, the approximation in Eq. 11-14 improves asymptotically.

Values for $\hat{x}_j$ can be obtained from appropriate subroutines on many computers or can be generated from a set of random numbers.

Since the probability of failure for structures is low, the process is clearly inefficient. It can be improved using a variance reduction technique such as importance sampling, Latin hypercube sampling, antithetic variables, stratified sampling, and others. In importance sampling Eq. 11-14 is replaced by

$$p_f \approx J_2 = \frac{1}{N} \sum_{j=1}^{N} \left\{ I[G(\hat{x}_j) \leq 0] \frac{f_X(\hat{v}_j)}{h_V(\hat{v}_j)} \right\} \qquad (11\text{-}15)$$

where $h_V(\ )$ is termed the importance sampling probability density function. The samples are now $\hat{v}_j$ taken from $h_V(\ )$ rather than from $f_x(\ )$. This has the advantage that $h_V(\ )$ can be selected by the investigator. Good choices of $h_V(\ )$ can considerably reduce the number of sample vectors $\hat{v}_j$ required for a good estimate of p_f, thereby significantly reducing computation times.

A robust choice for $h_V(\ )$ is a multinormal distribution with independent components and with standard deviations about 1 to 2 times those corresponding to X. Ideally the mean is placed at the point of maximum likelihood in the failure domain. This point may have to be found by trial and error. Fortunately, there is a reasonable degree of latitude in estimating this point without having much effect on the estimate of p_f.

11.5.4 Approximate Solution Methods—First Order Second Moment and First Order Reliability

An alternative approach to solving Eq. 11-12 is to simplify it. This can be done through (1) simplifying the failure domain D_f to be a linear function (i.e., a first order approximation) and (2) simplifying $f_x(\ )$, usually to a normal (or lognormal) form. Because the normal distribution can be described completely by its first two moments (mean and standard deviation), this approach is called the First Order Second Moment (FOSM) method. It is easy to use and is very popular. However, because of these two assumptions, it is approximate.

The basic concepts of FOSM are easy to understand. Consider again the bar shown Figure 11-7a. Its safety margin or limit state function is a linear function given by

$$Z = R - Q. \qquad (11\text{-}16)$$

If the load and resistance are defined by their second moments only, that is, by their means μ and variances σ^2, it follows from probability theory rules that (since R and Q are independent):

$$\mu_Z = \mu_R - \mu_Q , \qquad (11\text{-}17)$$

$$\sigma_Z^2 = \sigma_R^2 + \sigma_Q^2 . \qquad (11\text{-}18)$$

The probability of failure is then

$$p_f = \text{Prob}(R - Q < 0) = \text{Prob}(Z < 0)$$
$$= \Phi\left(\frac{0 - \mu_Z}{\sigma_Z}\right) = \Phi(-\beta) \qquad (11\text{-}19)$$

where $\Phi(\ )$ is the standard normal distribution function (with zero mean and unit standard deviation or variance). It is extensively tabulated in statistics texts, at least for higher probability levels. For the low values of probability usually associated with structural failure more detailed tables are required (e.g., Melchers 1999).

The simplicity of Eq. 11-19 is one of FOSM's major attractions. But clearly it cannot accurately represent situations in which the distributions of either or both the resistance and the load random variables is not normal. For this reason, the parameter β, known as the *safety index* or the *reliability index*, is often used to refer to the safety measure obtained from Eq. 11-19. Its relation to probability is shown in Figure 11-8. The failure region is the negative region. Evidently, $\beta = \mu_Z / \sigma_Z$ measures the distance from the mean of the safety margin to the failure condition in terms of the uncertainty σ_Z of the safety margin.

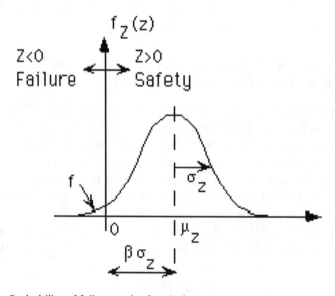

Figure 11-8. Probability of failure and safety index.

For situations involving more random variables, the standard approach is to transform each into the standard normal space $\mathbf{y}$ (with zero mean, unit variance). This can be done using $Y_i = x_i - \mu_{xi}/\sigma_{xi}$ (or a more general transformation, such as the Nataf or Rosenblatt, when there is statistical dependence between the variables).

The limit state function $G(\mathbf{x}) = 0$ must be transformed also and becomes $g(\mathbf{y}) = 0$. If it is then no longer linear, a first-order Taylor series expansion is applied to obtain a linear approximation. Without going into detail, it can be shown that the most appropriate expansion point to be used for this is the design or checking point (and not the mean value point as used in some algorithms). Of course, at this stage the checking point is not yet known so that an iterative procedure must be employed for cases with nonlinear $g(\mathbf{y}) = 0$.

When $g(\mathbf{y}) = 0$ is linearized the problem in $\mathbf{y}$ space is as sketched in Figure 11-9 in two dimensions. The contours of the joint probability density function $f_Y(\mathbf{y})$ describe a probability hill of all the (transformed) random variables $\mathbf{Y}$ in $\mathbf{y}$ space. The probability of failure p_f is represented by the volume under this hill in the failure region, which is the region where $g(\mathbf{y}) < 0$. As before, because of the simplifying assumptions it is appropriate to refer to the safety index β rather than to p_f.

With $g(\mathbf{y}) = 0$ linearized, the integration in the direction v, parallel to the linearized limit state $g(\mathbf{y}) = 0$, will produce exactly the probability density function shown in Figure 11-8. Also, $\beta\sigma_z$ in Figure 11-8 has now become just β (since all the standard deviations were made unity by transforming to the $\mathbf{y}$ space). As before, β represents the shortest distance from the checking point to the origin of $\mathbf{y}$ space. It is given by

$$\beta = \min\left(\mathbf{y}^T \cdot \mathbf{y}\right)^{1/2} = \min\left(\sum_{i=1}^{n} y_i^2\right)^{1/2} \tag{11-20}$$

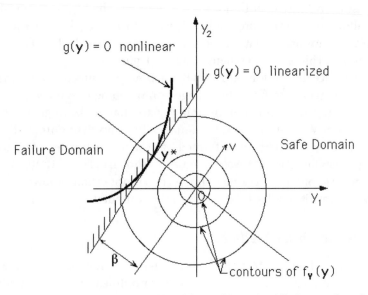

Figure 11-9. Space of standard normal variables and linearized limit state function.

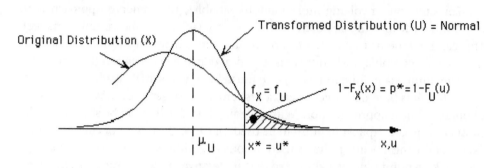

Figure 11-10. Original and transformed normal distributions for a tail region.

where the y_i represents the coordinates of any point on the limit state surface $g(\mathbf{y})=0$. Note that $\mathbf{y}^*$ is the design or checking point.

Alternatively, β can be obtained from an iterative algorithm that searches for the minimum Eq. 11-20 by searching for the design point $\mathbf{y}^*$ according to

$$\mathbf{y}^{(m+1)} = -\alpha^{(m)}\left[\beta^{(m)} + \frac{g\left(\mathbf{y}^{(m)}\right)}{l}\right] \tag{11-21}$$

where $\alpha^{(m)} = \mathbf{g}_y^{(m)}/l$, $l = (\mathbf{g}_y^{(m)T} \cdot \mathbf{g}_y^{(m)})^{1/2}$ and $g_{Yi} = \partial g\,(y^{(m)})/\partial y_i$. This is essentially a specialized minimization algorithm. It follows that other algorithms for minimization of Eq. 11-20 can be used.

The FOSM method can be refined to remove the limitation of all the random variables being represented only by their second moments. The key observation is that the probability of interest lies a long way from the origin in space. Only the tails of most probability distributions are of interest. The idea is to transform the relevant tail of a nonnormal probability density function to the corresponding tail of an equivalent normal probability density function (using the Nataf or Rosenblatt transformation). This is shown schematically in Figure 11-10.

Note that since the random variables in the limit state function are transformed, the limit state function itself will be changed on mapping it to the $\mathbf{y}$ space. Usually this results in a function more nonlinear in $\mathbf{y}$ space than in the original problem.

Once the transformations have been made, the equivalent normal distribution and a linearized version of the transformed limit state function are used to estimate the failure probability, using FOSM techniques. As shown in Figure 11-10, the transformation is made about the checking point. As this is not usually known beforehand, an iterative process is involved. This adds to the computational requirements.

11.5.5 Second Order Methods

The central pillar of the FOSM method and its first order reliability (FOR) extension is the linear (or planar) limit state function. For nonlinear limit state functions the probability content in the failure region can be seriously under- or overestimated. This can be seen in Figure 11-9. Eq. 11-19 and 11-20 can still be used to estimate the checking point and to evaluate β, but the meaning of β as a measure of probability now is not clear-cut since the link $p_{fi} = \Phi(-\beta_i)$ is no longer valid. This suggests that for nonlinear limit state surfaces it is better to employ probabilities rather than β.

FOSM/FOR methods can be modified to allow for the nonlinear limit state surfaces. This can be done by letting the actual nonlinear limit state surface be approximated by a second order function (a quadratic or parabolic surface). It is possible to develop a relationship between the curvature of this function at the checking point and the correction to be made for the probability content between the second order limit and the linear limit state functions (Hohenbichler et al. 1987; Breitung 1994). This is termed a second order reliability method (or SOR method).

It is evident that in attempting to refine the original FOSM approach through FOR and SOR to allow respectively for nonnormal probability distributions and nonlinear limit state functions, the computational effort increases.

11.6 Complex Structures and Reliability Analysis

11.6.1 Introduction

For many realistic structures the information about the structural response, behavior and stress states can be obtained only for particular selected values of x, such as from a finite element run. This means that the limit state function $G(x)$ itself is not available explicitly. $G(x)$ is available only on a point-by-point basis. Its derivatives also are not available unless the finite element code provides them (unusual). However, at extra computational cost the derivatives could be estimated from finite differences.

Three approaches have been identified for solving the structural reliability problem under these conditions (e.g., Lemaire and Mohamed 2000). These are:

1. The "direct" procedure using the iterative FOSM/FOR procedure in Eq. 11-21 directly. It is clearly expensive in terms of the number of $G(x)$ evaluations required both for $G(x)$ itself and for estimation of derivatives. However, any form of complex analysis can be involved in $G(x)$, including nonlinear behavior.
2. Using the FOSM/FOR minimization formulation Eq. 11-20 is a routine in a general purpose minimization algorithm. This is not unlike the direct procedure except that the efficiency of some optimization solvers can be utilized.
3. Generating a continuous and differentiable surrogate surface for $G(x)$ and that can be used in the usual manner for FOSM, FOR, or SOR. Such a surface is known as a *response surface*. It is the technique most commonly applied. It is discussed in the next section.

For use with Monte Carlo analysis the direct procedure consists only of $G(x)$ evaluations. It is very expensive in computational requirements. Response surfaces typically are somewhat more economical. However, it is fair to say that a significant research challenge remains in tying Monte Carlo-based reliability analysis with structural analyses for which evaluation of $G(x)$ is computationally intensive. The main reason is that for each evaluation of $G(x)$ the structural response is required for the particular set of deterministic parameter values in vector x.

11.6.2 Response Surfaces

A response surface may be viewed as a surrogate limit state surface constructed from whatever information is available from the analysis of complex structural systems. Typically the information is obtained through point-by-point discovery such as through repeated numerical analysis with different input values. These values could be random, as in Monte Carlo analysis, or specifically ordered. To apply FOSM and related methods, a closed, and preferably differentiable, form for the (surrogate) limit state function $G(x) = 0$ is required.

Another way of looking at this is as an extension of the polynomial fitting approach used in SOR. In the previous section the concept of using a quadratic surface to approximate a nonlinear limit state surface was mentioned. Such a surface is a special case of a response surface.

A function $\bar{G}(x)$ is sought that best fits the discrete set of values of $G(\bar{x})$, where $\bar{x}$ represents a set of discrete points in x space for which $G(x)$ is evaluated. The best surface will be that which minimizes the error of approximation, particularly in the region around the design point. However, usually this point is not known a priori, so trial and error will be involved.

Although in principle a higher order polynomial is preferred for $\bar{G}(x)$, usually a second order polynomial is employed (Der Kiureghian and Ke 1988; Faravelli 1989; Bucher and Bourgund 1990; El-Tawil et al. 1992; Rajashekhar and Ellingwood 1993; Maymon 1993):

$$\bar{G}(X) = A + X^T B + X^T C X \qquad (11\text{-}22)$$

with $(n^2 + n + 1)$ undetermined (regression) coefficients defined by A, $B^T = [B_1, B_2, \ldots, B_n]$ and

$$C = \begin{bmatrix} C_{11} & \cdots & C_{1n} \\ \vdots & & \vdots \\ sym & \cdots & C_{nn} \end{bmatrix}.$$

The (regression) coefficients may be obtained by conducting a series of numerical experiments, that is, a series of structural analyses with input variables selected according to some experimental design. Evidently, the computational effort increases sharply with n.

An appropriate experimental design takes into account that for estimating probability the main region of interest is in the neighborhood of the point of maximum likelihood within the failure domain, that is, the design point. However, as noted, this will not be known initially. One simple approach is to select input variables for the experimental design around the mean value of the variables. A simple experimental design is shown in Figure 11-11. More complex designs have been discussed (e.g., Myers 1971; Faravelli 1989; Rajashekhar and Ellingwood 1993).

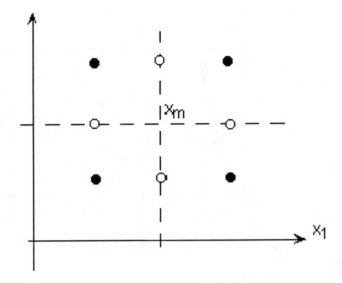

Figure 11-11. Simple experimental design for a two-variable problem with an x_m mean point.

There will be a difference between the actual response $G(\bar{x})$ for the values $X = \bar{x}$ and the evaluation of Eq. 11-21 for the same values of the random variables. This difference is due to intrinsic randomness and due to lack of fit between the surrogate function and the actual (but implicit) limit state surface. It is not possible to separate them without more refined analysis. The important step, however, is to try to select the (regression) coefficients A,B,C such that the total error is minimized. A least squares fit approach can be used to minimize the error.

The best points for fitting the response surface are not known a priori. An iterative search technique may be used to locate such points (Bucher and Bourgund 1990; Liu and Moses 1994; Kim and Na 1997). Particularly for large systems such points cannot always be identified without subjective interference, that is, they require some degree of input from the analyst.

11.6.3 Reliability Analysis with Response Surfaces

For reliability analysis, the response surface $\bar{G}(x)$ is treated as if it is the actual limit state function. Thus the same issues with approximation of limit state functions in FOSM and FOR arise, since in these methods the nonlinear response surface at the design point is approximated by a linear function.

If the transformed (y) space is of other than trivial size (dimensionality) and the limit state function $\bar{G}(x) \rightarrow \bar{g}(y)$ is even moderately nonlinear, the results can be inaccurate compared to estimates obtained from a Monte Carlo analysis. This is illustrated in Table 11-2 for the reliability analysis under uniaxial stress of a plate with a central hole (Fig. 11-12) (Guan and Melchers 1999). In this example the elastic modulus of the plate was represented as a stochastic field, with 132 finite elements. As usual, the

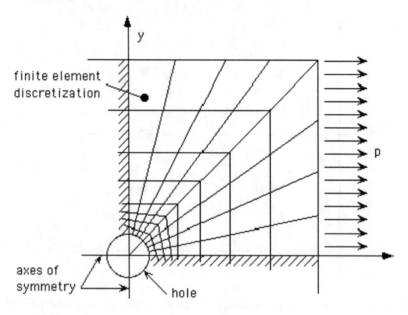

Figure 11-12. Plate with hole and stochastic elastic properties.

Table 11-2. Failure Probability Estimates for "Stochastic" Plate (Guan and Melchers 1999)

N_R	FOSM $p_f (\times 10^{-3})$	CPU Time (min)	Monte Carlo $p_f (\times 10^{-3})$	CPU Time (min)
6	2.77	0.81	3.37 [3.1, 3.6]	2,033
26	1.02	1.38	1.50 [1.3, 1.6]	2,710
132	0.61	3.69	0.93 [0.8, 1.0]	5,136

Note: N_R = number of random field elements.

representation of the random field also required discretization, indicated by N_R, into a number of random field elements.

In principle, more accurate results can be obtained by using the second order SOSM/SOR approaches. However, this requires more fitting points, more structural analyses and hence increase computation demands.

Other applications of the integration of response surfaces based on finite element analysis and structural reliability have been reported in the literature. Some involve stochastic material properties with a resultant finite element mesh of high dimension and a high level of stochastic discretization (Fig. 11-12). More commonly, finite element analyses includes relatively few parameters that are random variables. An example of the latter is shown in Figure 11-13 (Faravelli 1989). It shows a nuclear reactor pressure vessel. In the reliability analysis the random

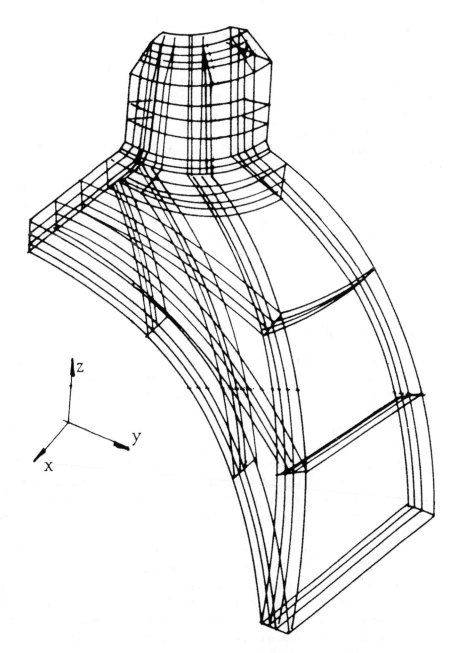

Figure 11-13. Part mesh for finite element analysis of section of pressure vessel. From Faravelli (1989). [Reproduced with permission from ASCE.]

variables were the following six quantities: elastic moduli of material and of cladding, yield stress of cladding, material hardening ratio, residual stress, and shell thickness. In addition, the elastic moduli were represented as random vectors. The geometry, apart from thickness, was assumed deterministic, as was the applied pressure loading. Similar examples have been discussed, for example, by Lemaire and Mohamed (2000).

11.6.4 Simplifications for Large Systems

When the number of random variables involved in a problem is very large, the direct application of the response surface technique may be impractical. One approach is to reduce the number of random variables such as through replacing random variables of low uncertainty with deterministic equivalents. This is a standard technique in probabilistic analysis generally.

Another approach is to reduce the set of random variables X to the smaller set X_A describing their spatial averages. An example, for a plate under stress, is to use the same (i.e., the spatial average) yield strength at every point, rather than specifically allowing for the variation of yield strength from point to point (or from finite element to finite element). A third approach is to simplify the error effect of random variables (or spatial averages) as having only an additive effect on the response, rather than some more complex relationship (Faravelli 1989). A further approach is to use intuitive input from the analysts about the location of the best response surface.

11.7 Practical Implications

Sections 11.4 and 11.5 have given a broad overview of the methods and tools available for estimation of the probability of structural failure. These are general within their own assumptions. The results they produce depend much on the data used. When generic data is used, the outcomes will be generic probabilities. This is satisfactory for the development of detailed load combination rules and partial factors for design codes. However, it is not sufficient for wider purposes such as for checking of compliance with regulatory requirements for potentially hazardous facilities and for economic evaluation. For these purposes, the probabilities must relate to the realistic behavior and the response of individual in situ structures as actually constructed and perhaps after years of operation. This in turn demands data for the actual, in situ structure.

Data, including probabilistic data, for a structure as it is expected to be on completion involves prediction for the individual structure but also reference back to experience with other structures. The data base, therefore, will be similar to the database used in code calibration work (Melchers 1999) but without the invariable trend in such work to generalize the data and to add some conservatism to the values adopted. Best estimates rather than conservative best estimates are required (Stewart and Melchers 1997).

For existing structures the same requirement holds except that detailed investigation of the possibility of deterioration (fatigue, corrosion, etc.) becomes increasingly more desirable with older structures. Again, best estimates are needed (Faber 2000).

The physical modeling of the structure should be consistent with the desired aim of the analysis. Typically it should not be a generic modeling effort but instead be sufficiently detailed to capture subtleties of the behavior of the structure under study. This is likely to require computationally demanding techniques such as finite element analysis. As discussed earlier, this adds considerable complexity to structural reliability analyses, irrespective of the precise approach adopted.

Realistic probability estimation also requires that the tools used for probability estimation are not so simplified as to preclude close approximation of the result that would be obtained with a fundamental technique such as ordinary Monte Carlo analysis. Its use has tended to be shunned because of the extreme computational effort required for reasonably realistic structures. However, the availability of so-called variance reduction techniques have produced tools such as importance sampling and directional simulation, which have considerably lower computational demand and which provide faster convergence. Coupled with the continued lowering of computer and computing costs this means that Monte Carlo simulation approaches are beginning to become viable even for moderately complex structural systems.

In principle, time-variant reliability analysis should be used whenever there is more than one load process acting on the structure. Few analytical results are available and numerical solutions demand much computation time. Some examples have been given, including efforts to simplify the formulation but these have still been for relatively simple structural systems (e.g., Moarefzadeh and Melchers 1996). Much remains to be done in this area.

The alternative approach, through time-invariant reliability analysis, has been exploited mainly through FOSM when finite element analysis was involved for modeling the structure. It is useful for notional probabilities but, except in very special cases, not usually adequate for anything more. Refinements, such as FOR and SOR, produce better results but at increased computing costs. It appears that even with these methods the gap between nominal and realistic can be closed only by resorting to Monte Carlo integration. Nevertheless, these methods of reliability analysis can be extremely useful for some types of application (particularly for code calibration work) and for comparative purposes. They have also provided some unique insights into the problem of structural reliability estimation and have led to advances in understanding and in Monte Carlo analysis.

Finally, it must be recognized that even the so-called realistic probability estimates cannot avoid relying on imperfect and subjective information. Not all (and probably very few) factors in a reliability analysis can be objectively evaluated. Subjective estimates must be made. In this structural reliability is no different from other probabilistic risk assessment techniques and from application of decision-theoretic tools. However, the aim always must be to obtain the best possible estimate, consistent with the weight of the decision to be made.

11.8 Conclusion

This chapter has been concerned with future trends in safety evaluation of complex structures. It was argued that there is increasing emphasis on the assessment of the safety of individual existing structures. There is also the need to consider the safety of structures that are part of larger infrastructure systems. Such systems tend to be assessed against regulatory and economic criteria and for these realistic estimates of failure probabilities are required. This requirement therefore carries over to structural safety and performance evaluation.

An overview of the basic principles of structural reliability analysis using probabilistic methods was given and the main techniques described. It was noted that the Monte Carlo methods should produce accurate results in the limit, although sometimes at extremely high computational cost. On the other hand, the FOSM method is approximate by definition. The various developments to improve its accuracy, such as the FOR methods based on mapping from nonnormal distributions to equivalent normals and the second order methods, involve additional computation for (successively) lower (but still undefinable) levels of approximation.

For earlier applications, such as calibration of structural design codes, these issues have not been of major consequence. Comparative measures rather than absolutes were sufficient. This is very unlikely to be the case in the future with increased emphasis on the safety assessment of individual structures and structures as part of larger infrastructure systems. Both require estimates of the probability of structural failure to be as realistic as possible.

References

Breitung, K. (1994). *Asymptotic approximations for probability integrals,* Springer-Verlag, Berlin.

Bucher, C. G., and Bourgund, U. (1990). "A fast and efficient response surface approach for structural reliability problems." *Struct. Safety,* 7, 57–66.

Conrad, J. ed. (1980). *Society, technology and risk assessment,* Academic Press, London.

Der Kiureghian, A., and Ke, J.-B. (1988). "The stochastic finite element method in structural reliability." *Prob. Engrg. Mech.,* 3(2), 83–91.

Ditlevsen, O., and Bjerager, P. (1984). "Reliability of highly redundant plastic structures." *J. Engrg. Mech.,* 110(5), 671–693.

Ditlevsen, O., and Madsen, H. O. (1996). *Structural reliability methods,* Wiley, Chichester.

Ditlevsen, O. (1983). "Fundamental postulate in structural engineering." *J. of Engrg. Mech.,* 109, 1096–1102.

Ditlevsen, O., Olesen, R., and Mohr, G. (1987). "Solution of a class of load combination problems by directional simulation." *Struct. Safety,* 4, 95–109.

Ellingwood, B. (1999). "Probability-based structural design: Prospects for acceptable risk bases." *Applications of statistics and probability,* R. E. Melchers and M. G. Stewart, eds., Balkema, Rotterdam.

Ellingwood, B., Galambos, T. V., MacGregor, J. C., and Cornell C. A. (1980). "Development of a probability based load criteria for American National Standard A58." *NBS Special Publication No. 577,* National Bureau of Standards, U.S. Department of Commerce, Washington, DC.

Elms, D. G., and Turkstra, C. J. (1992). "A critique of reliability theory." *Engrg. Safety,* D. Blockley, ed., McGraw-Hill, New York, 427–445.

El-Tawil, K., Lemaire, M., and Muzeau, J.-P. (1992). "Reliability method to solve mechanical problems with implicit limit functions." *Reliability and optimization of struct. systems,* R. Rackwitz and P. Thoft-Christensen, eds., Springer, Berlin, 181–190.

Faber, M. H. (2000). "Reliability based assessment of existing structures." *Progress in Struct. Engrg. and Mech.,* 2(2), 247–253.

Faravelli, L. (1989). "Response-surface approach for reliability analysis." *J. Engrg. Mech.,* 115(12), 2763–2781.

Freudenthal, A. M. (1961). "Fatigue sensitivity and reliability of mechanical systems, especially aircraft structures." *WADD Technical Report 61–53*, Wright-Patterson AFB, Dayton, Ohio.

Grandiori, S. G. (1991). "Paradigms and falsefication in earthquake engineering." *Meccanica*, 26, 17–21.

Guan, X. L., and Melchers, R. E. (1999). "A comparison of some FOSM and Monte Carlo results." *Applications of Statistics and Probability*, R. E. Melchers and M. G. Stewart, eds., Balkema, Rotterdam, 65–71.

Heyman, J. (1988). *Structural analysis: A historical approach*, Cambridge University Press, Cambridge.

Hohenbichler, M., Gollwitzer, S., Kruse, W., and Rackwitz, R. (1987). "New light on first- and second-order reliability methods." *Struct. Safety*, 4, 267–284.

HSE. (1999). "Reducing risks, protecting people." *Discussion Document*, Health and Safety Executive, London.

Julian, O. G. (1957). "Synopsis of the first progress report of committee on safety factors." *J. Struct. Engrg.*, 83(ST4), 1316.1–1316.22.

Kim, S.-H., and Na, S.-W. (1997). "Response surface method using vector projected sampling points." *Structural Safety*, 19(1), 3–19.

Lemaire, M., and Mohamed, A. (2000). "Finite element and reliability: a happy marriage?" *Proc., 9th IFIP Working Conf., Reliability and Optimization of Struct. Systems*, A. S. Nowak and M. M. Szerszen, eds., University of Michigan, Ann Arbor, Mich., 3–14.

Liu, Y. W., and Moses, F. (1994). "A sequential response surface method and its application in the reliability analysis of aircraft structural systems." *Struct. Safety*, 16(1,2), 39–46.

Madsen, H. O., Krenk, S., and Lind, N. C. (1986). *Methods of struct. safety*, Prentice-Hall, Englewood Cliffs, N.J.

Matousek, M., Schneider, J. (1976). "Untersuchungen zur Struktur des Sicherheitsproblems bei Bauwerken [Reaserches in the format of the safety problem in structural engineering]." *Bericht No. 59*. Institut fur Baustatik und Konstruktion, Eidgenossiche Technische Hochschule, Zurich.

Maymon, G. (1993). "Probability of failure of structures without a closed-form failure function." *Comp. and Struct.*, 49(2), 301–313.

Melchers, R. E. (1995). "Load space reliability formulation for Poisson pulse processes." *J. Engrg. Mech.*, 121(7), 779–784.

Melchers, R. E. (1999). *Structural reliability analysis and prediction*, 2nd Ed., Wiley, New York.

Melchers, R. E. (2000). "Discussion on 'The strategies and value of risk based structural safety analysis'." *Struct. Safety*, 22(3), 281–286.

Melchers R. E. (2002). "Safety and risk in structural engineering." *Progress in Structural Engineering and Mechanics*, 4(2), 193–202.

Moarefzadeh, M. R., and Melchers, R. E. (1996). "Sample-specific linearization in reliability analysis of off-shore structures." *Struct. Safety*, 18(2,3), 101–122.

Mori, Y., and Ellingwood, B. R. (1993). "Time dependent system reliability analysis by adaptive importance sampling." *Struct. Safety*. 12(1), 59–73.

Myers, R. H. (1971). *Response surface methodology*, Allyn and Bacon, New York.

Nowak, A. S. ed. (1986). *Modeling human error in structural design and construction*, ASCE, New York.

Pugsley, A. (1955). "Report in structural safety." *Struct. Engr.*, 33(5), 141–149.

Raiffa, H., and Schlaifer, R. (1961). *Applied statistical decision theory*, Harvard University Press, Cambridge, Mass.

Rajashekhar, M. R., and Ellingwood, B. R. (1993). "A new look at the response surface approach for reliability analysis." *Struct. Safety,* 12(3), 205–220 (see also Discussion (1994), *Struct. Safety,* 16(3), 227–230).

Rusch, H., and Rackwitz, R. (1972). "The significance of the concept of probability of failure as applied to the theory of structural safety." *The Significance . . . Development—Design—Construction,* Held und Francke Bauaktiengesellschaft, Munich.

Sibley, P. G., and Walker, A. C. (1977). "Structural accidents and their causes." *Proc., Inst. Civ. Engrs.,* 62(1), 191–208.

Stewart, M. G., and Melchers, R. E. (1997). *Probabilistic risk assessment of engineering systems,* Chapman & Hall, London.

Turner, B. A. (1978). *Man-made disasters,* Wykeham Press, London.

12

Concluding Reflections

Robert E. Melchers

As this collection of essays on aspects of the computational analysis of what we have called complex structures began to develop, there were some interesting discussions about what we actually had in mind. We hope that the meaning is now clear and that complexity in our sense relates to complexity of the system itself, complexity of the structural analysis of that system, and the degree of detail required in that structural analysis. One could well argue that even a two-span beam can present a highly complex analysis situation if it requires very detailed understanding of the stress states, even under a straightforward loading environment. It follows that complexity relates to the system being considered and the degree of detail being sought. This is evident also in the discussions in the various chapters. Longspan roofs, for example (Chapter 3), are topographically essentially simple systems—what makes them complex for analysis is the need to investigate in detail the effects that can be induce in them by the applied loadings. The discussion of the analysis of high-rise buildings (Chapter 2) shows this also, with details of requirements producing interesting issues for analysts, even though the overall system could be considered to be relatively simple.

Chapter 4 dealing with bridge structures and the increasing need to reassess their performance under gradually higher load requirements shows that alternative analysis techniques based on plastic theory may well allow a better assessment of capacity compared with traditional methods. For many practitioners such analyses could be unfamiliar and hence considered too complex. A similar assessment may be made for the possibility of analyzing the structural behavior of automobiles (Chapter 8), geotechnical structures (Chapter 5), offshore structures (Chapter 7), aircraft structures (Chapter 9), and ship structures (Chapter 10). In each case the authors show that the modeling required to capture the essential behavior of the structural system is itself already a major exercise. When this is coupled with sufficiently detailed models for the applied loads acting on these systems, very significant computational power is required to produce estimates of structural response. As Chapters 6 and 11 on aging nuclear structures and reliability estimation indicate, the computational demands are even greater when good quality structural modeling and adequate estimates of structural safety are involved. Taken together, all chapters demonstrate the tremendous structural analysis power now available both for new design and for assessment of existing structures. Moreover, given the trends in computer capabilities, there can be little doubt that the analyses once considered difficult and time-consuming will soon become within reach of all engineers.

The question for the future, therefore, is not whether capability for structural analysis can be available but rather how and when it should be used and for what purpose. The freedom to explore, quickly, highly complex structural scenarios should allow greater attention to be given to exploring structural concepts and different structural systems and detailed investigations of potentially difficult details or components. In a sense it should free the engineer from what has long been perceived as the mainstay of structural engineering effort—the structural analysis. More than ever before, with powerful computational tools at hand, the structural engineer should be able to concentrate on the creative or "art" side of a project.

Projecting this scenario into the future suggests, however, some caveats. One is that the definition of acceptability of a structure has become more complex. There are many more issues surrounding the safety, serviceability, and constructability of the structure to be considered—by the engineers and by the owners and also perhaps by the architects. Already the introduction of Limit States Design or Performance Based Design has brought about the need to consider variability and uncertainty of materials, of structural systems, and, importantly, of loadings. For many years this has been couched in output terms such as "structural risk." However, the recent increase in acts of terrorism has opened a new dimension to structural risk, and, more importantly the dimension of consequences of structural failure. These are not easy matters with which to deal, and there seems to be no alternative but to place structural engineering projects in the same category as all other projects with potentially hazardous outcomes if failure occurs. This will mean that for such structures risk assessments will need to be done, involving powerful tools such as Probabilistic Risk Assessment long used already in the nuclear and petrochemical areas.

The assessment of the remaining life of deteriorating or aging structures also opens up areas where some adaptation will be required. Not only is there an issue with how the safety and performance of such structures might be assessed compared to new designs, but the assessment of the degree of deterioration and the condition of the structure now become key issues. Thus the understanding of the behavior of materials and of their degradation with time becomes of considerable and of important interest.

These trends suggest that it will be beneficial in future to have much closer cooperation between analysts, designers and material scientists, and engineers. Such cooperation will be of major benefit to helping to improve the reliability of structural modeling and of practical structural outcomes since understanding of material performance with (long-term exposure) time is certain to become even more important, particularly for understanding and prediction of the performance of older structures. Moreover, this will be the case also if lifetime structural performance criteria are introduced as design norms, as has been proposed in various jurisdictions. For engineering educators these trends probably signal a need to return to subject areas such as chemistry, materials science, and thermodynamics, even in structural engineering courses. Interestingly, one might well imagine that this will eventually lead to both a demand for and a supply of better models for materials performance, increasing even more the demand for high-capacity computing power and increasing the profession's capability in predicting structural behavior under a variety of conditions and over long periods of time. An exciting prospect indeed.

Index